Handbook of Image Quality

OPTICAL ENGINEERING

Founding Editor

Brian J. Thompson

Distinguished University Professor
Professor of Optics
Provost Emeritus
University of Rochester
Rochester, New York

Editorial Board

Toshimitsu Asakura
Hokkai-Gakuen University
Sapporo, Hokkaido, Japan

Nicholas F. Borrelli
Corning, Inc.
Corning, New York

Chris Dainty
Imperial College of Science,
Technology, and Medicine
London, England

Bahram Javidi
University of Connecticut
Storrs, Connecticut

Mark Kuzyk
Washington State University
Pullman, Washington

Hiroshi Murata
The Furukawa Electric Co., Ltd.
Yokohama, Japan

Edmond J. Murphy
JDS/Uniphase
Bloomfield, Connecticut

Dennis R. Pape
Photonic Systems Inc.
Melbourne, Florida

Joseph Shamir
Technion–Israel Institute
of Technology
Hafai, Israel

David S. Weiss
Heidelberg Digital L.L.C.
Rochester, New York

1. Electron and Ion Microscopy and Microanalysis: Principles and Applications, *Lawrence E. Murr*
2. Acousto-Optic Signal Processing: Theory and Implementation, *edited by Norman J. Berg and John N. Lee*
3. Electro-Optic and Acousto-Optic Scanning and Deflection, *Milton Gottlieb, Clive L. M. Ireland, and John Martin Ley*
4. Single-Mode Fiber Optics: Principles and Applications, *Luc B. Jeunhomme*
5. Pulse Code Formats for Fiber Optical Data Communication: Basic Principles and Applications, *David J. Morris*
6. Optical Materials: An Introduction to Selection and Application, *Solomon Musikant*
7. Infrared Methods for Gaseous Measurements: Theory and Practice, *edited by Joda Wormhoudt*
8. Laser Beam Scanning: Opto-Mechanical Devices, Systems, and Data Storage Optics, *edited by Gerald F. Marshall*
9. Opto-Mechanical Systems Design, *Paul R. Yoder, Jr.*
10. Optical Fiber Splices and Connectors: Theory and Methods, *Calvin M. Miller with Stephen C. Mettler and Ian A. White*
11. Laser Spectroscopy and Its Applications, *edited by Leon J. Radziemski, Richard W. Solarz, and Jeffrey A. Paisner*
12. Infrared Optoelectronics: Devices and Applications, *William Nunley and J. Scott Bechtel*
13. Integrated Optical Circuits and Components: Design and Applications, *edited by Lynn D. Hutcheson*
14. Handbook of Molecular Lasers, *edited by Peter K. Cheo*
15. Handbook of Optical Fibers and Cables, *Hiroshi Murata*
16. Acousto-Optics, *Adrian Korpel*
17. Procedures in Applied Optics, *John Strong*
18. Handbook of Solid-State Lasers, *edited by Peter K. Cheo*
19. Optical Computing: Digital and Symbolic, *edited by Raymond Arrathoon*
20. Laser Applications in Physical Chemistry, *edited by D. K. Evans*
21. Laser-Induced Plasmas and Applications, *edited by Leon J. Radziemski and David A. Cremers*
22. Infrared Technology Fundamentals, *Irving J. Spiro and Monroe Schlessinger*
23. Single-Mode Fiber Optics: Principles and Applications, Second Edition, Revised and Expanded, *Luc B. Jeunhomme*
24. Image Analysis Applications, *edited by Rangachar Kasturi and Mohan M. Trivedi*
25. Photoconductivity: Art, Science, and Technology, *N. V. Joshi*
26. Principles of Optical Circuit Engineering, *Mark A. Mentzer*
27. Lens Design, *Milton Laikin*
28. Optical Components, Systems, and Measurement Techniques, *Rajpal S. Sirohi and M. P. Kothiyal*
29. Electron and Ion Microscopy and Microanalysis: Principles and Applications, Second Edition, Revised and Expanded, *Lawrence E. Murr*

30. Handbook of Infrared Optical Materials, *edited by Paul Klocek*
31. Optical Scanning, *edited by Gerald F. Marshall*
32. Polymers for Lightwave and Integrated Optics: Technology and Applications, *edited by Lawrence A. Hornak*
33. Electro-Optical Displays, *edited by Mohammad A. Karim*
34. Mathematical Morphology in Image Processing, *edited by Edward R. Dougherty*
35. Opto-Mechanical Systems Design: Second Edition, Revised and Expanded, *Paul R. Yoder, Jr.*
36. Polarized Light: Fundamentals and Applications, *Edward Collett*
37. Rare Earth Doped Fiber Lasers and Amplifiers, *edited by Michel J. F. Digonnet*
38. Speckle Metrology, *edited by Rajpal S. Sirohi*
39. Organic Photoreceptors for Imaging Systems, *Paul M. Borsenberger and David S. Weiss*
40. Photonic Switching and Interconnects, *edited by Abdellatif Marrakchi*
41. Design and Fabrication of Acousto-Optic Devices, *edited by Akis P. Goutzoulis and Dennis R. Pape*
42. Digital Image Processing Methods, *edited by Edward R. Dougherty*
43. Visual Science and Engineering: Models and Applications, *edited by D. H. Kelly*
44. Handbook of Lens Design, *Daniel Malacara and Zacarias Malacara*
45. Photonic Devices and Systems, *edited by Robert G. Hunsperger*
46. Infrared Technology Fundamentals: Second Edition, Revised and Expanded, *edited by Monroe Schlessinger*
47. Spatial Light Modulator Technology: Materials, Devices, and Applications, *edited by Uzi Efron*
48. Lens Design: Second Edition, Revised and Expanded, *Milton Laikin*
49. Thin Films for Optical Systems, *edited by François R. Flory*
50. Tunable Laser Applications, *edited by F. J. Duarte*
51. Acousto-Optic Signal Processing: Theory and Implementation, Second Edition, *edited by Norman J. Berg and John M. Pellegrino*
52. Handbook of Nonlinear Optics, *Richard L. Sutherland*
53. Handbook of Optical Fibers and Cables: Second Edition, *Hiroshi Murata*
54. Optical Storage and Retrieval: Memory, Neural Networks, and Fractals, *edited by Francis T. S. Yu and Suganda Jutamulia*
55. Devices for Optoelectronics, *Wallace B. Leigh*
56. Practical Design and Production of Optical Thin Films, *Ronald R. Willey*
57. Acousto-Optics: Second Edition, *Adrian Korpel*
58. Diffraction Gratings and Applications, *Erwin G. Loewen and Evgeny Popov*
59. Organic Photoreceptors for Xerography, *Paul M. Borsenberger and David S. Weiss*
60. Characterization Techniques and Tabulations for Organic Nonlinear Optical Materials, *edited by Mark Kuzyk and Carl Dirk*

61. Interferogram Analysis for Optical Testing, *Daniel Malacara, Manuel Servín, and Zacarias Malacara*
62. Computational Modeling of Vision: The Role of Combination, *William R. Uttal, Ramakrishna Kakarala, Sriram Dayanand, Thomas Shepherd, Jagadeesh Kalki, Charles F. Lunskis, Jr., and Ning Liu*
63. Microoptics Technology: Fabrication and Applications of Lens Arrays and Devices, *Nicholas F. Borrelli*
64. Visual Information Representation, Communication, and Image Processing, *Chang Wen Chen and Ya-Qin Zhang*
65. Optical Methods of Measurement: Wholefield Techniques, *Rajpal S. Sirohi and Fook Siong Chau*
66. Integrated Optical Circuits and Components: Design and Applications, *edited by Edmond J. Murphy*
67. Adaptive Optics Engineering Handbook, *edited by Robert K. Tyson*
68. Entropy and Information Optics, *Francis T. S. Yu*
69. Computational Methods for Electromagnetic and Optical Systems, *John M. Jarem and Partha P. Banerjee*
70. Laser Beam Shaping: Theory and Techniques, *edited by Fred M. Dickey and Scott C. Holswade*
71. Rare-Earth-Doped Fiber Lasers and Amplifiers: Second Edition, Revised and Expanded, *edited by Michel J. F. Digonnet*
72. Lens Design: Third Edition, Revised and Expanded, *Milton Laikin*
73. Handbook of Optical Engineering, *edited by Daniel Malacara and Brian J. Thompson*
74. Handbook of Imaging Materials, *edited by Arthur S. Diamond and David S. Weiss*
75. Handbook of Image Quality: Characterization and Prediction, *Brian W. Keelan*
76. Fiber Optic Sensors, *edited by Francis T. S. Yu and Shizhuo Yin*

Additional Volumes in Preparation

Optical Switching/Networking and Computing for Multimedia Systems, *edited by Moshen Guizani and Abdella Battou*

Handbook of Image Quality
Characterization and Prediction

Brian W. Keelan
Eastman Kodak Company
Rochester, New York

with contributions by
Robert E. Cookingham
Paul J. Kane
Karin Töpfer
Richard B. Wheeler
Eastman Kodak Company
Rochester, New York

MARCEL DEKKER, INC.   NEW YORK · BASEL

Some material described in this book may be protected by one or more U.S. and/or foreign patents; its description herein should not be construed as an implied license to use such patented inventions.

The authors and publisher have taken care in the preparation of this book, but make no expressed or implied warranty of any kind and assume neither responsibility for errors or omissions nor liability for incidental or consequential damage arising from the use of the information contained herein.

ISBN: 0-8247-0770-2

This book is printed on acid-free paper.

Headquarters
Marcel Dekker, Inc.
270 Madison Avenue, New York, NY 10016
tel: 212-696-9000; fax: 212-685-4540

Eastern Hemisphere Distribution
Marcel Dekker AG
Hutgasse 4, Postfach 812, CH-4001 Basel, Switzerland
tel: 41-61-261-8482; fax: 41-61-261-8896

World Wide Web
http://www.dekker.com

The publisher offers discounts on this book when ordered in bulk quantities. For more information, write to Special Sales/Professional Marketing at the headquarters address above.

Copyright © 2002 by Marcel Dekker, Inc. All Rights Reserved.

Neither this book nor any part may be reproduced or transmitted in any form or by any means, electronic or mechanical, including photocopying, microfilming, and recording, or by any information storage and retrieval system, without permission in writing from the publisher.

Current printing (last digit):
10 9 8 7 6 5 4 3 2 1

PRINTED IN THE UNITED STATES OF AMERICA

This volume is dedicated to my wife, Eileen, for her constant support and encouragement throughout its preparation, and to my colleagues Paul, Dick, Karin, Bob, Scott, and Jack, with whom I have spent so many stimulating hours pondering the mysteries of image quality.

Preface

The nature and scope of imaging are undergoing dramatic change as it enters the digital era. Portions of the formerly distinct photographic, electronic, software, television, computer, and printing industries are converging into a more generic imaging industry. The ways in which images are used are increasing in number and diversity, and the flexibility associated with digital imaging is leading to an increasingly complex field of opportunity. The rapid product cycle time of the electronics industry sets the standard for the new imaging industry, leading to an urgent need to streamline strategic, design, and development processes. In this more horizontal industry, the ability to effectively exchange specifications and evaluations based upon a common framework will become critical to the success of supplier-manufacturer and partnering relationships.

Each of these industry trends is leading to an increasingly acute need for methods of quantifying, communicating, and predicting perceived image quality. Consider the following two cases that exemplify these needs.

1. The development of the Advanced Photo System (APS) was carried out by a consortium of five companies. Prediction of image quality through computer modeling was used to select the optimal format size to meet the image quality aims while enabling such new features as smaller cameras and panoramic prints. It was quickly recognized that the modeling could simulate the results of the truly definitive experiments that engineers and analysts would like to run, but which would take far more time than was available. It was also soon appreciated that simplified experiments could yield misleading results, which, in fact, could be predicted from and explained by image quality modeling.

Subsequent to the critical step of format standardization, computer predictions were used for many purposes during component design, including: determining optimal film speed, specifying camera and printer lens and film performance requirements in terms of modulation transfer functions, setting tolerances for positioning of optical assembly subcomponents, establishing aims for film noise, and identifying the most valuable information to record on the magnetic layer for improved photofinishing. Trade trials prior to system introduction proved that the image quality distributions predicted by the modeling closely matched those produced in the marketplace.

2. A new charge-coupled device (CCD) electronic sensor architecture was proposed for professional digital still camera applications. This proposal was demonstrably advantageous to sensor manufacture. Based on engineering rules of thumb, there was some concern that adoption of the architecture might place greater demands on other system components, some of which were manufactured by other companies. Construction of prototype devices and subsequent testing and analysis would have required excessive expense and time. Image quality modeling was used to predict the impact that the change in the sensor architecture would have on requirements for each of the other system components if the image quality were to remain unaffected. The revised tolerances and performance specifications were compared to achievable levels and found to be impractical based on current technology. As a result, plans for new manufacturing lines were canceled, and the detrimental effects arising from a complex set of system interactions were avoided.

These cases exemplify some of the applications of image quality modeling techniques and suggest why the industry trends mentioned earlier are making such capabilities ever more critical. In general, the benefits of computer modeling are at least threefold:

1. cycle time compression and cost savings through reduced prototyping and experimentation;

2. identification of unexpected solutions that might be missed by empirical testing over a restricted range; and

3. education and training of practitioners through virtual experimentation.

At Eastman Kodak Company, prediction of image quality through computer modeling has proved to be of great value in all three regards and has been regularly used in formulating business strategies, guiding design decisions, establishing product aims, budgeting system tolerances, supporting advertising claims, and benchmarking competitors' offerings.

Despite such local successes, it is widely assumed that image quality, being a subjective attribute, is not amenable to quantitative analysis. This misconception is difficult to overcome because of several factors:

1. the infrequency of coverage of pertinent topics, such as psychometrics, in academic curricula;

2. the absence of a published, integrated approach to image quality characterization and prediction; and

3. the scarcity of non-proprietary examples of image quality modeling that could be shared among the industrial and academic communities.

The present volume addresses these issues through a review of needed background material, a description of an integrated and comprehensive approach to image quality modeling, and the provision of a number of examples of applications. This book is intended particularly for image scientists and product engineers, but portions of it should prove useful to individuals involved in project management, manufacturing quality control, marketing, business research, systems performance analysis, human factors and usability assessment, trade journal evaluations, and standards definition. It is hoped that this publication will focus new attention on, and stimulate further advances in, the fascinating field of image quality.

Brian W. Keelan

Acknowledgments

The majority of the results described in this book are drawn from investigations carried out by Karin Töpfer, Paul J. Kane, Robert E. Cookingham, Richard B. Wheeler, John E. Kaufman, Scott F. O'Dell, and the author, with assistance from Andrew D. Thompson, Donna L. Hofstra, James L. Miller, Stacey L. Mayo, and Sharon M. Skelly. The first four named individuals each co-authored three chapters in this volume and also assisted in its preparation in many additional respects.

This manuscript benefited from the constructive comments of Paul W. Jones, Scott F. O'Dell, Katherine S. Marsh, Eileen L. Keelan, John V. Nelson, Edward J. Giorgianni, David M. Woods, and the chapter co-authors. R. Brian Porter and J. Monty Wright provided valuable assistance regarding electronic document organization and page layout. Margaret L. Mauer consulted in the preparation of the index and Richard H. Repka wrote software for organizing the electronic graphics files.

Without the advice, encouragement and support offered by Brian J. Thompson, John V. Nelson, and James C. Weaver, it is doubtful whether this work could have been brought to fruition. Finally, I would like to thank my production editor, Eric F. Stannard, and the other staff of Marcel Dekker, Inc., who contributed to the production of this volume.

Note: Figures 14.2, 14.3, 22.2, 26.1, 26.8, 30.1, 30.4, and 31.2 have been reprinted from the proceedings of IS&T's PICS 2000 conference (Portland, Oregon) with the permission of the Society for Imaging Science and Technology (Springfield, Virginia).

Introduction

To create a computer model capable of predicting the image quality that would be produced by a hypothetical imaging system, researchers at Eastman Kodak Company have taken the following steps:

1. establishment of a numerical scale of image quality that is anchored to a set of physical standards (images) and is calibrated in perceptually useful terms that facilitate its interpretation (just noticeable differences);

2. development of a psychometric measurement technique efficiently yielding reproducible results that are calibrated in terms of the standard scale of image quality from Step #1;

3. elucidation of a theory for the prediction of the overall (multivariate) quality of an image from a knowledge of its individual quality attribute levels (e.g., sharpness, graininess, etc.);

4. investigation of a selected set of subjective image quality attributes (as in Step #3) using the psychometric technique from Step #2, leading to the definition of objective metrics (e.g., granularity) bearing a known relationship to calibrated assessments of the subjective attributes;

5. implementation of propagation models (e.g., linear systems theory) that, from key properties of system components, predict the corresponding properties of final images, in support of computation of the objective metrics from Step #4;

6. definition of measurement protocols for determining the key component properties of Step #5, and identification of engineering models that allow estimation of the same from basic design parameters; and

7. integration of the above, as well as system usage information derived from customer intercept studies, into a comprehensive Monte Carlo simulation for prediction of image quality distributions.

The outline of this book closely parallels the steps listed above. Part I describes the quantification of image quality. Although there is an extensive literature regarding psychometric methods, little attention has been devoted to obtaining calibrated perceptual measurements (Steps #1 and #2 above). This capability is critical in the construction of image quality models because the results from so many different experiments must be rigorously integrated. A novel multivariate theory (Step #3) is needed to avoid intractably large factorial experiments involving many combinations of multiple perceptual attributes.

Part II describes the derivation of objective metrics that correlate with individual attributes contributing to perceived quality (Step #4). Most readers will have some experience using such metrics; the intent here is to survey various aspects of their design that might prove useful in the development of new metrics or the generalization and extension of existing metrics. Advanced topics that are covered include: (1) treating attributes varying within an image (e.g., that depend on density); (2) accounting for attribute interactions; and (3) designing metrics that reflect preferences, as in color and tone reproduction.

Part III addresses the prediction of image quality distributions based on imaging system properties. Techniques for measuring, estimating, and propagating key properties of imaging system components (Steps #5 and #6) are well known and so are reviewed only briefly. Although Monte Carlo simulation (Step #7) is a standard computational technique, its application to image quality modeling and the interpretation of quality distributions will be unfamiliar to most readers, and so is considered in greater detail. The emphasis in the latter half of Part III is on practical examples and verification of image quality modeling analyses.

References have been collected in a single section following the epilogue. Acronyms have been kept to a minimum and are defined in the text where first used; those used more than once are listed in Appendix 1. Notation employed in the mathematical equations is also tabulated in Appendix 1.

Because of the breadth of scope of this book, the discussion of some topics is quite superficial. This is in no way meant to trivialize the complexity and richness of the associated fields of study. Instead, the intent is to provide just enough background to motivate and support the strategy we have followed in characterizing and modeling image quality. Although there are undoubtedly viable alternatives, we believe this approach, which has proven to be of great value in practical application, to be the most comprehensive available.

Contents

Preface   *v*
Introduction   *ix*

Part I: Characterization of Quality   1

Chapter 1: Can Image Quality Be Usefully Quantified?   3

    1.1 Introduction   3
    1.2 Classification of Image Quality Attributes   3
    1.3 Working Definition of Image Quality   8
    1.4 Historical Overview   10
    1.5 Areas of Current Research Emphasis   14
    1.6 Placing This Work in Perspective   16
    1.7 Summary   17

Chapter 2: The Probabilistic Nature of Perception   19

    2.1 Introduction   19
    2.2 A Probabilistic Model of Perception   19
    2.3 Properties of the Normal Distribution   23
    2.4 Predicting the Outcome of Paired Comparisons   26
    2.5 The Angular Distribution   29
    2.6 Summary   32

Chapter 3: Just Noticeable Differences — 35

    3.1 Introduction — 35
    3.2 Utility of JNDs — 35
    3.3 Certainty of Detection — 38
    3.4 Relationship of JNDs and Deviates — 39
    3.5 Determining JND Increments — 41
    3.6 Summary — 45

Chapter 4: Quantifying Preference — 47

    4.1 Introduction — 47
    4.2 JNDs of Preference — 50
    4.3 Preference Distributions and Quality Loss Functions — 51
    4.4 Analytical Approximation of Mean Quality Loss — 53
    4.5 Example of a Preference Analysis — 55
    4.6 Segmentation and Customization — 58
    4.7 Summary — 60

Chapter 5: Properties of Ideal Interval and Ratio Scales — 61

    5.1 Introduction — 61
    5.2 Ideal Interval Scales — 62
    5.3 Ideal Ratio Scales — 64
    5.4 Relationship of Ideal Interval and Ratio Scales — 67
    5.5 Calibrating Interval and Ratio Scales — 69
    5.6 Summary — 71

Chapter 6: Establishing Image Quality Standards — 73

    6.1 Introduction — 73
    6.2 Procedure for Constructing Calibrated Standards — 74
    6.3 Univariate and Multivariate Standards and JNDs — 78
    6.4 Adjectival Descriptors — 81
    6.5 Summary — 84

Chapter 7: Calibrated Psychometrics Using Quality Rulers — 87

    7.1 Introduction — 87
    7.2 Paired Comparison and Rank Order Methods — 88
    7.3 Categorical Sort Method — 89

Contents

7.4	Magnitude and Difference Estimation Methods	90
7.5	The Quality Ruler Concept	93
7.6	Attributes Varied in Quality Rulers	95
7.7	Using Quality Rulers to Assess Individual Attributes	98
7.8	Summary	99

Chapter 8: Practical Implementation of Quality Rulers — 101

8.1	Introduction	101
8.2	Hardcopy Quality Ruler	101
8.3	Softcopy Quality Ruler	111
8.4	Instructions to the Observer	113
8.5	Performance Characteristics of Quality Rulers	115
8.6	Summary	118

Chapter 9: A General Equation to Fit Quality Loss Functions — 119

9.1	Introduction	119
9.2	Dependence of JND Increment on Objective Metric	121
9.3	The Integrated Hyperbolic Increment Function (IHIF)	123
9.4	Effect of Fit Parameters on IHIF Shape	124
9.5	Summary	128

Chapter 10: Scene and Observer Variability — 129

10.1	Introduction	129
10.2	Scene Susceptibility and Observer Sensitivity	130
10.3	Selection of Scenes and Observers	139
10.4	Fitting Variability Data	142
10.5	Usefulness of Variability Data	145
10.6	Summary	147

Chapter 11: Predicting Overall Quality from Image Attributes — 149

11.1	Introduction	149
11.2	Attribute Interactions	150
11.3	Multivariate Formalism Assumptions	152
11.4	Distance and Minkowski Metrics	160
11.5	Predictions and Measurements of Multivariate Quality	162
11.6	Summary	167

Part II: Design of Objective Metrics — 169

Chapter 12: Overview of Objective Metric Properties — 171

12.1 Introduction — 171
12.2 Usefulness of Objective Metrics — 172
12.3 Determination of Objective Metric Values — 173
12.4 Other Types of Objective Quantities — 174
12.5 Example of a Benchmark Metric — 177
12.6 Summary — 179

Chapter 13: Testing Objective Metrics Using Psychometric Data — 181

13.1 Introduction — 181
13.2 Comparisons of Isotropic Noise, Streaking, and Banding — 186
13.3 Establishing a Primary Dimension and Reference Regression — 188
13.4 Investigating Variations in Secondary Attribute Dimensions — 190
13.5 Testing the Limitations of Objective Metrics — 194
13.6 Summary — 196

Chapter 14: A Detailed Example of Objective Metric Design — 197

14.1 Introduction — 197
14.2 Experimental Considerations — 197
14.3 Design of an Objective Metric — 198
14.4 Verification of an Objective Metric — 202
14.5 Fitting Scene and Observer Variability — 203
14.6 Summary — 206

Chapter 15: Weighting Attributes that Vary Across an Image — 207

15.1 Introduction — 207
15.2 Weighting in Objective versus Perceptual Space — 209
15.3 Location and Orientation Weighting — 211
15.4 Tonal Distribution and Importance Functions (TDF and TIF) — 212
15.5 The Detail Visibility Function (DVF) — 215
15.6 Determination of the Tonal Importance Function (TIF) — 217
15.7 General Applicability of the Tonal Weighting Scheme — 221
15.8 Summary — 224

Chapter 16: Analysis of Multi-Attribute Experiments — 227

16.1 Introduction — 227
16.2 Multivariate Decomposition — 228

Contents

| | 16.3 | A Practical Example: Oversharpening | 230 |
| | 16.4 | Summary | 237 |

Chapter 17: Attribute Interaction Terms in Objective Metrics — 239

	17.1	Introduction	239
	17.2	The Weak Interaction of Streaking and Noise	239
	17.3	The Strong Interaction of Contouring and Noise	241
	17.4	A Reconsideration of Perceptual Independence	249
	17.5	Summary	252

Chapter 18: Attributes Having Multiple Perceptual Facets — 253

	18.1	Introduction	253
	18.2	Sampling, Aliasing and Reconstruction	254
	18.3	Reconstruction Artifacts	258
	18.4	Perceptual Attributes Associated with Sampling Artifacts	263
	18.5	Summary	273

Chapter 19: Image-Specific Factors in Objective Metrics — 275

	19.1	Introduction	275
	19.2	Origin of Redeye and Factors Affecting Its Severity	276
	19.3	Design of a Scene-Specific Objective Metric of Redeye	278
	19.4	Summary	283

Chapter 20: Preference in Color and Tone Reproduction — 285

	20.1	Introduction	285
	20.2	Definition of Color and Tone Attributes	287
	20.3	Experimental Design Considerations	289
	20.4	General Form of Color/Tone Objective Metrics	292
	20.5	Fitting Psychometric Data	295
	20.6	Quality Contours of Preference	298
	20.7	Summary	303

Chapter 21: Quantifying Color/Tone Effects in Perceptual Space — 305

	21.1	Introduction	305
	21.2	Impact versus Compositional Importance	307
	21.3	Reference Images and Discrimination	311
	21.4	Multivariate Color/Tone Quality	315
	21.5	Summary	319

Part III: Modeling System Quality — 321

Chapter 22: Propagating Key Measures through Imaging Systems — 323

- 22.1 Introduction — 323
- 22.2 Systems, Subsystems, and Components — 324
- 22.3 Key Objective Measures in Imaging Systems — 325
- 22.4 Propagation of Mean Channel Signal — 329
- 22.5 Propagation of Image Wavelength Spectra — 331
- 22.6 Propagation of Modulation Transfer Functions — 337
- 22.7 Propagation of Noise Power Spectra — 339
- 22.8 Summary — 342

Chapter 23: Parametric Estimation of Key Measures — 343

- 23.1 Introduction — 343
- 23.2 Evolution of Modeling During a Product Cycle — 344
- 23.3 Examples of Parametric Estimation Applications — 346
- 23.4 Summary — 354

Chapter 24: Development of Measurement Protocols — 355

- 24.1 Introduction — 355
- 24.2 Definition of Measurement Protocols — 356
- 24.3 Verification of Measurement Protocols — 359
- 24.4 Summary — 364

Chapter 25: Integrated System Modeling Software — 367

- 25.1 Introduction — 367
- 25.2 User Interface Design — 368
- 25.3 Supporting Resources — 372
- 25.4 Program Output — 373
- 25.5 A Sample Modeling Session — 373
- 25.6 Summary — 376

Chapter 26: Examples of Capability Analyses — 379

- 26.1 Introduction — 379
- 26.2 Unsharp Masking Gain Selection — 380
- 26.3 Scanner Optical Positioning Tolerances — 384
- 26.4 Autofocus Ranging and Lens Aperture Specifications — 386
- 26.5 Spot Separation in Optical Anti-Aliasing Filters — 388
- 26.6 Capture and Display Resolution — 390

Contents

26.7 Digital Image Compression and File Size	393
26.8 Summary	396

Chapter 27: Photospace Coverage Metrics — 397

27.1 Introduction	397
27.2 Photospace Distributions	398
27.3 Photospace Coverage Requirements	399
27.4 Depth-of-Field Constraints	401
27.5 Exposure Constraints	403
27.6 Example of Photospace Coverage	407
27.7 Limitations of Photospace Coverage Metrics	410
27.8 Summary	412

Chapter 28: Monte Carlo Simulation of System Performance — 413

28.1 Introduction	413
28.2 Performance Modeling	414
28.3 Organization of Monte Carlo Calculations	415
28.4 Sampling Complex Distributions	419
28.5 Summary	422

Chapter 29: Interpreting Quality Distributions — 423

29.1 Introduction	423
29.2 Describing Quality Distributions Mathematically	424
29.3 Quality Distribution Shapes	426
29.4 Summary	431

Chapter 30: Examples of Performance Calculations — 433

30.1 Introduction	433
30.2 Photospace-Specific Camera Design	433
30.3 Digital Still Camera Sensor Size	435
30.4 Camera Metering	437
30.5 Output Color and Density Balancing Methods	439
30.6 Film Scanner Noise	443
30.7 Summary	447

Chapter 31: Verification of Performance Predictions — 449

31.1 Introduction	449
31.2 Performance Modeling	450
31.3 Advanced Photo System Aims	453
31.4 Verification of Predictions	456
31.5 Summary	458

Conclusion   **459**

Appendix 1: Definition of Acronyms and Symbols   **461**

    A1.1 Acronym Definitions   461
    A1.2 Symbol Definitions   462

Appendix 2: Sample Quality Ruler Instructions   **467**

    A2.1 Introduction   467
    A2.2 Instructions for Misregistration Psychophysical Test   467

Appendix 3: The Integrated Hyperbolic Increment Function   **471**

    A3.1 Continuity and Curvature   471
    A3.2 Derivatives   472

Appendix 4: Sample Help Screen   **475**

    A4.1 Introduction   475
    A4.2 Linear Sensor Help Screen Text   475

Appendix 5: Useful Optical Formulas and Photospace Coverage   **479**

    A5.1 Depth of Field and Focus   479
    A5.2 Angular Magnification   483
    A5.3 Photospace Coverage   488

References   **493**

Index   *501*

Part I

Characterization of Quality

Part I begins with a discussion of the nature of image quality and the degree to which it is amenable to quantification (Ch. 1). The just noticeable difference (JND), which provides a natural unit for quality scale calibration and making multivariate quality predictions, is mathematically defined (Chs. 2–3), and the properties of scales resulting from common rating procedures in terms of JNDs are then considered (Ch. 4). Recommendations are given for creation of physical standards and calibration of associated numerical scales (Chs. 5–6). A detailed description of an experimental method for obtaining assessments calibrated to such a standard scale is provided (Chs. 7–8). A response function suitable for fitting psychometric data obtained from such experiments is derived (Ch. 9) and the equation is used in the characterization of the variability that results from different sensitivities of observers and different susceptibilities of scenes to particular attributes (Ch. 10). Finally, a multivariate formalism for the prediction of overall image quality, given knowledge of the individual attribute levels, is discussed (Ch. 11).

1
Can Image Quality Be Usefully Quantified?

1.1 Introduction

One challenge faced by research scientists in the field of image quality is convincing product engineers, marketing personnel, management, and other scientists that image quality can be characterized in a quantitative fashion. We will address this challenge from two different viewpoints, one analytical, and the other historical. First, in Sects. 1.2 and 1.3, three schemes for classifying perceptual attributes are developed, and a working definition of image quality is developed from consideration of the correlations between the resulting categories. Second, in Sects. 1.4 and 1.5, very brief overviews of historical and more current research are presented, showing how investigators in this field have sought to quantify image quality. The final section of the chapter is intended to place the work described in this volume into perspective by comparing it with previous efforts in the field of image quality research.

1.2 Classification of Image Quality Attributes

In this and the following section, attributes contributing to perceived image quality, defined in a broad sense, are classified according to three criteria:

1. the nature of the attribute (personal, aesthetic, artifactual, or preferential), which affects its amenability to objective description;

2. the impact of the attribute in different types of assessment (first-, second-, third-party), which influences the difficulty of studying it; and

3. the extent to which the attribute is affected by imaging system properties, which largely determines the degree to which it concerns system designers.

It will be shown that there is a correlation between those attributes that are: (1) amenable to objective description; (2) readily studied; and (3) of concern to imaging system designers. This correlation suggests a more restricted definition of image quality that is tractable while still being useful.

Upon hearing a claim that image quality can be quantified, a skeptic is likely to cite certain factors that influence the satisfaction that a photographer derives from a particular image. These factors would seem to be distinctly personal in nature, and therefore unlikely to yield to objective description. For example, it can be difficult to assess whether a snapshot of a group of people will prove satisfying to the original photographer. An image that is technically deficient, perhaps being unsharp because of misfocus, noisy (grainy) because of underexposure, and dark from poor printing, may still be a treasured image because it preserves a cherished memory. Conversely, a technically sound image may still be a disappointment to the photographer because some aspect of the event that he or she wished to preserve was not successfully captured. The group of people may have been laughing at a joke and the picture taken a moment too late, when the expressions had faded to merely pleasant smiles; or the photographer may not have been able to approach closely enough to isolate the intended subject, whose facial details are hard to see in the resulting image. Furthermore, even if the photographer were pleased with the image, some of the subjects in the photograph might not be, because of certain aspects of their appearance.

The evaluation of images by the photographer who took them is referred to as first-party assessment. When the subject of a photograph renders an opinion concerning the image, the term second-party assessment may be applied. Finally, evaluation by individuals not involved in the picture taking (either as the photographer or the subject) is described as third-party assessment. Traditional consumer satisfaction primarily reflects first- and second-party evaluation, although there may be a contributing element of third-party assessment if images are often shared with acquaintances, friends, or relatives who were not involved with the events depicted. Advanced amateurs and professional photographers may be quite strongly influenced by third-party impressions, because their images are frequently displayed in exhibits, shown during presentations, entered into competitions, and sold to individuals not directly involved in the photography.

Personal attributes of image quality influence both first-party and second-party assessment, but would not be expected to affect third-party evaluation. Because first-party and second-party assessments are similar in this regard, they need not be distinguished for purposes of this discussion, and so subsequently the term first-party evaluation will be used more inclusively to refer to assessment by any party involved in the making of the image.

Compared to third-party assessment, first-party evaluation is resource-intensive, difficult to stage, and challenging to interpret. Images must be obtained from photographers, suitably modified to study the attribute of interest (usually through digital image manipulation), and returned to photographers for evaluation. The assessments of disparate images by different observers must then somehow be analyzed in a fashion that permits rigorous inter-comparison of the data. In third-party assessment, a single set of convenient images may be modified and shown to all observers, who need not meet particular requirements regarding photographic activity. First-party evaluation is a powerful tool for identifying features of importance and opportunities for improvement, but it is currently too cumbersome and expensive to use for routine image quality research. As digital imaging becomes more pervasive in consumer markets, modification and exchange of images will be facilitated; however, the other drawbacks of first-party assessment will persist, so third-party evaluation will likely remain the preferred method for image quality characterization.

In addition to personal attributes of quality, the aesthetic aspects of an image may be very subjective in nature. A skeptic might cite this as a second category of traits that affect photographer satisfaction, but would be difficult to quantify. For example, many books have been written about photographic composition and lighting, attesting to both the importance of the subject and the difficulty of mastering it. Although there are a range of opinions regarding what is aesthetically pleasing in photography, just as in other art forms such as music and painting, there actually is a good deal of agreement about desirable aspects of composition and lighting. In comparing otherwise similar images, the majority of people will prefer images in which one or more of the following are true: (1) the main subject is somewhat off-center; (2) the camera orientation reflects the shape of the subject; and (3) the main subject is sufficiently large to display good detail. Although many powerful images have been made that violate these guidelines, they still provide excellent default starting compositions that may be modified as desired by the photographer. It is difficult to argue that the prototypical snapshot of one standing person horizontally framed, with his or her face perfectly centered, and the image so small that it is hard to tell whether his or her eyes are open, would not be substantially improved by turning the camera to a vertical orientation, moving (or zooming) in closer, and shifting the person a bit off-center. Aesthetic attributes influence both first-party and third-party assessments, in a correlated but not necessarily identical fashion, because the photographer or subject in first-party evaluation has one particular opinion about what is aesthetically pleasing, which may deviate from that of the average third-party evaluator.

Although the objections raised by our skeptic have some validity, they focus only on a subset of the attributes (personal and aesthetic) that influence image quality. To obtain a more balanced perspective, two other types of attributes, artifactual and preferential, must also be considered.

There are many defects introduced by imaging systems that nearly always lead to a loss of image quality when they are detected by an observer. Examples of such problems include unsharpness, noisiness, redeye, and a variety of digital artifacts. We define an attribute as being artifactual if it is not always evident in an image, but generally leads to a degradation of quality when it is apparent. If an objective metric can be defined that is positively correlated with an artifactual attribute, the following behavior can be expected.

1. At low values of the metric, the attribute may not be readily detectable by the human visual system in pictorial images, in which case the attribute is described as being subthreshold.

2. Above the threshold for detection, quality should monotonically decrease with increasing values of the metric.

Thus, if the threshold point can be identified and the rate of quality loss above threshold can be characterized as a function of metric values, the impact of the attribute on image quality can be adequately quantified. Certainly, ranges in the threshold point and the rate of quality loss are to be anticipated as a function of scene content and observer sensitivity, but these can be described in a statistical sense by characterizing the distributions of variations. Evidently, then, artifactual attributes are amenable to quantification, and they may be expected to have a strongly correlated effect in first-party and third-party evaluations.

Preferential attributes are nearly always evident in an image and have an optimal position that usually depends on both the tastes of the observer and the content of the scene. At positions farther from the optimum, the quality loss relative to the optimum increases. One such example is contrast, with some people preferring higher contrast ("snappier") pictures, and others favoring lower contrast images with enhanced shadow and highlight detail. In general, aspects of color and tone reproduction correspond to preferential attributes. To characterize such attributes, the optimal position must first be described in objective terms. Contrast might be correlated with some average gradient of the system tone scale. Because preferences vary, the optimal position must be described as a statistical distribution. This distribution may contain information that is valuable for marketing purposes, particularly if criteria for segregation may be identified (see Sect. 4.6). At progressively greater objective distances from the optimum position, increasingly large quality loss will be experienced. This quality loss can be described in much the same way as in the case of artifactual attributes. Thus, preferential attributes are amenable to quantification through a combination of a distribution of optima, and a relationship between quality loss and the distance from the optimum. Like aesthetic attributes, preferential attributes influence both first-party and third-party assessments, in correlated but not identical ways.

We have now completed two of three distinct classification schemes of attributes of image quality, based on: (1) their nature (artifactual, preferential, aesthetic, or personal); and (2) how they influence first-party assessments compared to third-party evaluations. Our third and final classification scheme reflects whether the attributes are significantly affected by imaging system properties and so fall under the control of system designers. The term imaging system here refers to components of the imaging chain beginning with capture devices and ending with display devices or media, including all intermediate steps such as standardized processing algorithms, printing operations, etc. An example of an attribute affected by imaging system properties is noisiness, which is determined by the level of noise added to the system by certain components (such as film and the electronic sensors in scanners and digital still cameras), and the way in which these noise sources add together and are modified by other components. An example of an attribute of image quality that is not directly affected by imaging system properties is the extent to which an image captures a treasured memory. The utility of this classification scheme is that it identifies those attributes of image quality that are of greatest interest to companies and academic institutions involved in the design or manufacturing of imaging systems and their components. This is not to say that attributes unaffected or only weakly affected by imaging system properties are uninteresting. In addition to their intrinsic scientific value, such aspects may indicate significant opportunities in the imaging industry for new applications and services to enhance customer satisfaction; however, they are of lesser concern from the perspective of imaging system design.

1.3 Working Definition of Image Quality

Table 1.1 lists selected attributes of image quality of each of the four types (artifactual, preferential, aesthetic, or personal) and scores them on a −1, 0, and +1 scale for three characteristics:

1. amenability to objective description (+1 = straightforward, 0 = difficult, −1 = nearly intractable);

2. similarity of first-party and third-party assessments (+1 = strongly correlated, 0 = partially correlated, −1 = nearly uncorrelated,); and

Attribute by Type	Objective Tractability	$1^{st}/3^{rd}$ Party Correlation	System Dependence	Total Score
Artifactual				
Unsharpness	+1	+1	+1	+3
Graininess	+1	+1	+1	+3
Redeye	+1	+1	+1	+3
Digital artifacts	+1	+1	+1	+3
Preferential				
Color balance	+1	+1	+1	+3
Contrast	+1	0	+1	+2
Colorfulness (saturation)	+1	0	+1	+2
Memory color reproduction	+1	0	+1	+2
Aesthetic				
Lighting quality	0	0	0	0
Composition	0	0	0	0
Personal				
Preserving a cherished memory	−1	−1	−1	−3
Conveying a subject's essence	−1	−1	−1	−3

Table 1.1 Categorization of selected image quality attributes. There is a strong correlation between those attributes that are most amenable to objective description, those most easily investigated experimentally, and those influenced by imaging product design.

3. degree of dependence on imaging system properties (+1 = strongly influenced, 0 = somewhat influenced, −1 = minimally influenced).

An attribute might be a particularly attractive choice for study if it had +1 scores in each category, making it objectively tractable, experimentally accessible, and pertinent to imaging system design. As an overall measure of suitability for study, the three scores for each attribute are summed in the final column.

While the scoring in Table 1.1 is coarsely quantized and may be subject to debate, the overall trend evident in the table is quite compelling, and would only be made stronger by the inclusion of a more extensive list of image quality attributes. There is a strong correlation between the attributes that are artifactual or preferential in nature, those that are amenable to objective description, those that may be studied through third-party evaluation, and those that are significantly influenced by imaging system design. The attributes meeting these criteria cluster near the maximum total score possible. Personal attributes exhibit the opposite behavior, and have total scores at the low end of the scale, while aesthetic attributes fall in a middle tier. This segregation of attributes into three tiers suggests that a slightly restricted working definition of image quality, based on third-party assessment, be adopted. Specifically,

> the quality of an image is defined to be an impression of its merit or excellence, as perceived by an observer neither associated with the act of photography, nor closely involved with the subject matter depicted.

This narrower definition of image quality, which is based on third-party assessment, captures the artifactual, preferential, and aesthetic attributes, but excludes personal attributes. The included attributes, which constitute the upper and middle tiers in Table 1.1, are mostly tractable, experimentally accessible, and important in system design. Image quality so defined will certainly correlate well with broader concepts of image quality and customer satisfaction, but will be largely free of the ambiguities associated with personal attributes. While these omitted attributes can produce significant discrepancies in individual evaluations, distributions of assessments for collections of images and observers should be similar.

In summary, third-party image quality, while not including all attributes affecting a photographer's satisfaction with an image, is well-defined and of practical utility. With this definition in hand, we may confidently answer in the affirmative the question posed in this chapter's title, namely, whether image quality may be usefully quantified.

1.4 Historical Overview

The purpose of this section is to survey briefly the history of research related to image quality to show how scientists and engineers have sought to quantify its various aspects. Because the intent of the present section is to provide an overall perspective, the approach will be to highlight trends rather than provide a detailed review of the literature. Readers interested in pursuing certain topics in greater depth may wish to consult one of more of the following textbooks and monographs, which provide excellent accounts of the work accomplished through their publication dates:

1. James (1977), a comprehensive survey of all aspects of conventional photographic systems;

2. Dainty and Shaw (1974), a clear exposition of signal and noise analysis of imaging systems;

3. Bartleson and Grum (1984), a monograph on psychovisual methods and their application to the study of image quality; and

4. Hunt (1976), a treatise on the reproduction of color and tone in imaging systems, with both theoretical discussions and practical applications.

More specialized works providing greater detail are cited in the discussion that follows, which consists of chronological surveys of efforts related to: (1) perceptual measurement; (2) objective image structure metrics; and (3) color and tone reproduction. These three areas have, until recently, evolved fairly independently, in part because most workers have specialized in just one of the associated disciplines. To predict the quality produced by imaging systems, advances from each of these fields must be integrated into a common framework, such as that described in this volume.

Our first chronological survey is concerned with the measurement of perception or sensation. The field of psychophysics had its origins in the nineteenth century, when experimentation by Weber and others sought to relate discriminable differences in sensation to continuous physical properties such as weight. In 1860, Fechner proposed that such discriminable differences could be accumulated to form a quantitative scale of sensation, allowing perceptions to be mapped to numerical values. This concept met with considerable controversy, especially from physical scientists. Nonetheless, a variety of experimental techniques were developed for measuring perception of fundamental physical quantities.

Because image quality is not a single sensation that is fully correlated with a single physical continuum, useful application of psychophysics to image quality awaited the development of new scaling techniques and methods of analysis. The term psychometrics is used instead of psychophysics to describe experiments in which the stimuli do not vary along a single objectively measurable physical dimension. Thurstone (1927) placed psychometrics on a firm quantitative basis through his law of comparative judgments, which related the outcome of paired comparison experiments to the perceptual differences between the stimuli and the uncertainty of perception, without reference to the physical origins of the differences. The perceptual uncertainty, reflected in the lack of a deterministic outcome in comparisons of very similar stimuli, can be related to the extremely useful concept of a just noticeable difference (JND).

Paired comparisons are effective for measuring very small differences between stimuli, but perceptual scaling techniques are needed to study larger differences efficiently. Stevens (1946) defined the properties of several types of measurement scales that are of utility in image quality assessment, including ordinal, interval, and ratio scales. Such scales can in principle be obtained from various simple rating tasks, including rank ordering, categorical sorting, and magnitude estimation, although in many cases questionable assumptions and/or rather involved data analysis is required, as discussed in any of a number of available treatises on psychometrics, e.g., Guilford (1954), Torgerson (1958), Gescheider (1985), Nunnaly and Bernstein (1994), and Engeldrum (2000). Each of these scaling methods has proven to be of utility in image quality research (Bartleson and Grum, 1984). Unfortunately, the results of different rating experiments cannot readily be compared unless the scales are calibrated to some common standard, which has rarely been done. The present volume describes methods by which such calibrated results may be obtained.

Studies relating to image quality have mostly been carried out by image scientists and engineers, rather than psychologists, using the scaling methods mentioned above; a number of examples of such work are cited later. The greatest challenge has been the prediction of the overall quality of an image from a knowledge of its individual attributes. Prosser, Allnatt, and Lewis (1964) found impairments, which were harmonically related to the complement of a 1–5 quality scale, to sum in monochrome television images. Bartleson (1982) modeled overall quality as a Minkowski sum (n^{th} root of the sum of nth powers) of sharpness and the complement of graininess, each expressed on a 1–9 interval scale. De Ridder (1992) combined digital encoding impairments expressed as fractions of maximum quality loss using a re-normalized Minkowski metric. None of these methods, nor others previously proposed, have proven to be extensible and generally applicable (Engeldrum, 1999). This book describes a multivariate formalism that successfully predicts the results of a number of experiments, involving combinations of a variety of attributes.

Our second chronological survey pertains to investigations of image structure. The earliest objective measures relating to an aspect of image quality were probably the criteria developed around the turn of the century by Rayleigh and Strehl to describe the ability to discriminate fine detail in images produced by optical instruments. In the case of astronomical observations, resolving power was useful because it accurately described the ability of a telescope to distinguish stars having small angular separations as seen from the earth. Image noise was first characterized in terms of blending distance (the viewing distance at which the noise became visually imperceptible), and later by root-mean-square (RMS) granularity. Early analyses of photographic granularity were published in the 1920s and 1930s, by Silberstein, Siedentopf, and others.

Fourier theory began to permeate the field of optics in the 1940s, leading to the generalization of resolving power, a single-frequency metric, to the modulation transfer function (MTF), and of RMS granularity, an integrated quantity, to the noise power spectrum (NPS). Many successful applications of linear systems theory to imaging were made in the 1950s and 1960s. Linfoot (1964) described cascading of optical and photographic material MTFs to predict the metrics of image fidelity, relative structural content, and correlation quality. R. C. Jones (1955) explained how to measure the NPS of photographic materials, made the first such determinations, and related the NPS to RMS granularity measurements. Doerner (1962) derived an equation for propagation of NPS through an imaging system that is in common use today.

During this same timeframe, a number of investigations of the fundamental limitations of image capture were carried out. Definition and application of detective quantum efficiency (DQE) allowed the signal-to-noise performance of a detector to be related to that of an ideal detector, and facilitated the comparison of different image capture technologies, such as silver halide emulsions and electronic sensors.

Objective metrics reflecting properties of the human visual system began to appear in the 1950s and 1960s. Stultz and Zweig (1959) published a relationship between the magnification at which an image was viewed and the scanning aperture size yielding RMS granularity values best correlated with perceived noisiness. The usefulness of the RMS granularity metric was enhanced by the determination of its just noticeable difference (JND) increment by Zwick and Brothers (1975), and the quantification of the dependence of perceived graininess on density by Bartleson (1985). Similar progress was made in the definition of objective correlates of perceived sharpness. Crane (1964) proposed the system modulation transfer (SMT) metric, which was based upon system component MTFs and an assumed visual frequency response, and was stated to have JND units. Crane approximated the component MTFs and visual response by Gaussian functions for computational simplicity; with advances in computing power, subsequent workers were less constrained and proposed more generally applicable metrics.

Following its introduction by Shannon (1948), information theory was used to develop new objective metrics correlating with image sharpness. The theory was applied to optical systems in the 1950s and to photographic systems subsequently (e.g., R. C. Jones, 1961; Frieser, 1975). This approach led to the definition of a number of sharpness correlates based upon frequency-dependent signal-to-noise ratio (Görgens, 1987; Hultgren, 1990). The nonlinear response of the human visual system to modulation changes was modeled by Carlson and Cohen (1980) in their study of display quality, and this approach was improved and extended by Barten (1990, 1999) in his square root integral (SQRI) metric. Although signal-to-noise metrics account for the impact of noise on perceived sharpness, they do not necessarily predict perceived graininess quantitatively, either in terms of threshold location or JND increments above threshold, without empirical adjustment (Töpfer and Jacobson, 1993). They do, however, appear to be well suited for predicting interpretability (rather than quality) of images, based on their extensive application in reconnaissance systems analyses (Biberman, 2000). Bartleson (1982) successfully modeled the overall quality associated with simultaneous variations in both sharpness and noisiness by relating each attribute to its own objective metric, and combining their effects in a perceptual, rather than objective, space. As described later, subsequent workers have often adopted a similar, perceptually oriented perspective.

Our third and final chronological survey relates to the study of color and tone reproduction. The response of a photosensitive material to varying levels of exposure was first characterized by Hurter and Driffield in 1890, allowing its tone reproduction (and speed) to be quantified. L. A. Jones (1920) analyzed the propagation of tone reproduction through an imaging system by use of the diagrams that still bear his name. Preferred tonal rendition, particularly in terms of various gradients correlating with perceived contrast, was experimentally investigated for reflection prints (Jones and Nelson, 1942; Simonds, 1961) and television images viewed in differing surround conditions (DeMarsh, 1972).

There was a great deal of work done in the 1960s involving scaling of perceived brightness as a function of light level (controlling direct adaptation) and type of surround (controlling lateral adaptation), of which the work of Bartleson and Breneman (1967a) was particularly pertinent because of their use of complex images. From the results of these studies it was possible to predict, to first order, pleasing tone reproduction aims for different viewing conditions (Bartleson and Breneman, 1967b).

The trichromatic nature of color vision was well known by the mid-1800s, when color mixture experiments were performed by Grassman, and color images were created by Maxwell by projecting red, green, and blue "separations" in register. Maxwell's Principle, elucidated late in that century, provided a basic understanding of how to achieve metameric reproduction of original scene colors in additive color systems. Color matching functions and tristimulus values were standardized by the CIE in 1931, and subsequently more perceptually uniform color spaces were sought, leading to the definition of CIE L* u* v* and CIE L* a* b* coordinate systems (CIE Publication 15.2, 1986).

Following the introduction of practical subtractive color photographic systems, improvements in accuracy of color reproduction were made through the application of chemical interlayer interimage (Hanson and Horton, 1952) and masking (Hanson and Vittum, 1947) effects. As tolerably accurate reproduction of color became possible, at least within the constraints imposed by the display media, the question of whether accurate color reproduction was in fact preferred, assumed a greater importance. Studies of the color reproduction of Caucasian skin-tones, green foliage, and blue sky revealed that their preferred reproduction differed somewhat from the original colors (Hunt, Pitt, and Winter, 1974, and references therein).

The investigation of preference in color and tone reproduction, as well as the study of other aspects of image quality, has been greatly facilitated by digital image simulation techniques, which have become widespread during the last decade. Using such methods, recent studies have found a preference for overall color and tone reproduction that differs systematically from accurate reproduction in having higher contrast and colorfulness (e.g., Buhr and Franchino, 1994, 1995; de Ridder, 1996; and Janssen, 2001).

1.5 Areas of Current Research Emphasis

In addition to the investigation of preference in color and tone reproduction, several other broad areas of research are currently particularly active, and so we shall consider them very briefly. These areas are: (1) visual appearance modeling; (2) characterization of digital imaging system components; and (3) establishment of frameworks for understanding image quality.

Visual modeling refers to the prediction of the appearance of an image, or the difference in appearance of two images, accounting for as many known properties of the human visual system as possible, including those associated with neural processing. If a reference image were modified by an image processing operation, the difference in appearance between the modified and original images could be estimated from such a model. Furthermore, if the operation emulated a change in some attribute of image quality, the computed image difference might serve as a type of objective measure of that attribute (Jin, Feng, and Newell, 1998; Johnson and Fairchild, 2000). This approach is most likely to be successful with artifacts (which are generally detrimental if detected), and near threshold (where visual phenomena are best understood, and simple difference measures are most likely to prove predictive). Greater challenges are anticipated in the suprathreshold regime, particularly with preferential attributes because images that differ substantially in appearance from variation in such attributes may, nonetheless, have equal perceived quality.

With the emergence of digital imaging, a great deal of attention has been focused on the properties of digital components, including image processing algorithms and their effects on image quality. Areas of current emphasis include:

1. sampling, resampling, and reconstruction of images (Wolberg, 1990);

2. compression of digital images for reduced storage space and faster transmission (Rabbani and Jones, 1991; Taubman and Marcellin, 2001);

3. development of algorithms for correcting defects or enhancing images (Gonzalez and Woods, 1992; Bovick, 2000);

4. digital encoding of color information (Giorgianni and Madden, 1998);

5. interpretation (Kane et al., 2000) and standardization (ISO 14524, 1997; ISO 12233, 1998) of measurements of digital devices; and

6. rendering of images for half-tone output (Kang, 1999; Spaulding, Miller, and Schildkraut, 1997).

Although most image quality research has historically focused on the impact of one or a few factors, which often are peculiar to a particular imaging technology, recently some efforts have been made to develop more integrated approaches based upon a general, perceptually relevant framework. In our view, the best early example of this sort of work, which is still frequently cited, is the quality contour study of Bartleson (1982). His approach involved: (1)

identification of perceptual dimensions (attributes) of quality; (2) determination of relationships between scale values of the attributes and objective measures correlating with them; and (3) combination of attribute scale values to predict overall image quality. This basic framework, the structure of which is dictated by perceptual considerations, rather than being driven by objective criteria, has been adopted and extended by other workers. Examples of current image quality frameworks include: (1) the Image Quality Circle of Engeldrum (2000); (2) the Document Appearance Characterization (DAC) system used at Xerox (Dalal et al., 1998); (3) that of the European Adonis project (Nijenhuis et al., 1997) and other work carried out at the Institute for Perception Research (IPO); and (4) the approach described in the present volume. Efforts are now underway to define international standards partially reflecting a perceptual framework (Burningham and Dalal, 2000).

1.6 Placing This Work in Perspective

The research described in this volume is distinguished from that published previously in several ways. All psychometric experiments have been calibrated to a common numerical scale that is anchored to physical standards, allowing rigorous integration of results from different experiments. For maximum relevance to customer perception, the standard scale units are JNDs of overall quality, as determined in paired comparisons of samples varying in multiple attributes, assessed by representative observers. With attributes characterized individually in terms of JNDs of impact on overall quality, a general multivariate combination rule based on a variable-power Minkowski metric has been employed to avoid the necessity for large factorial experiments. Dozens of studies have been carried out within this framework, leading to the characterization of many perceptual attributes and the definition of a number of new objective metrics for both artifactual and preferential attributes.

The following features have been integrated into a single software application that uses Monte Carlo techniques and the results of the psychometric experiments to predict the quality distribution produced by imaging systems:

1. databases of component measurements and customer usage data;

2. engineering models for the estimation of component properties;

3. propagation models permitting prediction of a number of objective metrics from imaging system component properties;

4. transformations of the objective metrics to JNDs of quality; and

5. the multivariate combination rule.

Finally, this software has been applied to a wide range of imaging system and component design problems and its predictions have been rigorously verified against the independent assessment of the performance of imaging systems in the hands of customers.

1.7 Summary

For the purposes of this work,

> the quality of an image is defined to be an impression of its merit or excellence, as perceived by an observer neither associated with the act of photography, nor closely involved with the subject matter depicted.

This definition includes image quality attributes the nature of which are artifactual (degrading quality when detectable), preferential (always evident, and possessing an optimum position), or aesthetic (related to artistic merit). Also included are those attributes most readily correlated with objective metrics and those of particular importance in imaging system design. Excluded are those attributes that can only be assessed through first-party evaluation (by those involved in the photography), which is resource-intensive and can be difficult to interpret. This slightly restricted definition is well correlated with broader concepts of image quality and customer satisfaction, but is more clearly defined, while still maintaining a high level of practical utility.

There is a rich history of research pertinent to the quantification of image quality, starting in the nineteenth century, with investigations into the measurement of perception, the trichromatic nature of the human visual system, the characterization of photographic material response, and the definition of objective criteria of image resolution. During the twentieth century a great body of literature accumulated, with progress in areas such as psychometrics, signal and noise analysis, and preference in color and tone reproduction, being of particular importance in understanding image quality. In the last decade, particular effort has been expended in characterizing and optimizing the quality of digital imaging systems, and assembling integrated frameworks within which image quality may be organized, studied, and understood. The present volume, which describes the construction and application of general image quality models, reflects both of these recent trends.

2
The Probabilistic Nature of Perception

2.1 Introduction

The units of quality used throughout this work are just noticeable differences (JNDs), the merits of which are discussed in some detail in subsequent chapters. To define and interpret JNDs, an underlying model of perception that is probabilistic in nature is required. In this chapter, a very simple model of this type is presented in Sect. 2.2. Properties of the normal distribution and definitions of probability density functions (PDFs) and cumulative distribution functions (CDFs) are given in Sect. 2.3. This permits the outcome of paired comparison experiments to be related to perceptual distribution properties in Sect. 2.4. Finally, in Sect. 2.5, the properties of the angular distribution, a convenient alternative to the normal distribution, are reviewed.

2.2 A Probabilistic Model of Perception

Suppose that a number of pairs of images are prepared so that the members of each pair are identical except that one is very slightly blurred relative to the other, perhaps from digital spatial filtering or optical defocus. These pairs are presented to an observer who is asked to identify which image in each pair is sharper; the observer must make a guess even if they are not sure they see a difference. Under these circumstances, if the degree of blur is sufficiently small, the observer will sometimes select the blurred sample as being the sharper of the two. This choice may be considered an error, in the sense that it disagrees both with the known objective properties of the two samples and with the outcome of the majority of perceptual assessments made on the samples. As the degree of blur approaches zero, the responses of the observer will appear to become random, with the fraction of correct answers approaching a value of one-half.

These observations may be explained if it is assumed that perception is a probabilistic phenomenon because it is based on an inherently noisy process. A particular instance of perception is viewed as being drawn from a probability distribution of possible perceptions along a continuum. To represent this continuum in our example, let us assume that a numerical "blur value" may be associated with the perception of unsharpness of a single image, with greater positive values indicating increasing levels of blur. Suppose that Sample #1 of a

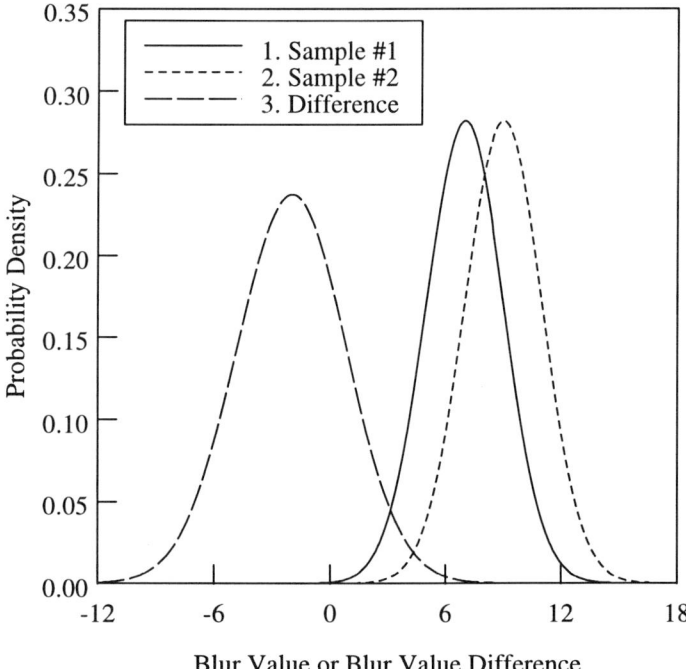

Fig. 2.1 Distributions of perceived degree of blur of each of two samples, and the distribution of perceived differences between them. Because the sample distributions overlap, the sharper sample will not always be correctly identified in a paired comparison.

The Probabilistic Nature of Perception

pair had a blur value of 7.0 and Sample #2 a blur value of 9.0. Sample #2 would be more blurred and, if perception were deterministic in nature, would always be identified as the less sharp sample. In contrast, Fig. 2.1 depicts a probabilistic model of how the observer makes the assessment of which image is sharper. Consider first the two probability distributions centered at x-axis blur values of 7.0 and 9.0, corresponding to Samples #1 and #2, respectively. These distributions show the relative likelihood of individual perceptions corresponding to different blur values for each sample. In the case of Sample #1, the observer is most likely to perceive the sample as having a blur value near 7.0, the mean perceived value, but in about one-third of the cases, will perceive it as being more than two blur value units different from the true position (either higher or lower), and rarely (about five percent of the time), more than four units different. Similar behavior applies in the perception of Sample #2, so there is considerable overlap between the perceptual distributions of the two samples.

The act of choosing which of the two samples is sharper is modeled as involving the assessment of each sample against the perceptual continuum of sharpness, followed by a comparison of the values obtained. Table 2.1 sketches one possible scenario. Sample #1 is perceived as having a sharpness equivalent to a blur value of 7.8, meaning that this single instance of perception resembles the average perception that would result from many assessments of samples having actual blur values of 7.8. This perception is a bit high in blur value compared to the mean perception, but is hardly unlikely, given the breadth of the perceptual distribution. Sample #2 is perceived as having an equivalent blur value of 7.4, somewhat low, but still reasonably probable. A comparison of the equivalent blur values of the samples leads to the conclusion that Sample #2 is sharper, an error. As shown in Table 2.1, the combination of perceptual discrepancies of +0.8 and −1.6 blur value units in the assessments of Samples #1 and #2, respectively, yields a discrepancy difference of 2.4 units, which is of sufficient magnitude to overcome the actual sample difference of 2.0 units.

	Sample #1	Sample #2	Difference
Individual perceived blur value	7.8	7.4	+0.4
Mean perceived blur value	7.0	9.0	−2.0
Perceptual discrepancy	+0.8	−1.6	+2.4
Standard deviation	2.0	2.0	2.8

Table 2.1 One possible assessment of the relative sharpness of two samples. Although Sample #2 is blurred relative to Sample #1, noise in the perception of each sample caused discrepancies that more than offset the actual sample difference, leading to an erroneous assessment.

If a numerical simulation were performed, in which perceived blur values for Samples #1 and #2 were randomly drawn from their distributions in Fig. 2.1, the distribution of blur value differences between Sample #1 and Sample #2 would be that shown at the left of Fig. 2.1. The mean and peak value of the distribution lies at a blur value difference of $7.0 - 9.0 = -2.0$ as expected; the mean of the difference distribution is equal to the difference of the means of the individual perceptual distributions. The difference distribution is considerably wider than the individual distributions because both of them contribute to the variability of the difference.

In Table 2.1, an erroneous assessment resulted when the discrepancy difference, +2.4 blur value units, was greater in magnitude than, and opposite in sign to, the actual sample difference (−2.0 units). Restated, an incorrect choice occurred when the sum of the discrepancy and sample differences had a sign opposite that of the sample difference. Analogously, in Fig. 2.1, the probability of an erroneous choice being made is equal to the fraction of the area under the difference distribution that is to the right of $x = 0$, because those are the assessments in which the difference is positive, whereas the sign of the actual sample difference (−2.0 units) is negative. It is evident that the likelihood of an erroneous choice depends strongly on the difference in degree of blur; e.g., shifting the difference distribution to the left, corresponding to a larger blur value separation, would rapidly reduce the fractional area of the distribution lying to the right of the origin. This result is of note, and will now be expressed in a more convenient form, which will be useful in the following section.

Let us define a modified difference distribution that is shifted by the mean difference so that it is centered on the origin. The previous result now may be restated as follows: the probability of a correct response is equal to the fraction of the area under the zero-mean difference distribution that lies to the left of the absolute value of the mean sample difference. The zero-mean difference distribution is a useful concept because, if the perceptual distributions had the same shape for different samples, as is the case in Fig. 2.1, then the modified difference distribution would be identical for all sample pairs, and would reflect the fundamental nature of the perception, rather than being a property of individual samples.

The simple probabilistic model of Fig. 2.1 was developed by Thurstone (1927), who assumed that the individual distributions were normal (Gaussian). He outlined a number of limiting cases, of which our simple model corresponds to his Case V, in which the two sample distributions are equal and uncorrelated. To proceed further it is necessary to understand some of the mathematical properties of normal distributions, which are described in the following section.

2.3 Properties of the Normal Distribution

The normal distribution, which describes the familiar bell curve, is ubiquitous in nature because the sums of the outcomes of independent random events tend towards a normal distribution even if the distribution of outcomes of the individual events are not normally distributed, This remarkable result is embodied in the central limit theorem, which is discussed in all elementary statistics texts. As a simple example, consider a single roll of a die. The probability of rolling 1, 2, 3, 4, 5, or 6 is equal, which corresponds to a uniform probability distribution having magnitude 1/6, as shown in Fig. 2.2, Curve #1. If two dice are rolled and their counts summed, the resulting probability

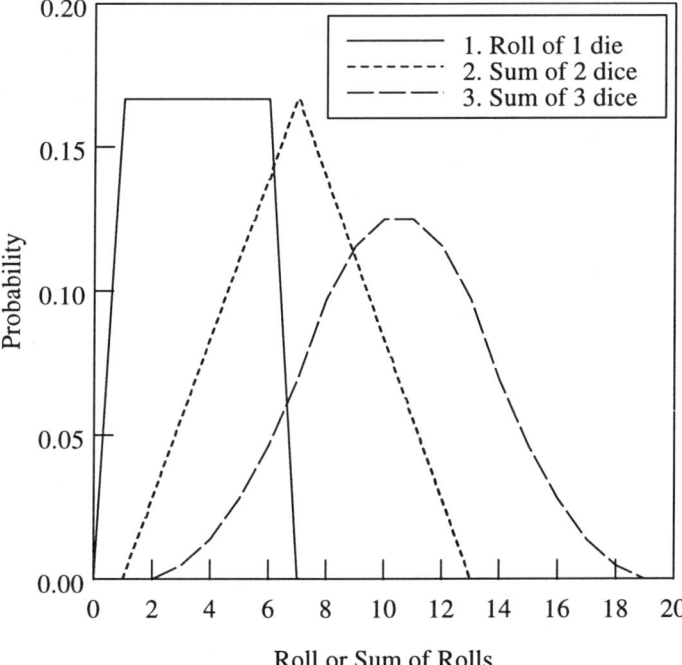

Fig. 2.2 Demonstration of the central limit theorem. Although the probability distribution of values from the roll of a single die is uniform (equal probabilities of one to six), that of the sum of values from multiple rolls rapidly approaches a normal distribution.

distribution is triangular (Curve #2), as can easily be confirmed by a simple tabulation of all 36 possible outcomes. The sum of three rolled dice is already rather bell-shaped (Curve #3), and as the number of rolls being summed increases, the shape of the distribution of the sum even more closely approaches that of a normal distribution. Thus, the sum of the outcomes of several random elements tends toward a normal distribution, even though the individual events possess a distribution of outcomes that may be very different in shape.

The normal distribution is mathematically described by a Gaussian function, which is a negative exponential of an independent variable squared. A normalized (unit area) Gaussian function $g(u)$, having mean $\bar{u}$, and standard deviation σ_g, is given by Eq. 2.1.

$$g(u) = \frac{e^{-(u-\bar{u})^2/(2\cdot\sigma_g^2)}}{\sigma_g \cdot \sqrt{2\cdot\pi}} \qquad (2.1)$$

This normal distribution has the property that 68% of its area lies within plus or minus one standard deviation of the mean, 95% within two, and > 99% within three. It is convenient to define a coordinate called the Gaussian deviate, denoted by z_g, which is the number of standard deviations from the mean.

$$z_g = \frac{u - \bar{u}}{\sigma_g} \qquad (2.2)$$

Substituting Eq. 2.2 into Eq. 2.1 and re-normalizing to unit area yields the simplified normal distribution as a function of deviate value.

$$g(z_g) = \frac{e^{-z_g^2/2}}{\sqrt{2\cdot\pi}} \qquad (2.3)$$

This function is plotted in Fig. 2.3 (Curve #1). We have been referring to such functions simply as probability distributions, but it is now necessary to refine our terminology to distinguish between two cases. The continuous distributions shown in Fig. 2.1 and Curve #1 of Fig. 2.3 are of the type known as probability density functions (PDFs). PDFs are functions having the property that the probability of occurrence of a value of an independent variable between two limits is the area under the PDF between those same two limits. Denoting an arbitrary PDF by $h(u)$, this can be expressed as:

The Probabilistic Nature of Perception

$$p(u_1 < u < u_2) = \int_{u_1}^{u_2} h(u) \cdot du \qquad (2.4)$$

where p is probability and u_1 and u_2 are the limits of the independent variable. As these limits extend towards ±∞, the probability approaches one, because the independent variable must have exactly one value. The integral on the right side of Eq. 2.4 must therefore equal one, and so a PDF must have unit area.

A second type of distribution is known as a cumulative distribution function (CDF). It describes the cumulative probability of occurrence of values of the independent variable that are less than or equal to a given value, and so it ranges

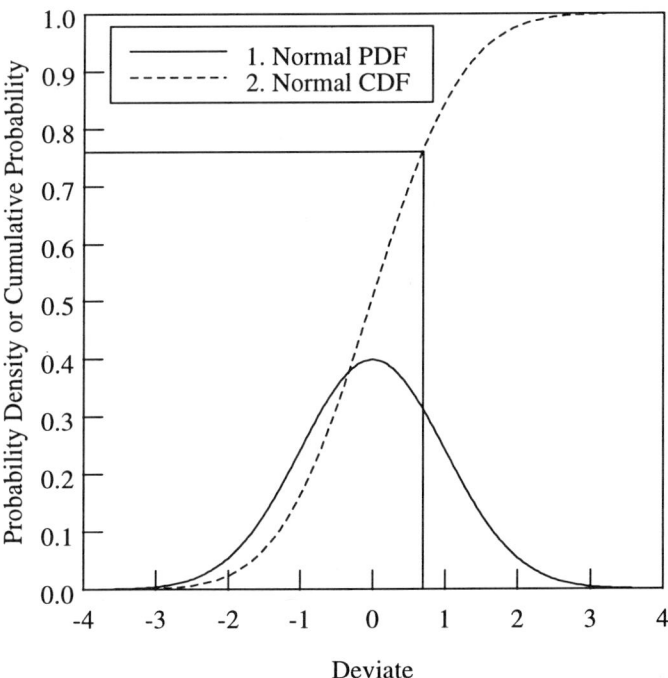

Fig. 2.3 Gaussian (normal) probability density function (PDF) and cumulative distribution function (CDF), plotted against the deviate z_g. The CDF is the integral of the PDF, which has unit area.

monotonically from zero to one in magnitude. The normal CDF p_g at deviate value z_g is computed by integrating the PDF from $-\infty$ to z_g, yielding:

$$p_g(z_g) = \frac{1}{\sqrt{2 \cdot \pi}} \int_{-\infty}^{z_g} e^{-z_g'^2/2} \cdot dz_g' \qquad (2.5)$$

where z_g' is a dummy variable of integration. This CDF is plotted in Fig. 2.3 (Curve #2). PDFs and CDFs can be derived from one another by integration (PDF $\rightarrow$ CDF) or differentiation (CDF $\rightarrow$ PDF) and so contain the same fundamental information; we will make use of both types of functions depending on which is more convenient in a given situation. For example, if the most probable value of the independent variable were sought, the PDF would be the natural choice because the PDF peak occurs at this value. If instead the median value were sought, the CDF would be preferred because that is the value at which the CDF is equal to one-half.

2.4 Predicting the Outcome of Paired Comparisons

Recall that in the discussion of Fig 2.1, the difference PDF was said to be computed numerically by randomly sampling from the individual perceptual PDFs and taking the difference between pairs of outcomes. Although this is a viable means of calculating the difference distribution, a more elegant approach is to convolve the two individual PDFs. The mathematical convolution of two PDFs, $h_1(u)$ and $h_2(u)$, is defined as follows:

$$h(\Delta u) = \int_{-\infty}^{+\infty} h_1(u) \cdot h_2(u - \Delta u) \cdot du \qquad (2.6)$$

where Δu is a difference between values of the independent variable. As can be intuitively appreciated in the context of the sample PDFs in Fig. 2.1, this equation expresses the probability density of a particular difference as the sum of the joint probability densities of all possible pairs of individual outcomes differing by that particular difference. Although Eq. 2.6 may still have to be evaluated numerically for arbitrary individual PDFs, the convolution can be done analytically in the case of two normal distributions. The well-known result is that the convolution of two Gaussian functions yields another Gaussian function, the variance (standard deviation squared) of which is equal to the sum of the variances of the two functions being convolved (Castleman, 1996). For example, the individual sample PDFs in Fig. 2.1 are each normal distributions with standard deviations equaling two blur value units. Consequently, the difference distribution is a Gaussian function with a variance of $2^2 + 2^2 = 8$ and a standard deviation of $8^{1/2} \approx 2.8$. These values are noted in Table 2.1 for use later. Although the individual sample perceptual distributions and their variances are certainly of theoretical interest, most psychometric experiments involve either a direct comparison of stimuli (as in a rank ordering) or a comparison of ratings of different samples during the data analysis for purposes of calibration. Consequently, references to perceptual distributions, PDFs, and assessment variances shall hereafter pertain to the case of sample differences unless otherwise noted.

Recall from the previous section that in a paired comparison, "the probability of a correct response is equal to the fraction of the area under the zero-mean difference distribution that lies to the left of the absolute value of the mean sample difference". If it is understood that the sample difference is to be positive, and the difference PDF is to have zero mean, then, with the concept of a CDF available, this result may be stated more succinctly as follows: the probability of a correct response is equal to the value of the CDF at the mean sample difference. In particular, if a difference distribution is Gaussian and the mean difference between two samples is expressed in positive deviate units z_g (ensuring a zero-mean PDF), then p_g in Eq. 2.5 must be exactly the probability of a correct response in a paired comparison, which will be denoted by p_c.

We can now predict the outcome of paired comparisons as a function of actual sample differences. For example, from Table 2.1, the difference PDF of Fig. 2.1 has a standard deviation of 2.8 blur value units, and Samples #1 and #2 are separated by 2.0 blur value units, so the mean sample difference is $z_g = 2.0/2.8 \approx$ 0.7 deviate units. As shown graphically in Fig. 2.3, $p_g(0.7) \approx 0.76$, so the sharper of the two samples should be correctly identified 76% of the time in paired comparisons.

In analyzing the data from a paired comparison experiment, the opposite transformation must be made, from the probability of a correct response, p_c, which is measured, to the sample difference, z_g, which is calculated. In the tails of the CDF, the slope is very low, so z_g cannot be determined precisely from p_c. For example, if in 39 of 40 comparisons the correct sample were identified, $p_c =$ 0.975, implying $z_g \approx 2$; therefore, the samples would be deduced to differ by two deviate units. If just the one dissenting choice were changed, then p_c would become one, and the sample difference would be deduced to be infinite. Evidently, it is not possible to obtain robust estimates of sample differences on the order of two deviate units without huge sample sizes. One might think of paired comparison experiments as saturating in response for larger sample differences. This behavior fundamentally limits what we might call the dynamic range of a paired comparison experiment, necessitating the use of other experimental methods when larger sample differences are to be characterized, as discussed in detail in subsequent chapters.

Not only is it difficult to obtain precise estimates of sample differences in the tails of the normal CDF, but also the Gaussian PDF may not accurately describe the perceptual difference distribution for larger sample differences. In many cases, more extended tails are observed experimentally, as shown in Fig. 2.4, which compares the upper half of the normal CDF with results from a large paired comparison study of sharpness differences. Curve #1 is simply a plot of

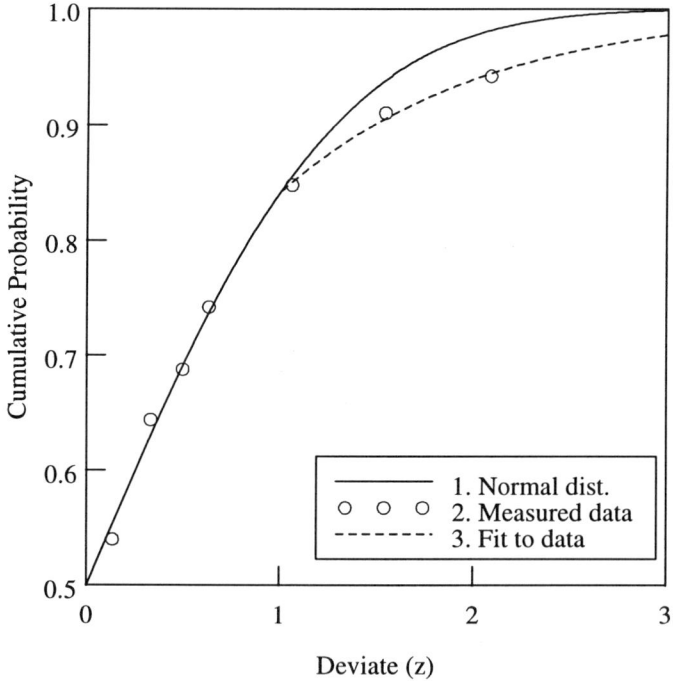

Fig. 2.4 Comparison of normal and experimentally measured CDFs. The measured distribution has a substantially more extended tail, indicating a greater frequency of errors at larger sample differences.

The Probabilistic Nature of Perception

the normal CDF given by Eq. 2.5. The measured data points have y-values equal to the fraction of correct responses for a particular stimulus difference. The determination of the measured x-values involves some subtleties (which are described in detail in subsequent chapters), but is essentially equivalent to the following procedure. The difference distribution is assumed to be shaped like a Gaussian function at least near its center, so that comparisons of very close samples, yielding relatively low proportions of correct responses, should produce accurate deviate values. These values should also be precise, because the region of saturation (high correct response rates) has been avoided. Deviate differences between more widely separated stimuli are then computed by adding up smaller deviate increments. For example, suppose that sharpness decreases from Sample #1 to Sample #3, and when Samples #1 and #2, or Samples #2 and #3 are compared, the correct response is obtained in 76% of cases. This implies that the difference between Samples #1 and #2, and between Samples #2 and #3, is 0.7 deviate units each. The difference between the more widely separated Samples #1 and #3 is therefore assigned a value of 0.7 + 0.7 = 1.4 deviate units.

As seen in Fig. 2.4, up to a difference of about one deviate unit, the measured data is well described by a normal CDF, but at larger sample differences, considerably more errors are observed than are expected based on a Gaussian distribution, so that a greater than anticipated stimulus difference is required to reach a given high level of correct responses. Given that determination of larger sample differences from paired comparison experiments is both imprecise because of saturation, and potentially inaccurate because the underlying distributions may not conform to the normal shape in their tails, it is tempting to somehow restrict the tails of the assumed difference distribution. Such a restriction could limit the possible deduced deviate values to a reasonable range, and contain the uncertainties of the resulting values within acceptable bounds. In fact, there is a function called the angular distribution, which does this in a particularly elegant fashion, as described in the next section.

2.5 The Angular Distribution

The angular CDF, $p_a(z_a)$, may be defined by the equation:

$$p_a(z_a) = \sin^2\left(\frac{z_a}{\sqrt{2 \cdot \pi}} + \frac{\pi}{4}\right) \qquad |z_a| \leq \sqrt{\pi^3/8} \qquad (2.7)$$

where z_a is the angular deviate and the argument to the sine function is in radians. Outside the indicated deviate range, the CDF is zero (at more negative values) or one (at more positive values). If p_a is set equal to p_p, the fraction of times one sample is chosen in a paired comparison, inversion of Eq. 2.7 yields the corresponding sample difference in angular deviate units:

$$z_a(p_p) = \sqrt{2 \cdot \pi} \cdot \left(\sin^{-1}\left(\sqrt{p_p}\right) - \frac{\pi}{4} \right) \tag{2.8}$$

The origin of the alternative name "arcsine distribution" is evident in this equation. Conveniently, compared to use of a normal distribution, this transformation can be done analytically, rather than requiring a numerical solution. Equation 2.8 is a principal result of this chapter, and it will be used in the following chapter to define a JND.

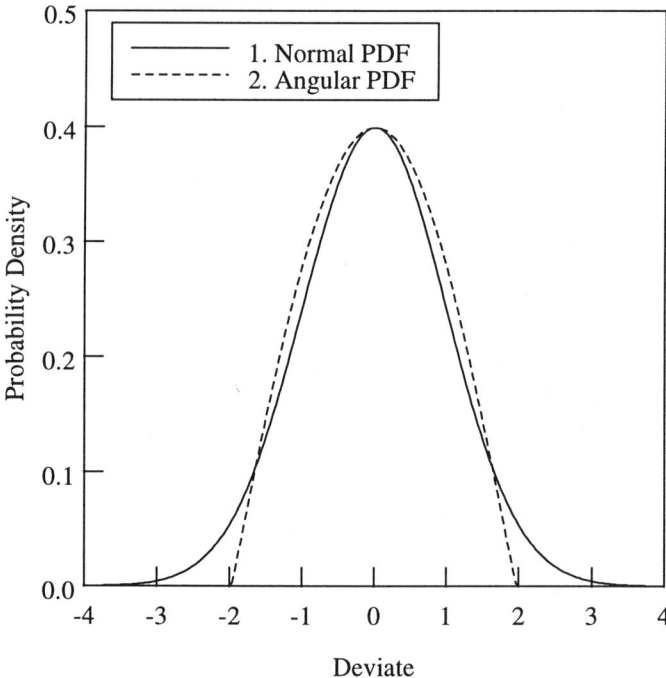

Fig. 2.5 Comparison of normal (Gaussian) and angular (arcsine) PDFs. The functions are similar except in the tails, where the angular distribution is truncated.

The Probabilistic Nature of Perception

The angular PDF is obtained by differentiating Eq. 2.7 with respect to the deviate z_a, which yields:

$$a(z_a) = \frac{1}{\sqrt{2 \cdot \pi}} \cdot \sin\left(\sqrt{\frac{2}{\pi}} \cdot z_a + \frac{\pi}{2}\right) \qquad |z_a| \leq \sqrt{\pi^3/8} \qquad (2.9)$$

The PDF is zero outside the indicated deviate range. Figure 2.5 compares the normal and angular PDFs, and Fig. 2.6 their CDFs. The angular and normal distributions are very similar except in their tails, which are truncated in the angular distribution. One consequence of this behavior is that the uncertainty of a deviate deduced from Eq. 2.8 is not a strong function of probability, unlike the previously discussed case of a normal distribution, where the uncertainty

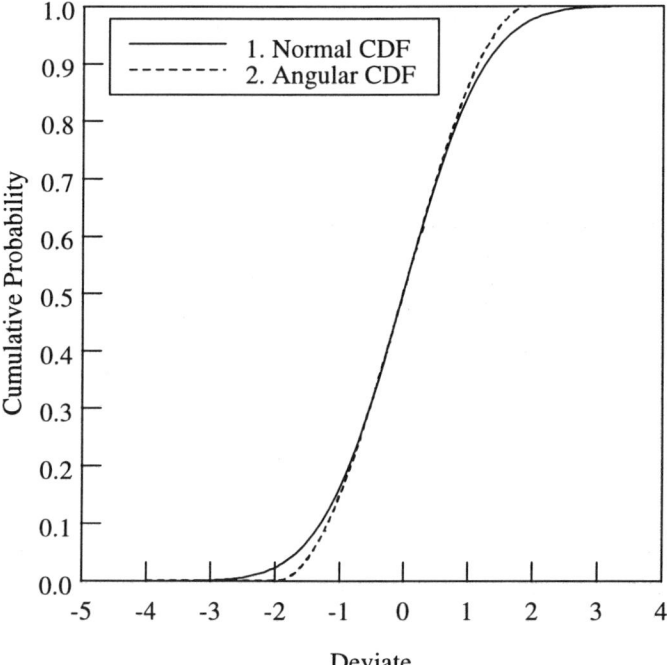

Fig. 2.6 Comparison of normal (Gaussian) and angular (arcsine) CDFs. Advantages of the angular distribution include bounded deviate values, uniform deviate uncertainties, and its analytical form.

diverges as the probability of a correct response approaches one. This leads to a statistical convenience when deviate values are regressed against objective metrics or psychometric scale values, a type of analysis that will later be shown to have utility in determining JND increments. If normal deviates are used, it is usually necessary to employ one of several techniques, such as weighted regression, to account for the variation in uncertainties of the deviates. In contrast, if angular deviates are used, a simple unweighted regression can be employed instead (Bock and Jones, 1968).

As seen in Fig. 2.6, the normal and angular CDFs start to diverge slightly near $|z| = 1$, which is where the uncertainty of the normal deviate begins to be inflated. This is also the region where the experimental data in Fig. 2.4 starts to depart from a normal CDF. These observations suggest that the boundary $|z| = 1$ may be regarded as the approximate onset of problematical behavior in paired comparison analysis.

In the next chapter, just noticeable differences (JNDs), the units of quality employed throughout this work, will be defined in terms of deviate values.

2.6 Summary

Paired comparison data is useful for determining the perceptual differences between similar stimuli. The data is interpreted by assuming that perception is probabilistic in nature and can be described by a distribution function over a subjective continuum. The normal distribution is a good candidate for quantifying perception because it describes many phenomena that depend on the sum of multiple independent events, as reflected in the central limit theorem. The probability of a correct response in a paired comparison is equal to the value of the cumulative distribution function (CDF) evaluated at the sample difference. The CDF is the integral of the zero-mean perceptual difference distribution, and sample differences are expressed in positive deviate units, i.e., separation in multiples of the standard deviation of perception.

In practice, sample differences in deviate units are deduced from the outcome of paired comparison experiments. Because the normal distribution has extensive tails of low slope, and because these tails may not accurately mimic those of actual perceptual distributions, inferred deviate values for larger sample differences may be both imprecise and inaccurate. The precision problem may be ameliorated through the use of the angular distribution, which produces deviates like those of the normal distribution, except in the tails, which are truncated. Equation 2.8 allows the sample difference in angular deviates to be conveniently calculated from the measured probability of a correct response in a paired comparison experiment. Nonetheless, even using the angular distribution, paired comparisons between samples differing by more than approximately one deviate unit should be interpreted with caution.

3

Just Noticeable Differences

3.1 Introduction

In this chapter the just noticeable difference (JND) and the JND increment are defined for arbitrary levels of detection certainty. The JND is a measure of the perceptual continuum; differences in JNDs are proportional to deviate values from a zero-mean difference distribution. A JND increment is the number of units of an objective metric or rating scale required to produce a sample difference of one JND. Both JNDs and JND increments are defined in terms of the results of paired comparison experiments.

This chapter is organized as follows. In Sect. 3.2, the reasons why JNDs are such useful measures of image quality and its attributes are discussed. The relationship between certainty of detection and the outcome of paired comparisons is derived in Sect. 3.3. The equation relating JNDs and deviate values is presented in Sect. 3.4. Finally, Sect. 3.5 describes how JND increments are determined from regressions of paired comparison data.

3.2 Utility of JNDs

The concept of a just noticeable difference is extremely useful in image quality characterization and prediction for several reasons, as listed on the next page.

1. Most image quality analyses involve the comparison of a test system with an analogous reference system, and it is important to understand the significance of the quality difference between them.

2. Although JNDs are small units of change, they may be used as the basis for constructing or calibrating numerical scales that quantify wide ranges of quality.

3. JNDs are natural units with which to perform certain image quality calculations, notably the prediction of overall quality from a knowledge of the impact, in isolation, of a collection of quality attributes.

Each of these reasons is now discussed in more detail.

Most well posed questions that can be addressed by image quality analyses involve a comparison of a test system with a reference system, the latter serving as a sort of experimental control. The test system may be hypothetical, in a breadboard or prototype stage, or in manufacture. A similar range of possibilities applies to the reference system, although most frequently a system currently available in the marketplace is chosen, because associated customer satisfaction is then likely to be known, providing a useful benchmark position. Proper choice of a reference system usually leads to quality differences between the test and reference systems that are relatively small, because the systems are likely to share a number of characteristics. It is, therefore, commonly of interest whether modest quality differences are of perceptual significance. Following are three examples of the types of questions encountered, with test and reference systems identified, and the application of JNDs noted.

1. Is an improvement in an imaging system component, which is known to be feasible, large enough to be advertised? The test system differs from the reference system only in possessing the improvement. A difference large enough to justify advertisement can be expressed as the number of JNDs required so that only a small, specified fraction of customers will not be able to see the difference.

2. Is the quality loss associated with a manufacturing cost reduction sufficiently small to be largely undetected in practice? The test system differs from the reference system by virtue of incorporation of the cost reduction feature. The difference can be expected to be largely undetected if it is smaller than one JND.

3. What specifications must an imaging system component meet to effectively match the performance of a product in the marketplace? The

reference system is the existing product; the test system may differ from it in a variety of ways, to be determined in the analysis. The goal is for the test system to produce quality or attributes of quality within one JND of that of the reference.

In each of these three cases, and in general, expressing quality differences in terms of JNDs greatly facilitates the interpretation of predictions or psychometric measurements in light of the question being addressed.

The second reason for the particular utility of JNDs is that they may be used to construct calibrated numerical scales that quantify wide ranges of quality, which are needed to describe the full distribution of quality produced by imaging systems under conditions of actual usage. Consider a simple example of constructing a quality scale using JNDs. Suppose that we have a series of samples A, B, C, etc., for which it is known that each sample is one JND better than the preceding sample. We could designate these samples as physical standards and assign them numerical values such as A = 10, B = 15, C = 20, etc., so that 5 scale units corresponds to one JND. This type of assignment method produces what is called an ideal interval scale, which is discussed in detail in Ch. 5. More commonly, a series of samples might be rated using some arbitrary scale in a psychometric experiment. Selected pairs of similarly rated samples could then be assessed by direct comparison to establish JNDs of difference between them. These results could be used to determine how the JND increment varies as a function of position on the scale, and if desired, the scale could be mathematically transformed so that the JND increment became a constant, thereby simplifying its interpretation, as described in Sects. 5.5 and 6.2.

The third and final reason why JNDs are so important in image quality analysis is that JNDs are natural units with which to perform certain image quality calculations. Suppose that the quality of an image were adversely affected by three artifactual attributes, each of which had been studied in isolation, so that their individual impact on some numerical scale of quality would be known, if each were the only defect present. It would be highly desirable to be able to predict the total change in numerical scale value arising from the three attributes in combination, without performing additional experimentation. Such a prediction may be made through the multivariate formalism described in Ch. 11. The first, crucial step in making a prediction using the multivariate formalism is to express the impact of each attribute in terms of JNDs of overall quality, which then allows a generally applicable and extensible multivariate combination rule to be applied.

3.3 Certainty of Detection

In the previous chapter we focused on the number of correct and incorrect responses in paired comparisons, but from an intuitive point of view it would also be of interest to know the fraction of assessments in which the observer genuinely detected the sample difference. Although the probabilistic model of Ch. 2 does not provide such a value, because it contains no absolute threshold, an intuitive argument can provide a simple relationship for estimation of the probability of genuine detection. We assume that there are two types of assessments: (1) those in which the sample difference is detected, which always lead to correct responses; and (2) those in which the sample difference is not detected, which lead to an equal number of correct and incorrect responses by chance. If this is the case, then the probability of a correct response, denoted p_c, is equal to the probability of detection, denoted p_d, plus one-half the complement of that probability.

$$p_c = p_d + \frac{1-p_d}{2} = \frac{1+p_d}{2} \qquad (3.1)$$

For example, if 75% of responses are correct ($p_c = 0.75$), then 25% were incorrect guesses and presumably an equal number, 25%, were correct guesses, leaving 50% that were not guesses at all ($p_d = 0.50$) and therefore represented genuine detection events. As a check of the behavior of this equation, if the samples in a pair were nearly identical, p_d should approach zero, and by chance, we would expect the fraction of correct responses to approach one-half. If instead the samples were extremely different, p_d and p_c both should approach one. Equation 3.1 is consistent with both these expectations.

Because most people find it more intuitive to think in terms of detection probability than probability of a correct response, we hereby define a JND of certainty level p_d to be a perceptual difference detectable in a fraction p_d of paired comparisons, and leading to a fraction p_c of correct responses, given by Eq. 3.1. We shall refer to such quantities as different percentage JNDs, e.g., 50% JNDs when $p_d = 0.5$. Being a perceptual quantity, the JND may be applicable to individual attributes such as sharpness or to overall image quality, depending upon the task defined in the paired comparison experiment. For example, if the observers were asked to select the sample that appeared sharper, the deduced JND difference would be one of sharpness. In contrast, if the observers were asked to select the sample that was of higher quality, the differences derived from the data would be JNDs of overall quality. The magnitude of a JND may depend upon the details of the paired comparison experiment, such as expertise

Just Noticeable Differences 39

of observers, image content, number of attributes varying, etc., which should be selected based on the intended application, as discussed in Ch. 6.

We further define a JND increment of certainty level p_d to be the sample difference, expressed in units of an objective metric or a rating scale value, that corresponds to one JND of the same certainty level. Recall the earlier example in which a series of samples known to be spaced apart by one JND were assigned numerical values of 10, 15, 20, etc. In that case, the JND increment was five scale value units. The distinction between JNDs and JND increments is important and so bears repeating. JNDs are units of perceptual attribute difference between stimuli, defined by the outcomes of paired comparison experiments, whereas JND increments are the objective metric or rating scale differences corresponding to one JND.

It will be convenient to choose a standard level of certainty for JNDs and JND increments. The most common choice, which we will adopt hereafter, is that the probability of detection be 50% ($p_d = 0.50$). In cases where a different level of certainty is desired, the results of the next section will allow the requirement to be stated in terms of multiples of 50% JNDs.

3.4 Relationship of JNDs and Deviates

In the example of Fig. 2.1 and Table 2.1, two samples differing by two blur value units are compared for sharpness. The standard deviation of the perceptual difference distribution is 2.8 blur value units, so the samples differ by 2.0/2.8 ≈ 0.714 standard deviations. This should correspond to a probability of correct response of $p_a(0.714) \approx 0.77$ from Eq. 2.7. From Eq. 3.1, $p_c \approx 0.77$ corresponds to $p_d \approx 0.54$. This information is summarized in the first row of Table 3.1 (next page), which will be filled out as we progress through this section. Based on this calculation, we may make the following two statements:

1. the two samples perceptually differ by one 54% JND of sharpness; and

2. a 54% JND increment of sharpness corresponds to two blur value units, at least in the region of the blur values of the two samples.

It would be more convenient if we could express these results using our adopted certainty level, i.e., in terms of 50% JNDs and JND increments. A fundamental assumption of the probabilistic model of Sect. 2.2 was the existence of an underlying perceptual continuum of the attribute being assessed (in our case, sharpness). This continuum was numerically represented by blur values, and differences along the continuum were expressed either in terms of blur value differences or deviates. The continuum can equally well be quantified in terms of JNDs if a suitable reference position is chosen, as discussed further in Sect. 5.5, but for now only sample differences are of concern. Because deviate values (as from Eq. 2.8) are derived from a zero-mean difference distribution, identical samples have $z = 0$. Obviously, identical samples differ by zero JNDs, and so sample differences expressed in deviates and JNDs are proportional to one another. One JND of certainty p_d corresponds with the deviate $z_a(p_c)$, where Eq. 3.1 relates p_d and p_c, so we may write:

$$\Delta Q_d = \frac{z_a(p_p)}{z_a(p_c)} \qquad (3.2)$$

where p_p is the fraction of times one sample is chosen in a paired comparison and ΔQ_d is the perceptual difference between the samples in JNDs of certainty p_d. The notation Q is meant to suggest quality, although in our present example it is instead the attribute of sharpness that is being assessed. In this book, all quantities denoted with a Q have units of JNDs. The notation ΔQ indicates that the quantity is a perceptual difference. Equations 3.1 and 3.2 serve to define JNDs of any level of certainty.

	p_d	p_c	z_a	50% JNDs
Fig. 2.1	0.54	0.77	0.714	1.09
One 50% JND	0.50	0.75	0.656	1.00
$z = 1$	0.72	0.86	1.000	1.52
Two 50% JNDs	0.86	0.93	1.312	2.00

Table 3.1 Relationships between angular deviates z_a, probabilities of detection (p_d) and correct response (p_c), and numbers of 50% JNDs for several cases. These quantities are related by Eqs. 3.1, 3.2, 2.7, and 2.8.

Equation 3.2 may now be applied to our example of Table 3.1. The second row of the table will list the characteristics of a 50% JND. In Eq. 3.1, setting $p_d = 0.50$ yields $p_c = 0.75$. From Eq. 2.8, this corresponds to an angular deviate of $z_a \approx 0.656$. It was already noted that a 54% JND corresponds to an angular deviate of $z_a \approx 0.714$, so by Eq. 3.2, a 54% JND must be $0.714/0.656 \approx 1.09$ 50% JNDs, completing the first row of the table. (A very slightly different result would be obtained if a normal distribution were used rather than an angular distribution, because of their slight shape differences even away from the tails.) These values seem plausible; a 54% JND should be a bit larger than a 50% JND, because slightly greater certainty should arise from a slightly larger sample difference.

Two other rows are filled out in Table 3.1. It was noted that precision and accuracy of deviates begins to be compromised above $z = 1$; by Eq. 3.2, this position corresponds to $1/0.656 \approx 1.5$ 50% JNDs. Discrepancies in the tails notwithstanding, the last row tabulates the characteristics of two 50% JNDs, with $z_a = 2 \times 0.656$, which corresponds to $p_c \approx 93\%$ by Eq. 2.7. Two JNDs seems like a reasonable requirement for an improvement to be advertised because so few customer assessments should lead to the contrary conclusion that the improved version is inferior.

If we substitute Eqs. 2.8 and 3.1 into Eq. 3.2, we obtain a single equation relating the fraction of time a particular sample is selected in a paired comparison, p, the desired probability of detection p_d, and the number of associated JNDs, ΔQ_d.

$$\Delta Q_d = \frac{\sin^{-1}\left(\sqrt{p_p}\right) - \frac{\pi}{4}}{\sin^{-1}\left(\sqrt{\frac{1+p_d}{2}}\right) - \frac{\pi}{4}} \quad (3.3)$$

The right-hand side of Eq. 3.3 reduces to $(\sin^{-1}(p_p^{1/2}) - 0.785)/0.262$ for 50% JNDs. Equation 3.3 is the principal result of this chapter. It defines the perceptual difference between two samples in JNDs of any level of certainty, in terms of the outcome of a paired comparison experiment.

3.5 Determining JND Increments

It would be convenient to have a formula like Eq. 3.3 that directly relates the JND increment, in objective metric units, to paired comparison data. Typically, in a paired comparison experiment, a number of pairs of samples differing by various amounts are assessed. If the angular deviates of each comparison, from Eq. 2.8, are plotted against the objective metric difference between the samples, the resulting data can usually be fit by a linear regression (and because we are using angular deviates, the regression can be unweighted). The slope of the regression is the change in deviate per unit change in objective metric difference. One JND increment produces a deviate change from zero to $z_a(p_c)$, so one JND increment must equal $z_a(p_c)$ divided by the regression slope, as shown in Eq. 3.4:

$$\Delta\Omega_J = \frac{z_a(p_c)}{\left(\frac{\partial z_a}{\partial \Delta\Omega}\right)} \tag{3.4}$$

where $\Delta\Omega$ is the objective metric difference, $\Delta\Omega_J$ is the JND increment in objective metric units, and the partial derivative in the denominator is the slope of a z_a versus $\Delta\Omega$ regression line. Throughout this book, objective metrics and quantities derived therefrom are denoted by an Ω, the pronunciation of which is suggestive of the acronym OM for objective metric. For 50% JNDs the right side of Eq. 3.4 reduces to 0.656 divided by the regression slope. Although this equation is written in terms of an objective metric, it applies equally well to a rating scale. The only properties required of Ω are that it be monotonically related to the perceptual attribute being tested and that its JND increment not vary dramatically over the range of a few JNDs. The latter condition permits a paired comparison experiment spanning such a range to yield a meaningful mean value of the JND increment. Almost any reasonable objective metric or rating scale easily meets these requirements.

Substitution of Eqs. 2.8 and 3.1 into Eq. 3.4 allows the results to be expressed in terms of the probability of detection p_d and the fraction of times a particular sample is chosen in a paired comparison, p_p, in a fashion analogous to Eq. 3.3:

$$\Delta\Omega_J = \frac{\sin^{-1}\left(\sqrt{\frac{1+p_d}{2}}\right) - \frac{\pi}{4}}{\left(\frac{\partial \sin^{-1}(\sqrt{p_p})}{\partial \Delta\Omega}\right)} \tag{3.5}$$

The phase shift of $\pi/4$ disappears when the derivative of z_a is taken, and all factors of $(2\cdot\pi)^{1/2}$ cancel, leading to this simpler form. For 50% JNDs, the right hand side of Eq. 3.5 reduces to 0.262 divided by the slope of a regression of $\sin^{-1}(p_p^{1/2})$ versus objective metric difference. Equation 3.5 permits determination of the JND increment of an objective metric from the results of a series of paired comparisons.

Just Noticeable Differences

An example of such an analysis based on Eq. 3.4 is shown in Fig. 3.1. A number of different samples were compared against a common sample and assessed for sharpness. Some test samples were sharper than the reference sample and some were less sharp. For simplicity, in previous discussions we have worked with probabilities of correct responses (always > 0.50) and necessarily positive deviates, but it is also possible to express results in terms of the fraction of times one sample is chosen over another (which may then be less than 0.5) and deviates that can be positive or negative, as is done in this figure. Specifically, the probability p_p was defined to be the fraction of times the test sample was chosen as being sharper, and the objective metric difference was defined to be positive when the test sample was predicted to be sharper than the reference.

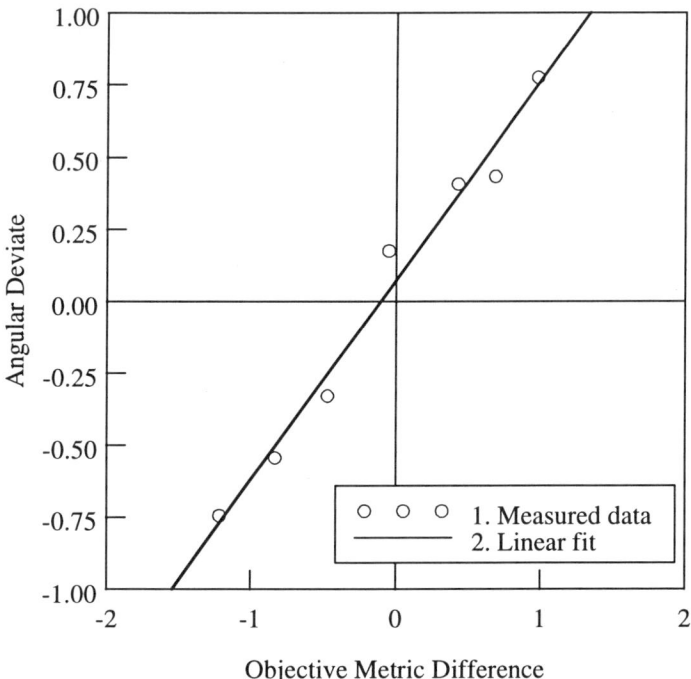

Fig. 3.1 Linear regression of angular deviate (from Eq. 2.8) against objective metric difference between samples in paired comparisons of sharpness. Equation 3.4 relates the JND increment of the objective metric to the slope of the regression line, which passes close to the origin, as it should.

With these choices, we expect a positively sloped deviate versus metric difference line that passes through the origin, corresponding to no objective or subjective sample difference. These expectations are approximately met in Fig. 3.1. The regression line does not exactly pass through the origin but is close. The degree of scatter of data is typical for the sample size employed, which is either 90 or 180 assessments per data point. The slope of the regression line is ≈ 0.693, and by Eq. 2.8, $z_a(0.75) \approx 0.656$, so, by Eq. 3.4, a 50% JND increment is equal to $0.656/0.693 \approx 0.95$ objective metric units.

Three brief notes regarding Fig. 3.1 are in order.

1. If $\sin^{-1}(p^{1/2})$ rather than deviates z_a had been regressed against objective metric differences, as in Eq. 3.5, the regression line would be expected to pass through $(0, \sin^{-1}(0.5^{1/2})) \approx (0, 0.785)$ rather than through the origin.

2. The use of signed deviates and probabilities of a particular sample being chosen, rather than positive deviates and probabilities of correct responses, is helpful when determining the point of subjective equality, which is the objective metric difference at which the stimuli are found to be equal in terms of the attribute being assessed. Although one might expect this point always to occur at or near an objective metric difference of zero, if the samples differ in certain ways not properly reflected in the metric, it may inaccurately predict the degree of the attribute being assessed. Determination of the point of subjective equality permits a very sensitive test for such effects.

3. In data of this type, it is not uncommon for the uncertainty in the x-values and y-values to be comparable in magnitude. The standard least-squares approach, which assumes that all error is in the y-value, can sometimes lead to poor estimates of slope in such cases, especially if the data is noisy because of inadequate sample size. In such cases it is recommended that the uncertainty in both the metric and deviate values be estimated independently and a linear regression method that accounts for precision in both coordinates be employed (Kendall and Stuart, 1967).

Our discussions in the last two chapters have implicitly assumed that the perceptual attribute being assessed is artifactual in nature, so that, if a difference between stimuli is genuinely detected, observers will agree in their choices. This behavior does not hold for preferential attributes, where the appearance of samples may be quite different, and so evident to all observers, yet there may

Just Noticeable Differences   45

not be agreement regarding which sample is of higher quality. In the next chapter, the application of JNDs is extended to include matters of preference.

3.6 Summary

JNDs are useful measures of image quality and its attributes for three reasons:

1. they are easily interpreted units for quantifying the smaller quality differences usually found between test and reference systems;

2. they provide a basis for calibrating rating scales spanning a wider range of quality; and

3. they are the natural units with which to perform certain perceptual transformations, notably the prediction of overall quality from the multivariate formalism.

Equation 3.1 defines the certainty of detection in terms of paired comparison outcomes. If two stimuli differ by one 50% JND, the difference between them will actually be detected 50% of the time, and in the remaining cases the observer must guess. By chance, half of these guesses will be correct, so a 75%:25% split in responses will result. Image quality positions within one 50% JND of one another are considered to be functionally equivalent; improvements exceeding two 50% JNDs may be deemed sufficient to justify advertising. A certain percentage JND increment is the change in objective metric or rating scale value that produces stimulus differences of one JND of the same certainty level.

Equation 3.2 generally relates JNDs of any certainty level to deviates from any zero-mean difference distribution. Equation 3.3 represents the special case of Eq. 3.2 in which angular deviates are assumed. This equation is very convenient because it directly transforms fractions of correct responses in a paired comparison to JNDs of an arbitrary certainty level. Equations 3.4–3.5 are the analogues of Eqs. 3.2–3.3, but for JND increments rather than JNDs.

JNDs and JND increments are fundamentally defined in terms of the outcomes of paired comparison experiments. Although these experiments have a dynamic range not greatly exceeding 1.5 50% JNDs, they may be used to calibrate rating scales covering much wider ranges of quality, as discussed in Chs. 5 and 6.

4
Quantifying Preference

4.1 Introduction

In the previous chapter, JNDs were defined in the context of detection. Unlike artifactual attributes, which degrade quality when detected, preferential attributes are essentially always visible, and there is a range of preferred degrees of the attribute depending upon observer and scene. It is, therefore, clear that the term "noticeable" in just noticeable difference is not appropriate for application to preferential attributes. This observation raises the question whether the results of the previous chapter may even be applied to preferential attributes.

As will be shown in this chapter, it is indeed possible to do so, and the meaning of JNDs of preference can be well understood through consideration of two functions:

1. the preference distribution, a probability density function (PDF) that characterizes the relative frequency of preference of different degrees of an attribute, for some set of observers and scenes; and

2. the quality loss function, which quantifies how rapidly quality falls off with distance from the scene- and observer-specific optimum.

These functions and their implications regarding product segmentation and customization are the principal topics of this chapter, which is organized as follows. The working definition of JNDs of preference is provided in Sect. 4.2.

48   Chapter 4

Fig. 4.1 Lower contrast rendition of a scene having prominent shaded and sunlit regions.

Quantifying Preference

Fig. 4.2 Higher contrast rendition, having a distinctly different appearance, but a roughly comparable degree of preference.

In the next section, the mean quality loss associated with the variability of preference over observer and scene is expressed as a convolution of the preference distribution and the quality loss function. In Sect. 4.4, analytical approximations of these functions are assumed and a simple equation for quality loss from preference variability is derived. The equations of Sect. 4.4 are applied to a practical example involving contrast preference in Sect. 4.5. Finally, Sect. 4.6 discusses two methods for reducing quality loss arising from preference variability, namely product segmentation and customization of images on a scene- and observer-specific basis.

4.2 JNDs of Preference

In this chapter, perceived contrast will be used to exemplify a preferential attribute. Contrast will be defined more precisely in Ch. 20, but for now it may be regarded as an impression of the degree of separation of different tones in an image. Figures 4.1 and 4.2 show two different tonal renditions of a scene, at lower and higher contrast levels, respectively. The difference in appearance of the two images is sufficient so that nearly all observers can readily see it, so in paired comparisons between the images, there would be few, if any, guesses. Despite this fact, some observers would prefer one image, and some the other, and so the responses of different observers would vary. If comparable numbers of observers preferred each image, the quality of the two images would have to be considered approximately equal, because the outcome would be the same as if no differences were apparent and observers simply had to guess.

Extending this notion, we will define a JND of preference by requiring that it be consistent with a JND of detection. Suppose that for each of a number of scenes, two pairs of images were made, the members of each pair differing from one another in only one attribute. In the first set of pairs one image has a higher sharpness level and the other a lower sharpness level, the two sharpness levels being the same for every pair. In the second set of pairs two levels of contrast are similarly arranged. Representative observers are asked to select the image in each pair in each set that is of higher quality. In the first set, the sharper samples are chosen 75% of the time, so the sharpness positions differ by one 50% JND. In the second set, the higher contrast position is chosen 75% of the time. If an otherwise well-designed imaging system were at the less sharp, lower contrast position, and could be shifted to the sharper position or the higher contrast position for equal cost, which change would be better to make? It doesn't appear to matter. Both changes would lead to an equal degree of selection of the improved product over the existing version.

Quantifying Preference 51

Although in this example one attribute is artifactual and one preferential, they are equivalent in terms of paired comparison outcome. Therefore, it seems natural to speak of the higher contrast position as being preferred by one 50% JND over the lower contrast position, even though every observer may have detected the difference in contrast in every image pair of the second set. By simply regarding a JND as a unit of image quality (or some other attribute), which is applicable for some particular selection of scenes and observers, there appears to be no conflict in using a JND to describe preference. The probability of certainty or detection, p_d, may readily be reinterpreted as a probability of preference. The term just noticeable difference might better be generalized to just significant difference or something similar, but the JND nomenclature is entrenched and little is to be gained by defining a new term. It is sufficient to regard a 50% JND as being a stimulus difference leading to a 75%:25% proportion in paired comparisons, regardless of the nature of the attribute or the origin of the sample differences.

4.3 Preference Distributions and Quality Loss Functions

Preferred contrast may be a function of both the scene and observer. For example, in an image taken on a bright, sunny day with strong shadows, most observers may prefer a lower contrast position because the shadows are lightened and more detail is thereby evident. The higher contrast position may appear too harsh by comparison. In comparison, an image taken on an overcast day with flat lighting may be preferred at the higher contrast position, which causes it to appear crisper. Similarly, some observers may tend, on average, to prefer the "snappy", eye-catching appearance of higher contrast prints, whereas others may favor a more muted and understated rendition.

Suppose an experiment were performed in which a representative group of observers rated the quality of a number of series of images, each series depicting a scene rendered at several different contrast levels. Examining the data from one observer assessing one scene, we would expect to find that there is a preferred contrast level, and that as contrast deviates farther from that scene- and observer-specific optimum, the rated quality drops off. Intuitively, we might further expect that the fall-off in quality would accelerate at greater deviations from the optimum, with a 10% deviation from the optimum causing more than twice as much quality loss as a 5% deviation. The function quantifying the change in quality in JNDs as a function of the deviation from the optimum is called the quality loss function. This function passes through the origin, because the quality loss at zero deviation from the optimum is defined to be zero.

Examining data from other observers and scenes, we might find a general similarity of shape between their quality loss functions. Certainly, the functions of observers more sensitive to contrast might exhibit more rapid fall-off, and some scenes, being more susceptible, might evoke stronger responses to changes in contrast. Nonetheless, the quality loss functions for each observer and scene could be viewed as members of a family, and could be averaged to produce a mean quality loss function for the collection of observers and scenes tested.

The quality loss function quantifies behavior relative to the scene- and observer-specific optimum. To fully characterize preference, a second function is needed that quantifies the distribution of the preferred positions, thereby accounting for variability in preference across different scenes and observers. This function, a PDF, is called the preference distribution. If the preferred attribute value is not correlated with the sensitivity of observers and susceptibility of scenes, then the mean quality loss $\overline{\Delta Q}$ associated with preference variability is the convolution of the preference PDF, denoted h_p, and the mean quality loss function ΔQ_p:

$$\overline{\Delta Q}(\Omega) = \int_{-\infty}^{+\infty} h_p(\Omega') \cdot \Delta Q_p(\Omega' - \Omega) \cdot d\Omega' \qquad (4.1)$$

where Ω is the value of an objective metric or rating scale correlated with the preferential attribute under consideration, Ω' is the analogous variable of integration, and the preference distribution, being a PDF, is normalized to unit area.

Equation 4.1 provides a basis for understanding preferences over a set of scenes and observers. If preferred positions, observer sensitivity, and scene susceptibility are correlated, then Eq. 4.1 must be reformulated as the properly normalized sum of a finite number of convolutions, each over a subset of scenes and observers chosen so that, within the subset, the correlation is sufficiently small to be neglected. Such cases of correlation appear to be rare, although one example is described in Ch. 20. In that case, observers preferring more saturated rendition of foliage are less sensitive to deviations from their preferred position.

Like preferential attributes, artifactual attributes have quality loss functions associated with them, as discussed in Ch. 9, and the shapes of these functions also change depending on observer and scene, as described in Ch. 10; however, the optimum position is replaced by the threshold of detection of the artifact. The preference distribution for an artifact may be regarded as a delta function at the detection threshold, with any variability of detection threshold being included in the quality loss function. This view is reasonable, because detection

Quantifying Preference

threshold is fundamentally related to observer sensitivity and scene susceptibility. One advantage of characterizing preference using these two functions is that shared behavior of artifactual and preferential attributes are all collected in the quality loss function, whereas aspects unique to preference are isolated in the preference distribution function.

4.4 Analytical Approximation of Mean Quality Loss

Equation 4.1 relates the mean perceived quality at some amount of a preferential attribute to the preference distribution and the quality loss function associated with that attribute. If these two functions are measured experimentally (as discussed in the next section), Eq. 4.1 can be solved numerically. To develop some intuition regarding the implications of Eq. 4.1, it is helpful to consider an analytical solution of the equation that might provide an approximate description of actual behavior. Such an analytical solution is derived in this section, and is tested against measured data in the next section.

First, an analytical expression for the preference distribution is needed. For the reasons outlined in Sect. 2.3, in connection with perceptual distributions, the normal distribution is a plausible choice. Rewriting Eq. 2.1, with the mean being equated to the preferred value of objective metric Ω_p, and the standard deviation of preference being denoted by σ_p, yields Eq. 4.2.

$$h_p(\Omega) = \frac{e^{-(\Omega-\overline{\Omega}_p)^2/(2\cdot\sigma_p^2)}}{\sigma_p \cdot \sqrt{2\cdot\pi}} \qquad (4.2)$$

A widely applicable form of a quality loss function will be derived in Ch. 9, but for purposes of this discussion, a simpler function that facilitates analytical integration of Eq. 4.1 will be used. It seems likely that as the attribute deviates from its optimal position, the quality loss is at first slow, but then accelerates with increasing difference of appearance. The simplest function having this sort of behavior is a quadratic function of the form:

$$\Delta Q_p(\Omega) = \Delta Q_1 \cdot (\Omega - \Omega_p)^2 \qquad (4.3)$$

where ΔQ_1 is the curvature of the parabola, which is negative and corresponds to the quality loss associated with a deviation of one objective metric unit from the preferred value. Substituting Eqs. 4.2 and 4.3 into Eq. 4.1, and employing the substitution:

$$u = \frac{\Omega' - \overline{\Omega}_p}{\sigma_p \cdot \sqrt{2}} \qquad (4.4)$$

yields the following expanded integral.

$$\overline{\Delta Q}(\Omega) = \frac{\Delta Q_1}{\sqrt{\pi}} \cdot \int_{-\infty}^{+\infty} (2 \cdot \sigma_p^2 \cdot u^2 + (\Omega - \overline{\Omega}_p)^2 - 2 \cdot (\Omega - \overline{\Omega}_p) \cdot \sqrt{2} \cdot \sigma_p \cdot u) \cdot e^{-u^2} \cdot du \qquad (4.5)$$

The first term yields an integrand of the form $u^2 \cdot \exp(-u^2)$, which is tabulated, and has a value of $\pi^{1/2}/2$. The second term, of the form $\exp(-u^2)$, is also tabulated, having a value of $\pi^{1/2}$. The third term produces an integrand of the form $u \cdot \exp(-u^2)$, which is antisymmetric with respect to the origin and so vanishes when integrated over symmetric limits. With these results, Eq. 4.5 reduces to:

$$\overline{\Delta Q}(\Omega) = \Delta Q_1 \cdot \left(\sigma_p^2 + (\Omega - \overline{\Omega}_p)^2 \right) \qquad (4.6)$$

a remarkably simple result.

Equation 4.6 expresses the mean quality loss over a collection of observers and scenes, at some particular value of objective metric, as the sum of two terms. The first term, which is proportional to the variance of the preference distribution, reflects the fact that the broader the distribution of preference, the less any single position can satisfy all observers in all scenes. The second term, which is proportional to the square of the deviation from the preferred position, in objective metric units, quantifies the quality loss associated with choosing a position that does not represent the best compromise for the particular collection of observers and scenes. The implications of Eq. 4.6 will be discussed in greater depth later in this chapter, but first a practical example of its application will be presented.

4.5 Example of a Preference Analysis

In this section, experimental results regarding the preference of different contrast positions will be analyzed using the equations just discussed. Contrast is correlated with the gradient of a reproduction (see Ch. 20), which may be defined as the derivative of reproduced lightness (as perceived by the observer viewing the image) with respect to original lightness (as perceived by the photographer viewing the scene). The average gradient over a range of tones will serve as our trial objective metric.

An experiment was conducted in a manner yielding assessments calibrated in JNDs of overall quality (using the method described in Ch. 8). Multiple images were prepared from each of twelve scenes, using seven different tonal renderings that produced a range of contrast positions. Nineteen observers rated the resulting images. To construct the preference PDF, the gradient of the highest rated image (of the seven contrast positions) was identified for each scene and observer, and the fraction of times each position was preferred, over all observers and scenes, was computed.

To convert each fraction to a probability density, it was divided by the width of a gradient interval including the sample position. These interval widths were assigned by splitting the spacing between adjacent samples and assuming symmetric intervals at the extreme positions. As an example, this approach would assume that samples with gradients of 0.9, 1.0, 1.2, and 1.5 would be rated highest if the preferred gradient fell in the ranges of 0.85–0.95, 0.95–1.1, 1.1–1.35, and 1.35–1.65, respectively, yielding interval widths of 0.1, 0.15, 0.25, and 0.3, respectively.

The resulting probability densities were plotted against the gradient values, and it was observed that the resulting PDF was noticeably skewed. If Eq. 4.1 were being evaluated numerically, this would be of no consequence, but to apply Eq. 4.6, the preference distribution must be normal, so that Eq. 4.2 is satisfied. To obtain this desired result, the sample gradients were nonlinearly transformed to create a new, arbitrary scale that compressed the expanded side of the gradient distribution and vice versa, thereby removing the skew. This arbitrary scale constitutes an objective metric correlating with perceived contrast and will be referred to as a contrast metric for brevity. The probability densities were recomputed using contrast metric intervals instead of gradient intervals, yielding a new preference distribution of the contrast metric, which was fit well by a normal distribution, as shown in Fig. 4.3.

Although the preference distribution of the contrast metric was approximately normal as desired, there was no guarantee that the quality loss function of contrast metric would be parabolic in shape. This behavior was tested by computing the mean JNDs of quality for each of the seven contrast levels, averaged over all observers and scenes. These mean JNDs were plotted against the contrast metric and fit with a quadratic function, as shown in Fig. 4.4. The fit was quite reasonable, so Eq. 4.3 was satisfied and the analysis could proceed. It is important to recognize that it would not have been acceptable to obtain a parabolic quality loss function by defining a different contrast metric; the objective metrics of Eqs. 4.2 and 4.3 must be identical.

For simplicity, liberties were taken in the above calculations of both the preference PDF and the quality loss function. A more rigorous and consistent

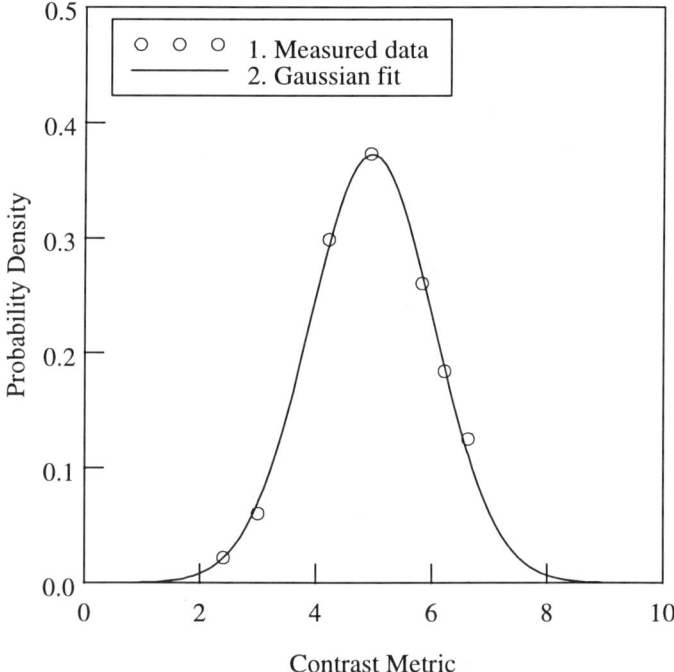

Fig. 4.3 Measured preference distribution of a transform of reproduction gradient, which correlates with perceived contrast. The data is well fit by a normal distribution, as assumed in Eq. 4.2.

Quantifying Preference

approach for calculating the two functions would involve: (1) fitting the JNDs of quality versus contrast metric data for each observer and scene with a peaked function (say, a parabola); (2) computing the preference PDF from the contrast metric values of the peak of each parabola; (3) shifting each data pair (contrast metric, JNDs) by the amounts required to map the peak of its associated parabola to the origin (so that the estimated preferred position coincides with the origin); and (4) computing the quality loss function from the mean shifted data, averaged over scene and observer. Because the data being fit in the first step are usually individual ratings, rather than averaged quantities, assessment noise may be significant, and some poor fits are to be expected.

The standard deviation of the Gaussian fit to the preference distribution of Fig. 4.3 is 1.07 contrast metric units. The curvature of the quadratic fit to the quality

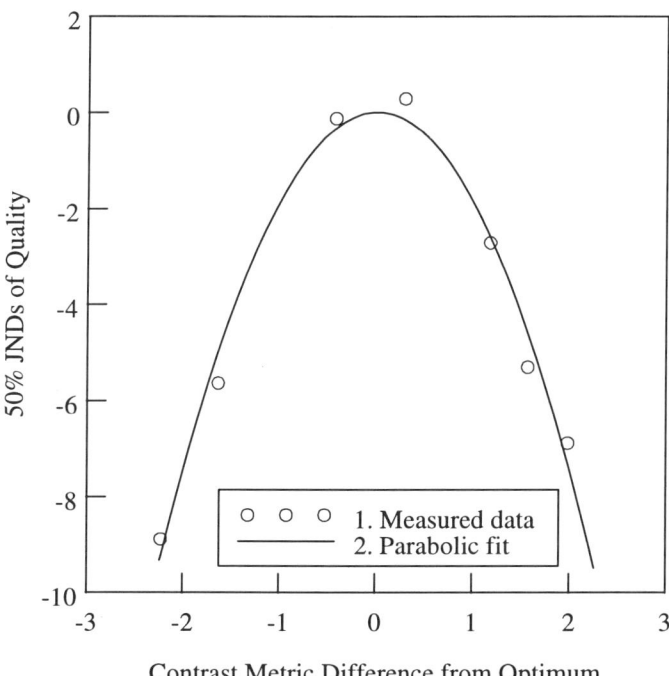

Fig. 4.4 Measured quality loss function of reproduction gradient transformed as in Fig. 4.3. The data is fit well by a quadratic function, as assumed in Eq. 4.3.

loss function in Fig. 4.4 is −1.86 JNDs per contrast metric unit squared. Therefore, from Eq. 4.6, if an imaging system produced a gradient having the optimal value ($\Omega = \Omega_p$), the quality loss arising from the variability of contrast preference among observers and between scenes would be $-1.86 \cdot 1.07^2 \approx -2.1$ JNDs. If the system deviated from the best compromise contrast, a further quality loss would be suffered. For example, if the contrast metric differed from the optimal value by one standard deviation of preference, an additional quality loss of −2.1 JNDs would be incurred.

It might seem that the quality loss arising from the variability of preference is unavoidable, and that the chief aim of the system designer should be to achieve the best compromise so that no quality is lost unnecessarily. However, several mechanisms exist for reducing the quality loss associated with preference variability, as described in the next section.

4.6 Segmentation and Customization

The first way to reduce quality loss from preference variability is to provide versions of a product having different degrees of the attribute in question. This strategy is called segmentation, and it will improve quality if, in practice, the proper version of the product can be identified and utilized sufficiently frequently. The maximum improvement possible from segmentation can be computed by breaking the convolution of Eq. 4.1 into multiple integrals. For example, suppose that two versions of a product are offered, one having $\Omega = \Omega_1$ and one $\Omega = \Omega_2$, where $\Omega_2 > \Omega_1$. Further suppose that, for $\Omega < \Omega_s$, the switch point, Ω_1 is preferred, but above the switch point Ω_2 is preferred. The switch point Ω_s might simply be the average of Ω_1 and Ω_2. If the better choice of version were always made, the mean quality of the segmented product would be:

$$\overline{\Delta Q(\Omega_1, \Omega_2)} = \int_{-\infty}^{\Omega_s} h_p(\Omega') \cdot \Delta Q(\Omega' - \Omega_1) \cdot d\Omega' + \int_{\Omega_s}^{+\infty} h_p(\Omega') \cdot \Delta Q(\Omega' - \Omega_2) \cdot d\Omega' \qquad (4.7)$$

Depending upon the breadth and shape of the preference distribution, the improvement from segregation may be significant, even if there are only two choices. In the example of the previous section, if two choices were provided at roughly ±0.8 contrast metric units from the optimum, with a switch point at the optimum, the quality loss from preference variability would be reduced from −2.2 JNDs to only −0.8 JNDs, according to Eq. 4.7. Even larger improvements might be anticipated if the preference distribution were flatter or bimodal.

In practice, the challenge in segregation is having a choice that is both easy to make and convenient to implement. For example, marketing of parallel films with two levels of contrast, rather than one position of compromise, might offer a significant potential improvement, particularly if preferred contrast depended more on observer than scene, and/or if the scene variations were largely predictable based on factors such as subject matter and lighting (e.g., a preference for lower contrast when photographing people, or higher contrast under dull lighting conditions). These criteria would allow the photographer to select the preferred product version with some accuracy, and to make adjustments accordingly, at least when loading a new roll of film.

There would be even greater leverage for improvement if there were a way to customize the attribute on an image-by-image basis. In a digital camera, the photographer might indicate at the time of capture what degree of contrast he or she expected to prefer in a particular image. This information could be attached to the digital image as metadata, and then could be used during output (e.g., in commercial photofinishing) to adjust the rendering accordingly. As an example of an even more automated embodiment, consider a digital printing algorithm that performed the following operations for each image to be printed:

1. compiled a histogram of scene tones present in the image to characterize its dynamic range and estimate the potential importance of highlights and shadows;

2. looked up, in the user profile of the customer who submitted the order, their average contrast preference;

3. predicted, from the results of the first two steps, the contrast that would be preferred by the customer in that particular image; and

4. modified the tonal rendition of the image accordingly.

The digital printing algorithm described above is an example of image-specific processing because the dynamic range and areas of highlights and shadows of the individual scene are taken into account. It is also an example of observer-specific customization because personal preference is encoded via a user profile. If the digital printing algorithm worked perfectly, each image would be rendered at its scene- and observer-specific optimum, and the quality loss associated with contrast preference variability would be completely eliminated. In practice, such an algorithm would sometimes misestimate the scene effect, because a simple histogram analysis could not reflect all the factors influencing the preferred rendering of an image. Furthermore, the user profile could hardly classify the customer's preferences so accurately and completely that discrepancies would not occur. Nonetheless, the potential leverage is considerable, and customization through image processing algorithms and the use of metadata is a promising area in which substantial research is currently being conducted by many parties.

4.7 Summary

A JND of preference is defined as a stimulus difference producing the same outcome in a paired comparison experiment as would a JND of an artifactual attribute. In this view, the probability of certainty or detection, p_d of Eq. 3.1, is reinterpreted as the probability of preference, and a JND is regarded more generally as representing a just significant difference of quality.

The dependence of quality on preferential attributes may be understood in terms of two functions, the quality loss function and the preference distribution. The latter is a probability density function characterizing the preferred degree of an attribute for a collection of observers and scenes. The quality loss function describes how quality, expressed in JNDs, falls off as the attribute deviates from its preferred scene- and observer-specific position. The mean quality loss associated with the variability of preference may be computed by convolving the preference distribution and quality loss function according to Eq. 4.1.

If the quality loss function is assumed to be quadratic, and the preference distribution to be normal, Eq. 4.1 may be solved analytically to yield the very simple result of Eq. 4.6. It states that the mean quality loss arising from a preferential attribute is the curvature of the quality loss function multiplied by the sum of the variance of the preference distribution and the squared distance from the position of best compromise. A well-designed imaging system should endeavor to minimize or eliminate deviation from this optimum, although a variety of constraints and interactions may render this goal infeasible. The second component of quality loss, which is proportional to the variance of the preference distribution, can be mitigated by segmentation or customization. Segmentation involves offering multiple versions of a product having different attribute values. Customization on an image-by-image basis is enabled by digital image processing algorithms, which may use image metadata, user profile information, and/or measures derived from the digital image itself to accomplish this task.

5
Properties of Ideal Interval and Ratio Scales

5.1 Introduction

As discussed in Sects. 2.4 and 3.4, although paired comparisons form the basis for determination of JNDs, they have a limited dynamic range, and so cannot be used directly to measure sample differences or to characterize quality ranges that exceed a few JNDs. To characterize the distribution of quality produced by practical imaging systems, much wider ranges of quality must be quantified. As discussed in the next chapter, an effective approach to this problem is to establish a standard numerical rating scale anchored to physical samples, calibrate the scale once through selected paired comparisons of the standard samples, and subsequently rate samples directly against the standard scale (how this might be done is described in Chs. 7 and 8). The current chapter describes the properties of two types of numerical rating scales that are particularly useful for this purpose, namely, interval and ratio scales.

The terms interval and ratio scales were coined by Stevens (1946), but there has been considerable disagreement regarding the properties ascribed to these scales and which statistical measures can be rigorously applied to each type of scale (Velleman and Wilkinson, 1993). The viewpoint adopted here is to define the properties of what shall be called ideal interval and ratio scales in terms of their JND increments, because such scales are of limited value unless scale value differences can be interpreted in terms of JNDs. Rating scales obtained in real experiments may approximate ideal interval or ratio scales, but for quantitative research should be calibrated against the results of paired comparisons.

This chapter is organized as follows. Sections 5.2 and 5.3 discuss properties of ideal interval and ratio scales, and Sect. 5.4 demonstrates the relationship between them. Finally, Sect. 5.5 describes how arbitrary rating scales from psychometric experiments can be transformed to produce ideal interval or ratio scales, based on JND increment determinations from paired comparisons.

5.2 Ideal Interval Scales

An ideal interval scale may be defined as a numerical scale in which pairs of samples having equal differences between their scale values appear to be equally different perceptually. For example, suppose that Samples #1–#4 have interval scale values of 10, 15, 30, and 35, respectively, and that higher scale values correspond to higher quality. Sample #2 must be as much better than Sample #1, as Sample #4 is compared to Sample #3, because Sample #2 is rated five units higher than Sample #1, and Sample #4 is also five units higher than Sample #3. Obviously, in an interval scale, a JND corresponds to a constant difference in scale values, because samples one JND apart would have to differ by the same number of scale value units regardless of their absolute scale value.

It is possible to define an infinite number of ideal interval scales that are equivalent to a given scale. For example, if the interval scale in the above example were simply doubled, the difference between Samples #1 and #2, and between Samples #3 and #4, would still be equal (although the difference would now be ten units, rather than five). Likewise, if the original scale were shifted by twenty units, the differences between Samples #1 and #2, and between Samples #3 and #4 would again remain equal (each difference would still be five units). Evidently, linear transforms of an ideal interval scale are still ideal interval scales. In Fig. 5.1 a selection of linear transforms are plotted against an original ideal ratio scale (shown as a solid, 45° line). All the new ideal interval scales are straight lines but the slopes and/or intercepts have been modified. This would hardly be worth plotting except that it provides an interesting contrast with the properties of equivalent ratio scales, described in the next section.

Although invariance of an ideal interval scale under a linear transformation is intuitively obvious, it is helpful to formalize this result to facilitate comparison with ratio scale behavior. Suppose t were an original, ideal interval scale. We seek a general mathematical transform that can generate a new ideal interval scale t'. If both t and t' are ideal interval scales, then sample pairs differing by equal amounts in one scale must also differ by equal amounts in the other scale, although the constant difference of the two scales need not equal each other. This criterion may be written as:

Properties of Ideal Interval and Ratio Scales

$$\frac{\iota_4 - \iota_3}{\iota_2 - \iota_1} = 1 \implies \frac{\iota'_4 - \iota'_3}{\iota'_2 - \iota'_1} = 1 \tag{5.1}$$

The first relationship of Eq. 5.1 states that the difference between Samples #4 and #3 is equal to the difference between Samples #2 and #1 on the original scale. This result implies that these same differences must be equal to each other on the new scale as well, which is embodied in the second relationship of Eq. 5.1. If the new scale were related to the original scale by a linear transform having an arbitrary slope c_1 and intercept c_2, as in Eq. 5.2:

$$\iota' = c_1 \cdot \iota + c_2 \tag{5.2}$$

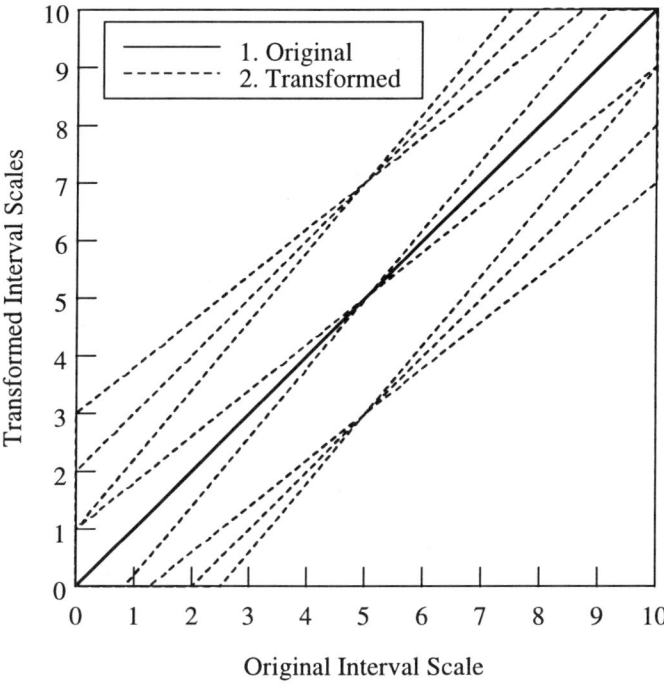

Fig. 5.1 Equivalent ideal interval scales. These scales are related by linear transforms (Eq. 5.2), and so may be shifted or scaled relative to one another.

then the criterion of Eq. 5.1 would be met, because:

$$\frac{t'_4 - t'_3}{t'_2 - t'_1} = \frac{(c_1 \cdot t_4 + c_2) - (c_1 \cdot t_3 + c_2)}{(c_1 \cdot t_2 + c_2) - (c_1 \cdot t_1 + c_2)} = \frac{t_4 - t_3}{t_2 - t_1} \quad (5.3)$$

where the last equality is based on the first relationship in Eq. 5.1, which causes Eq. 5.3 to agree with the second relationship in Eq. 5.1.

5.3 Ideal Ratio Scales

We will now consider a second useful type of numerical scale, the ratio scale, in a fashion analogous to that employed above with regard to interval scales. An ideal ratio scale is defined as a numerical scale in which pairs of samples having equal ratios (rather than differences) between their scale values appear to be equally different perceptually. For example, suppose that Samples #1 through #4 have ratio scale values of 10, 20, 50, and 100, respectively, and that higher scale values correspond to higher quality. Sample #2 must be as much better than Sample #1, as Sample #4 is compared to Sample #3, because Sample #2 has twice the scale value of Sample #1, and Sample #4 also has twice the scale value of Sample #3. In the case of a ratio scale, a JND "increment" evidently corresponds to a constant ratio or percentage of scale values, e.g., a factor of 1.1× if a 10% difference corresponds to one JND.

As with interval scales, an infinite number of ideal ratio scales may be defined so that they are equivalent to a given reference scale. For example, if the ratio scale in the above example were doubled, it would still be true that the ratios between Samples #1 and #2, and Samples #3 and #4, would be equal to each other (and would still equal a factor of two). However, if a constant of twenty were added to the original ratio scale, the ratios between Samples #1 and #2, and Samples #3 and #4 would become unequal (40/30 ≈ 1.3; 120/70 ≈ 1.7). However, there is a second operation that preserves ratio values, namely, raising the scale values to a constant power. For example, if the original ratio scale were squared, the ratios of Samples #2 to #1 and #4 to #3 would be 400/100 = 4 and 10000/2500 = 4, which are equal as required for a ratio scale. In Fig. 5.2, a sampling of power transforms are plotted against an original ideal ratio scale (shown as a solid, 45° line). A common intercept at the origin is maintained, but the slopes at the origin and the degree of curvature now vary.

Proceeding as we did in the previous section, let ρ be an original, ideal ratio scale. We seek a general mathematical transform that can generate a new ideal

Properties of Ideal Interval and Ratio Scales

ratio scale ρ'. If both ρ and ρ' are ideal ratio scales, then sample pairs having equal rating ratios in one scale must also have equal rating ratios in the other scale, although the constant rating ratios of the two scales need not equal each other. This criterion may be written as shown in Eq. 5.4.

$$\frac{\rho_4/\rho_3}{\rho_2/\rho_1} = 1 \Rightarrow \frac{\rho'_4/\rho'_3}{\rho'_2/\rho'_1} = 1 \qquad (5.4)$$

The first relationship of Eq. 5.4 states that the ratio of scale values of Samples #4 and #3 is equal to the ratio of values of Samples #2 and #1 on the original ratio scale. This result implies that these same sample ratios must also be equal on the new ratio scale, which is embodied in the second relationship of Eq. 5.4.

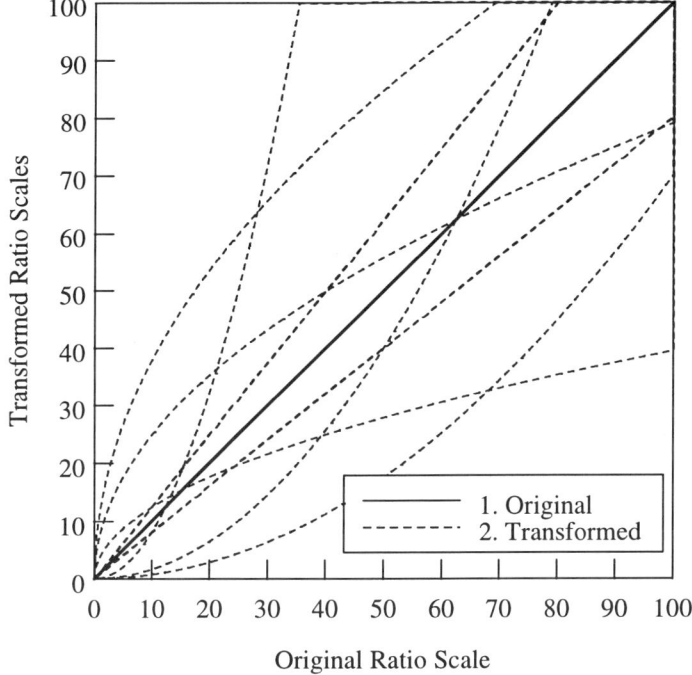

Fig. 5.2 Equivalent ideal ratio scales. These scales are related by power transforms (Eq. 5.5), and so may be scaled or "warped" relative to one another; however, they share a common intercept at the origin.

If the original and new ideal ratio scales were related by a power transform,

$$\rho' = c_3 \cdot \rho^{c_4} \tag{5.5}$$

then the criterion of Eq. 5.4 would be met, because:

$$\frac{\rho'_4/\rho'_3}{\rho'_2/\rho'_1} = \frac{(c_3 \cdot \rho_4^{c_4})/(c_3 \cdot \rho_3^{c_4})}{(c_3 \cdot \rho_2^{c_4})/(c_3 \cdot \rho_1^{c_4})} = \left(\frac{\rho_4/\rho_3}{\rho_2/\rho_1}\right)^{c_4} = 1^{c_4} = 1 \tag{5.6}$$

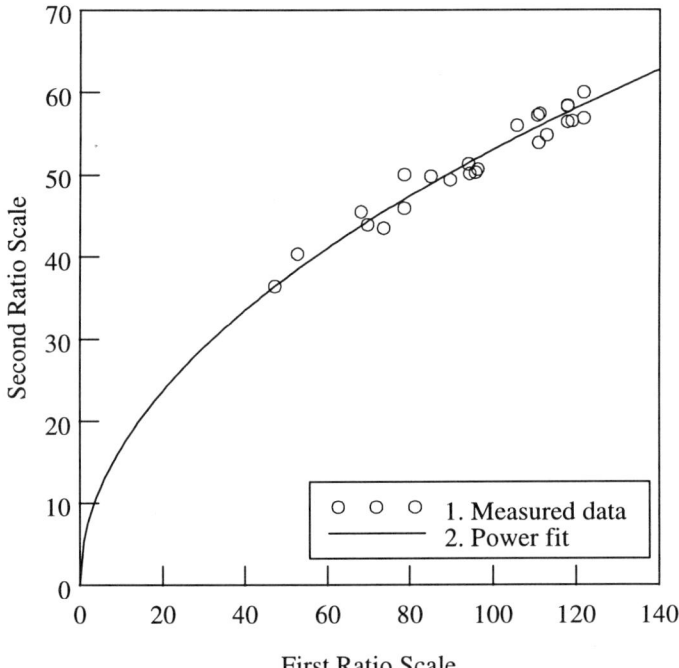

Fig. 5.3 Comparison of approximate ratio scales from two different magnitude estimation experiments, involving the same attributes but different levels, scenes, observers, viewing conditions, and instructions. The relationship between two ratio scales is well fit by the power transform of Eq. 5.5; note how this transform preserves the origin.

Properties of Ideal Interval and Ratio Scales 67

where the next to last equality is based on the first relationship in Eq. 5.4, which causes Eq. 5.6 to agree with the second relationship in Eq. 5.4.

Knowing that ideal interval scales should be related by linear transforms, and that ideal ratio scales should be related by power transforms, is useful when trying to compare or combine the results of different psychometric experiments. Figure 5.3 shows an example of such an analysis involving two approximate ratio scales derived from two different experiments. Each experiment used the method of magnitude estimation, discussed in Sect. 7.4, in which observers are asked to provide a rating proportional to quality, compared to a reference image. The two experiments, which examined the impact of sharpness and noisiness on overall image quality, involved different scenes, observers, viewing conditions, and instructions. The relationship between two approximate ratio scales is well fit by a power function of the form given in Eq. 5.5, shown as a solid line in Fig. 5.3. The extrapolation back to the origin is shown to emphasize that the power transform preserves the origin, as required for ideal ratio scales. Although a linear fit would have been acceptable in terms of residual error over the range of the experimental data, the resulting line would have had a large positive intercept, and extrapolations based on the straight line would have been unreliable.

5.4 Relationship of Ideal Interval and Ratio Scales

There is a simple relationship between ideal interval and ratio scales of a given attribute. In an ideal interval scale, a JND is a constant difference, whereas in an ideal ratio scale it is a constant fraction or percentage. Therefore, if we formed a new scale that was the logarithm of an ideal ratio scale, it would have a JND corresponding to a constant difference, and therefore would be an ideal interval scale. More formally, as shown in Eq. 5.7, by taking the logarithm of the first relationship in Eq. 5.4, we can recover the first relationship of Eq. 5.1 by identifying $\ln(\rho_i)$ with ι_i for all i, proving that the logarithm of an ideal ratio scale is an equivalent ideal interval scale.

$$\ln\left(\frac{\rho_4/\rho_3}{\rho_2/\rho_1} = 1\right) \Rightarrow \frac{\ln\rho_4 - \ln\rho_3}{\ln\rho_2 - \ln\rho_1} = 1 \Rightarrow \ln\rho = \iota \qquad (5.7)$$

As a check of this result, note that if the logarithm of Eq. 5.5 is taken, the resulting equation shows that equivalent ideal interval scales are related by linear transforms, as expected based on Eq. 5.2. Any ideal ratio scale and any ideal interval scale can be related by a transform involving only two unknown constants, because the scale can be logged and linearly transformed (ratio to interval) or exponentiated and power transformed (interval to ratio).

An example supporting this result is shown in Fig. 5.4. This figure compares an approximate ratio scale of quality (y-axis) from a magnitude estimation experiment, to an approximate interval scale (x-axis). The interval scale is from a categorical sort experiment, discussed in Sect. 7.3, in which observers sort samples into bins that are intended to be approximately uniformly spaced perceptually. The two experiments involved the same samples but different observers, viewing conditions, and instructions. The sample treatments involved emulation of the effects of different film exposures spanning a wide range of

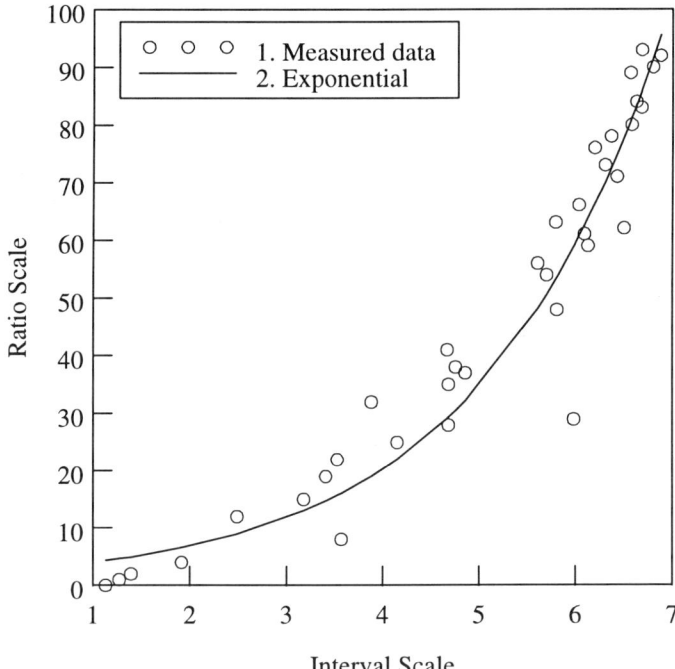

Fig. 5.4 Comparison of a ratio scale of quality (y-axis) to an interval scale (x-axis) from two different experiments, involving the same samples but different observers, viewing conditions, and instructions. The transform between the two scales is well fit by an exponential function, as predicted by Eq. 5.7.

Properties of Ideal Interval and Ratio Scales 69

quality. The transform between the two scales is fit reasonably well by an exponential function, as predicted by Eq. 5.7.

Equation 5.7 emphasizes the equivalence of ideal interval and ratio scales from a mathematical viewpoint. Which type of scale is used is largely a matter of preference. Because image quality position differences are usually expressed in JNDs, an interval scale might be preferred because differences on the scale can easily be converted to JNDs. In fact, the scale could be defined so that one unit was one JND for convenience. Some care is required so that scale values and scale value differences not be confused, the possibility of which can be avoided if a large offset is added to the interval scale. For example, if some high image quality position were defined to have an interval scale value of one hundred and the scale had a unit JND increment, ratings would usually be much larger numerically than rating differences.

5.5 Calibrating Interval and Ratio Scales

Although certain psychometric methods are intended to yield interval or ratio scales directly or indirectly (see Ch. 7), there is no guarantee that the resulting scales will have the properties here ascribed to ideal interval or ratio scales. Therefore, it is of interest how an arbitrary rating scale might be converted to an ideal interval or ratio scale with known JND increments or ratios. This may be accomplished by calibrating the rating scale against the results of selected paired comparisons. First, JND increments are measured at a number of scale values, using Eq. 3.4 or Eq. 3.5. Next, the JND increments are fit by a smooth function of the rating scale value. If the rating scale were an ideal interval scale, the plot would be a flat line; if it were instead an ideal ratio scale, the plot should be a sloped line passing through the origin. Therefore, if the rating scale has any sort of reasonable behavior, the dependence of JND increment on rating scale value is likely to be linear or fairly close thereto, so exotic mathematical functions should not be required to fit the relationship.

Denoting the arbitrary rating scale s, and the JND increment function by $\Delta s_J(s)$, an ideal interval scale ι having a constant JND increment of $\Delta \iota_J$, and a value of ι_r at the reference rating scale value s_r, can be constructed by the following equation.

$$\iota(s) = \iota_r + \Delta \iota_J \cdot \int_{s_r}^{s} \frac{ds'}{\Delta s_J(s')} \quad (5.8)$$

The definite integral is zero when $s = s_r$ so s_r maps to t_r. Starting from this position, an infinitesimal change in s', the variable of integration, is divided by the JND increment at s' to yield an infinitesimal portion of a JND. These infinitesimal portions of JNDs are accumulated up to the point s to yield the cumulative JNDs between s_r and s. This value is scaled by the chosen parameter Δt_J, which thus becomes the JND increment of the ideal interval scale. For example, if it were desired to have a scale where one JND corresponded to ten scale units, so that only integer values were needed to achieve a precision of 0.1 JNDs, Δt_J would be chosen to be ten. If a ratio scale were desired instead, this interval scale could be exponentiated. Then s_r would map to $\exp(t_r)$ and a JND would correspond to a factor of $\exp(\Delta t_J)$. Thus, if a JND were desired to

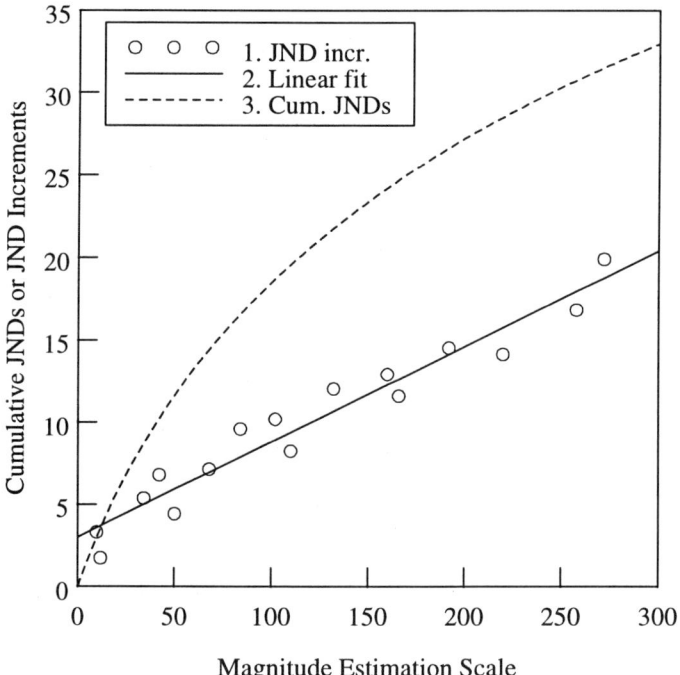

Fig. 5.5 Variation of JND increments within an approximate ratio scale, and cumulative JNDs from Eq. 5.8. The JND increments are linearly related to the magnitude estimation scale, but the intercept is not zero, so the scale is not an ideal ratio scale.

Properties of Ideal Interval and Ratio Scales

correspond to a 10% change, Δt_J would be chosen to be $\ln(1.10) \approx 0.095$.

Figure 5.5 shows an example of this procedure for an approximate ratio scale of overall quality obtained from a magnitude estimation experiment. Groups of samples having numerically close ratings, at different points along the approximate ratio scale, were assessed in a paired comparison experiment. Equation 3.5 was used to determine the JND increments of the rating scale at different positions along the scale. As seen in Fig. 5.5, the data is fit quite well by a straight line, but neither is the slope zero, as in an ideal interval scale, nor is the intercept zero, as in an ideal ratio scale. Evidently, the rating scale is intermediate between a ratio and an interval scale, albeit closer to the former, as expected from a magnitude estimation experiment. Substituting the regression line (Curve #2) into Eq. 5.8 and selecting $s_r = 0$, $t_r = 0$, and $\Delta t_J = 1$ yields the cumulative JNDs in Curve #3, with an original rating of zero corresponding to zero JNDs (this condition being arbitrary). Although a JND scale is of greatest utility, any other desired ideal interval scale could be obtained by linearly transforming the scale of Curve #3, and any desired ideal ratio scale could be generated by exponentiating the corresponding interval scale.

Calibration of a numerical rating scale using Eq. 5.8 may be particularly valuable if the scale is to be adopted as a standard, against which subsequent assessments will be made, as described in the next chapter.

5.6 Summary

Paired comparison experiments are the best method available for determining small sample differences in terms of JNDs; however, to characterize the distribution of quality produced by practical imaging systems, much wider ranges of quality must be quantified. This may conveniently be done through rating experiments of various types, which yield numerical scales of the attribute being tested. The most useful types of numerical rating scales are interval and ratio scales.

Ideal interval scales are characterized by constant JND increment differences; equivalent interval scales are related by linear transforms. In contrast, ideal ratio scales are characterized by constant JND increment ratios, and equivalent ratio scales are related by power transforms. The logarithm of an ideal ratio scale will form a corresponding ideal interval scale.

Most psychometric experiments are intended to produce an interval scale or a ratio scale, either directly, by observer estimation, or inferentially, through data analysis based on probabilistic models. Nonetheless, the actual rating scales obtained do not necessarily have the properties ascribed to ideal interval or ratio scales as defined above. Arbitrary rating scales may be converted into ideal interval or ratio scales with desired JND increments and reference quality positions through the use of Eq. 5.8, based upon JND increment data from paired comparison experiments.

6

Establishing Image Quality Standards

6.1 Introduction

Designation of physical standards and definition of associated numerical scales are requisites to conducting quantitative research on, and making predictions of, image quality. The establishment of a standardized scale of quality has the following benefits.

1. Such a scale promotes unambiguous communication of image quality levels between those involved in research, development, management, marketing, and manufacturing.

2. Associated physical standards anchor the numerical scale, preventing drift over time, and providing reference samples for incorporation into psychometric experiments.

3. The properties of a standard scale may be investigated once, and then applied to the results of any number of experiments that are calibrated against the scale.

4. The results of calibrated experiments can be compared and combined rigorously, even if they are done at different times and/or by different individuals.

In this chapter, a procedure for establishing image quality standards is described and a number of attendant considerations are discussed. The chapter is organized as follows. Section 6.2 provides an overview of the recommended process for establishing reference standards. The relative merits of standards that vary in a single attribute (univariate) compared to those that differ in multiple attributes (multivariate) are considered in Sect. 6.3, which also discusses the important distinction between univariate and multivariate JNDs. Finally, the utility and limitations of mapping numerical scale values to adjectives such as excellent, acceptable, etc. are explained in Sect. 6.4.

6.2 Procedure for Constructing Calibrated Standards

The intent of the process described in this section is to generate a physical standard and an associated numerical scale of quality meeting the following two criteria.

1. The physical standard should be largely free of ambiguous samples, so that it may be used either by experts or representative observers with consistent results.

2. The numerical scale should have a known and constant JND increment that is characteristic of the observers and images for which perceived image quality inferences are to be made, so that interpretation of scale differences is straightforward.

One procedure that meets these criteria and has proven to be practical consists of the seven steps below.

1. Samples from which physical standard will be constructed are prepared or collected.

2. Expert observers rate the samples in a fashion yielding an approximate interval or ratio scale.

3. Representative observers evaluate the samples, possibly by a categorical sort procedure, to provide mappings to adjectival descriptors such as excellent, acceptable, etc.

4. The experts' scale values are regressed against the representative observers' ratings to identify and eliminate ambiguous samples.

Establishing Image Quality Standards 75

5. Representative observers perform paired comparisons of similarly rated samples to determine JND increments at different scale positions.

6. Equation 5.8 is applied to convert the experts' rating scale to an ideal interval or ratio scale (the latter via subsequent exponentiation), which constitutes the standard numerical scale of image quality.

7. The physical standard is constructed by ordering and labeling selected samples according to their associated standard numerical scale values.

It may seem that this approach, which entails three psychometric experiments, is more complex than is necessary; however, to meet the stated criteria, such an involved process is often needed. One rating experiment spanning the full range of quality is required to generate a preliminary scale; a second rating experiment is needed to check expert and representative observer consistency; and the paired comparison experiment is required to determine the numerical scale JND increments for representative observers. One rating experiment (Step #2 or Step #3) and the regression analysis (Step #4) may be omitted if the consistency of expert and representative observers is not required or need not be confirmed (see Sect. 6.3). The expert and representative observer experiments may also be combined into a single larger experiment, although there are some advantages associated with maintaining their separation, as discussed later in regard to adjectival descriptors. These seven steps are now considered in greater detail.

In the first step, a range of physical reference samples is prepared for assessment. The number of samples should be sufficient so that, at different quality levels, some pairs of samples are very similar in quality, permitting paired comparison experimentation later in the process. It may be helpful for the reference samples to contain a variety of scenes so that when test samples are subsequently compared to the standard generated, it is likely that reasonably similar scenes will be found. The reference samples should physically resemble the types of test samples that are likely to be rated against them. For example, if both continuous-tone three-color and halftone four-color test samples will be assessed against the standard, it may be helpful to have both types of samples represented in the reference series. Furthermore, as new imaging materials and display methods become available, it may be desirable to augment existing physical standards with additional samples. Discussion of the advantages and disadvantages of using samples that vary in a single attribute, compared to those differing in a number of attributes, is deferred to Sect. 6.3.

The second step involves experts, or carefully trained observers, evaluating the reference samples in a fashion yielding an approximate interval or ratio scale. As mentioned in the preceding chapter, approximate interval scales can be inferred from categorical sort and rank order data, and approximate ratio scales should result from magnitude estimation. A discussion of the relative merits of these methods is deferred to the next chapter; the author's personal preference is to use magnitude estimation in this step. The primary reason for seeking an approximate interval or ratio scale is that its JND increment can be expected to vary in a simple fashion with the scale value. If the scale obtained were an ideal interval scale, the JND increment would be a constant; if it were instead an ideal ratio scale, the JND increment would be proportional to the scale value. Even if the scale obtained were intermediate between ideal interval and ratio scales, its JND increment could be fit by a linear function of the scale value. The simpler the function that can be used to fit the JND increment versus scale value relationship, the more robust the application of Eq. 5.8 in Step #6 in is likely to be, particularly near the extremes of the scale, where some extrapolation may be required.

The third step is to have the same reference samples rated by observers representative of the group or groups for which image quality inferences are to be made. For example, observers might be chosen from among individuals untrained in photography to ensure relevance of the standard scale for characterizing perceived quality among traditional consumers. As in Step #2, different experimental rating methods may be employed, but it is less critical whether an approximate interval or ratio scale is obtained, because the results may be transformed to the experts' scale if desired, usually without extrapolation. In this instance, a categorical sort technique may be slightly preferable to magnitude estimation because of the simplicity of the task and the potential for mapping of adjectival descriptors to scale values. To achieve the latter, the bins into which the samples are sorted are labeled with adjectives such as excellent, good, etc. and the data is analyzed to determine the scale values at the boundaries between the descriptors. A knowledge of adjectival descriptor boundaries can be useful in interpreting numerical scale values of image quality, but there are associated pitfalls, as discussed in Sect. 6.4.

The fourth step is to regress the numerical ratings from the representative observers in Step #3 with the approximate ratio or interval scale from the expert observers in Step #2. Samples falling unusually far from the regression curve should be excluded from the physical standard because they represent samples that may be perceived differently by different types of observers. It is desirable to identify the characteristics that might lead to such discrepancies, because such properties may provide interesting topics for future study. It also is advisable to eliminate samples having an unusually high degree of variability in ratings by either group of observers, for similar reasons. With these exclusions made, the remaining samples should be satisfactory candidates for inclusion within the final standard. The ratings of the representative observers may be converted to

Establishing Image Quality Standards

equivalent experts' scale values using the regression equation, and then the data from either or both of the groups may be used in subsequent analyses. If a categorical sort experiment was done with the representative observer group, the experts' scale values corresponding to boundaries between the adjectival descriptors may also be determined.

The fifth step is to perform paired comparison experiments between samples with very close scale values, from different regions of the numerical scale. Although scene and observer selection is addressed in greater detail in Ch. 10, pertinent considerations are summarized briefly here. Because the standardized scale of quality will be calibrated using the JND increments derived from this experiment, it is critical that both the observer group and scene set be as representative as is possible, to ensure maximum relevancy. Characteristics that might be considered in making a balanced selection of observers include level of experience, primary imaging applications, types of equipment used, frequency of use, demographic factors, etc. The selection of observers in the two rating experiments is less important, because their overall sensitivity to image quality is normalized out during the JND increment calibration process (see Sect. 10.3); however, variations of relative sensitivity to different attributes are preserved. The scenes selected should reflect the variety of subject matter, lighting, composition, etc. encountered in practice. Some scenes display particular attributes more noticeably than others; such scenes are said to be more susceptible to that attribute. For example, scenes with slowly varying areas, such as blue sky, exhibit noise to a greater degree than scenes with more high-frequency detail, which visually masks the noise. By choosing a representative selection of scenes, an appropriate mixture of susceptibilities to a variety of attributes may be achieved. The nature of variation of image quality attributes in the samples significantly affects JND increments and is discussed in Sect. 6.3.

The sixth step is the transformation of the experts' approximate interval or ratio scale to an ideal scale having the desired reference position and a constant JND increment (either as a difference in an interval scale or as a percentage in a ratio scale). This is accomplished using the results of the paired comparisons and Eq. 5.8, as in the example of Fig. 5.5. Equation 5.8 yields an ideal interval scale, which may be exponentiated to yield an ideal ratio scale if desired.

The seventh and final step is to assemble the physical standard from the reference samples found to have low judging variance and to be assessed consistently by experts and representative observers (Step #4). Although closely spaced samples were required for the paired comparisons, there is little advantage to including, in the final standard, samples of similar scenes that are so close in quality that individual observers would frequently disagree about their ordering. Display size restrictions may also limit the number of samples in the final standard, which should be convenient to use and easy to transport. The final samples can be arranged in many possible fashions, but at a minimum they should be sorted in an ascending or descending sequence (perhaps within subsets sorted by scene), and should be labeled with the associated numerical scale value of the ideal interval or ratio scale.

6.3 Univariate and Multivariate Standards and JNDs

The reference samples from which a standard is constructed may vary in many different attributes or may vary in a single attribute. One potential advantage of the multivariate approach is that, when assessing a test sample the quality of which is significantly affected by a particular attribute, reference samples exhibiting that same attribute are likely to be found in the standard, perhaps aiding in the evaluation. In practice, this benefit seems to be quite limited; we have extensive data showing that both the accuracy and precision of assessments made by experts using multivariate and univariate standards is similar. Furthermore, accuracy and precision approaching that of trained experts has been obtained from inexperienced observers using a univariate standard, the quality ruler, which is described in Ch. 8.

Three significant advantages associated with univariate standards are listed below.

1. An individual observer is unlikely to perceive the quality ordering of the reference samples to contain inversions of scale values unless the reference samples are extremely similar in appearance. Such inversions can easily happen with multivariate samples because different observers may have different sensitivities to the attributes represented.

2. For the same reasons that inversions are unlikely in univariate standards, discrepancies between experts and representative observers are not expected for univariate standards, except perhaps between different scenes because of their content. Consequently, one rating experiment and the regression analysis might be omitted from the seven-step procedure of Sect. 6.2 if the standard were chosen to be univariate.

3. After an initial calibration, the standard scale values of new reference images can be predicted based largely, or entirely, on a single objective measurement. This can facilitate convenient, calibrated production of new standards at any time, using routine image simulation techniques.

Establishing Image Quality Standards

The first point, regarding inversions, is worthy of additional discussion, because perception of strong inversions can undermine confidence in a physical standard. Consider two reference images in a standard, Sample #1 and Sample #2, with Sample #1 having a slightly higher scale value, indicating higher quality. Suppose that Samples #1 and #2 are each otherwise excellent images that have a single (but different) artifactual attribute degrading their quality, these being Artifacts #1 and #2, respectively. Because the scale values are based on the mean response of a group of observers, a majority of observers (but perhaps only a slight one) will perceive Sample #1 to be of higher quality. A minority (but, again, perhaps only a slight one) will regard the scale values as being inverted because they are relatively more sensitive to Artifact #1 than #2, compared to the mean observer. This minority of observers cannot assign a meaningful scale value based only on Samples #1 and #2 because of their inversion. This situation will not arise with univariate standards because sensitivity variation among observers does not change the perceived ordering of the samples, but merely the magnitude of the perceived differences between them. Of course, if two univariate reference samples were close enough together, their ordering could be confused because of overlap of the perceptual distributions, as discussed in Sect. 2.2. But in the multivariate case, additional confusion results from the variation of ratios of sensitivities to different attributes.

This analysis suggests that a JND increment in multivariate samples assessed for quality will be larger than that in univariate samples. This effect might be augmented because in the assessment of univariate samples, the attention of the observer is focused on the one attribute that varies, so that the measurement is one of detectability, rather than significance (in terms of impact on quality). The JND increment determined in a univariate experiment quantifies the ability to distinguish different degrees of the attribute, as measured by some rating scale or objective metric. In contrast, the JND increment determined in paired comparisons of multivariate sample quality measures the extent to which attribute differences matter to the observers. Because a difference must be detected before it can influence quality, the univariate JND increment must be less than or equal to the multivariate JND increment. In the cases of sharpness and noisiness, we have found JND increments from paired comparison assessments of overall quality of multivariate samples to be approximately twice as large as the JND increments found in univariate paired comparisons. The relative importance of the two effects mentioned above, namely, observer variability in relative attribute sensitivities, and detectability versus significance, is not known, but it seems likely that both contribute to the difference of a factor of two that is observed.

Which JND increment is of interest to us? The univariate JND increment is a measure of detectability, and so would be appropriate for critical comparisons of individual attributes in an analytical context, as might be done by scientists or occasionally professionals interested in a single property of the image. For example, researchers developing human visual system models might first try to explain univariate JND increments because they reflect primarily the detection properties of the system, with fewer complications arising from higher order cognitive effects. The multivariate JND increment is a measure of value, and is appropriate for comparisons of the overall quality of images, which are certainly of primary concern in most imaging applications. It might seem that univariate JNDs could be useful in evaluating advertising claims such as "improved sharpness", which direct the attention of the observer to a single attribute. But even in this case, a customer may not be receptive to an improvement that they can see, but do not care about because it is of so little significance in terms of their satisfaction.

Based on the above considerations, we have chosen to calibrate our standard scale of image quality in terms of multivariate JND increments. These have been determined from paired comparisons of samples depicting diverse subject matter and many combinations of attributes that span the variations of images encountered in consumer and professional photography. We have obtained similar JND increments from untrained and expert observers in these experiments. One positive outcome of the choice of multivariate JND increments is that it has been possible to develop a general equation for predicting how multiple attributes combine to determine overall quality (see Ch. 11). It is unclear whether such a prediction could be made directly from univariate JNDs.

The primary use of JND increments is in calibrating standardized scales of quality. Once such a scale exists, results are nearly always expressed in JNDs, so it is worth recasting the above results in these terms. JNDs and JND increments are related inversely (Eq. 5.8), so there will be fewer multivariate JNDs than univariate JNDs for some objective stimulus difference. Restated, it will take a larger objective stimulus change to produce a multivariate JND of quality difference than to produce one univariate JND of detectability. For example, a given noise difference between stimuli will produce approximately twice as many univariate JNDs of noisiness as multivariate JNDs of quality. Hereafter, in this volume, all JNDs will be multivariate JNDs of quality unless otherwise noted. In cases where a quality change is produced by a variation in a single attribute, the change will be referred to as being a certain number of JNDs of quality arising from (or caused by) the attribute; e.g., an improvement in MTF might yield an increase of two JNDs of quality arising from sharpness.

Establishing Image Quality Standards 81

Let us return now to the original question of this section, namely, whether standards should be univariate or multivariate. Because multivariate JNDs of quality have been adopted as our standard units, the paired comparison experiment (Step #5) must be performed with multivariate samples. Nonetheless, the final standards could still be univariate, because only a subset of the samples used in the paired comparisons need be included in the final standard, and those varying in a single attribute could be chosen exclusively. It seems a waste to calibrate samples exhibiting many attributes and then only make use of a small subset of them in the physical standard. Yet, the three advantages of univariate standards listed at the beginning of this section, which are of considerable significance, still are valid, notwithstanding the adoption of a multivariate JND increment calibration. There is no single correct choice in this matter; our approach represents something of a compromise. Our primary physical standard is multivariate and includes the majority of reference samples tested in the paired comparison experiment, and so is composed of a representative set of images from consumer and professional photography. The bulk of our image quality research, however, is carried out using secondary univariate standards that are traceable to the primary multivariate standard. These univariate standards are referred to as quality rulers, and their design and use are motivated and described in the two following chapters.

6.4 Adjectival Descriptors

Once a standard scale of quality is established, the numerical scale should be invariant with time, although as noted previously, physical samples in the standards may need to be augmented or updated occasionally. The time invariance of the scale is critical to enabling the results of experiments performed at different times, perhaps many years apart, to be accurately compared and combined. But temporal invariance of the numerical scale does not imply that the level of satisfaction associated with a particular scale value does not change over time.

Figure 6.1 shows the boundaries between a frequently used set of adjective descriptors and a static scale of JNDs of quality for three cases:

1. 110 format users in 1979, who were shown representative images for that time and format;

2. 35-mm format users in 1990, who again were shown representative images; and

3. the latter group in the same year, shown the same set of images plus additional images, comprising 19% of the total, and having one to six JNDs higher quality than the highest quality sample in the original set (denoted by "Enhanced Quality" in Fig. 6.1).

Looking first at the data from representative sample distributions (left and middle columns of data), the adjective boundaries expanded at higher quality with the later group, so that they required progressively greater scale values at higher quality to assign the same adjectival descriptor as their predecessors. This is exactly what would be anticipated if customer expectations increased as a result of using newer systems yielding higher image quality. Now comparing the

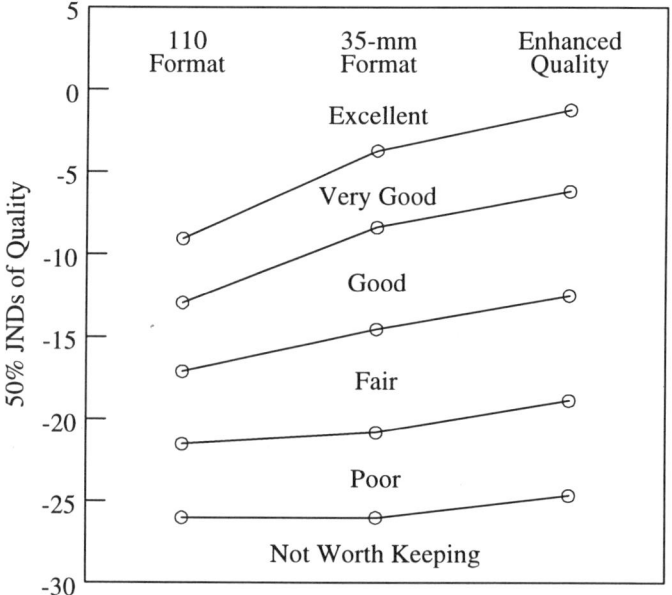

Fig. 6.1 Adjectival boundaries for: (1) 110-format users in 1979; (2) 35-mm format users in 1990; and (3) the same as in (2) but with exposure to samples of very high quality. Comparison of (1) and (2) suggests that user expectations increased as system performance improved; comparison of (2) with (3) exemplifies a range effect.

middle and right columns, it is seen that the addition of a high-quality tail on the distribution of samples presented shifted the application of all of the adjectives by a fairly constant amount, roughly two JNDs. Evidently, the sample distribution affects the fashion in which adjectives are applied, which is one form of a range effect. Range effects of this type plague categorical sort experiments, and make interpretation of adjectival descriptors tenuous. When range effects may be of significance, it is critical that sample set characteristics be as representative as possible, a condition that is not necessarily easy to meet or verify. The magnitude of such range effects limit the usefulness of data from categorical sort experiments unless subsequent calibrations are performed.

One might now reasonably ask whether the first comparison, between the representative distributions in 1979 and 1990, is not simply another example of a range effect. Restated, would the 110 format users in 1979 have given answers similar to those of the 35-mm format users in 1990, had they been shown the same set of samples? The best evidence that the 1979 to 1990 change is not a range effect is that the shifts in category boundaries ranged widely (0.0 JNDs at low quality to 5.4 JNDs at high quality), whereas the shifts from the enhanced quality range effect varied little (1.3 to 2.5 JNDs). This argument is plausible but hardly compelling, and demonstrates the difficulty of interpreting adjectival mappings.

Nonetheless, adjectival mappings are useful for developing some intuition regarding the level of satisfaction associated with different numerical scale values. They are perhaps even more helpful in envisioning the magnitude of differences represented by multiple JNDs of difference; for example, the average width of a bounded adjective category in the 1990 data in Fig. 6.1 is about 6 JNDs. The bounded categories vary slightly in their widths; more perceptually uniform, but less familiar, adjectives have been identified (Zwick, 1984). Another set of adjectival descriptors used in describing artifact severity is undetectable, detectable, objectionable, and not worth keeping. From data for a number of digital artifacts, the average boundaries between these descriptors correspond to about 4, 9, and 15 JNDs of quality loss, respectively. Some inconsistency between the application of adjectives to individual attributes and to overall quality is evident, as the 1990 excellent/very good to poor/not worth keeping boundary difference is about 22 JNDs (Fig. 6.1), whereas that between undetectable/detectable and objectionable/not worth keeping, which might be expected to be of similar magnitude, is only 11 JNDs.

Once a standardized scale of image quality is established, the next problem faced is that of performing psychometric experiments that are calibrated to the scale. This topic is the focus of the next two chapters.

6.5 Summary

Numerical scales of image quality associated with physical standards constitute a fundamental and important tool in image quality characterization. Such scales promote effective communication regarding image quality levels between different individuals and disciplines. Furthermore, results of different psychometric experiments can be calibrated against such a common scale, permitting them to be rigorously compared or combined.

One procedure for establishing a standardized scale of image quality involves the following steps:

1. selection or preparation of physical samples;

2. rating of quality of samples by expert observers in a fashion yielding an approximate interval or ratio scale;

3. rating of quality of samples by representative observers, often by a categorical sort to obtain adjectival mappings;

4. conversion of the representative observers' results to the experts' numerical scale to identify ambiguous or variably assessed samples, which are discarded;

5. paired comparison of samples with similar scale values by representative observers to determine JND increments as a function of scale value;

6. application of Eq. 5.8 to transform the experts' scale to an ideal interval or ratio scale; and

7. creation of a physical standard from a subset of the samples in the paired comparison experiment.

The physical standards may be multivariate or univariate, although the latter has several key advantages, including freedom from perceived inversion of scale values and simplicity of generation and calibration of new physical standards. The samples used in the paired comparison experiment, however, should be multivariate in nature if the JND increment calibration is to reflect assessments of overall quality, rather than detectability of an individual attribute.

Establishing Image Quality Standards

Although numerical scales of quality should be invariant with time if they are anchored to a physical standard, customer expectations may evolve with time. Such effects can be monitored by occasional categorical sort experiments involving selected adjective descriptors, although the interpretation of such results is not straightforward. As a rough measure of the significance of multiple JNDs of quality, the average width of the bounded adjectival categories in the sequence excellent, very good, good, fair, poor, and not worth keeping is approximately six JNDs.

7

Calibrated Psychometrics Using Quality Rulers

with Robert E. Cookingham
Eastman Kodak Company
Rochester, New York

7.1 Introduction

In the previous chapter, we saw that establishing a standard numerical scale of image quality that is tied to physical standards provides many advantages. Yet, the existence of a numerical scale and an associated physical standard does not solve the problem of how to perform experiments in such a manner that their results are calibrated to the standard scale. There are a variety of psychometric techniques available for assessing a series of samples and expressing the results on some type of rating scale, but none of these are fully satisfactory for producing assessments that are calibrated to a standardized scale.

In most cases, the best that can be done is to incorporate within the psychometric experiment, a substantial number of reference samples from a physical standard, which then reduces the number of test samples that can be investigated in a given time. In such cases, a regression is run between the experimental scale values of the included reference samples and their known standardized scale values, and the regression equation is applied to the experimental scale values of the test samples to map them to standardized scale values. This approach is inefficient, especially with a small number of test stimuli (which may then be outnumbered by reference samples), so a better method of obtaining calibrated psychometric data is clearly desirable, especially if a number of investigations are to be conducted.

This chapter reviews the reasoning leading to the development of a new psychometric method, the quality ruler, which is designed specifically to obtain results calibrated to a standardized scale in an efficient manner. The chapter is organized as follows. Sections 7.2–7.4 describe selected psychometric techniques (paired comparison, rank ordering, categorical sort, magnitude estimation, and difference estimation) and analyze their suitability for evaluating samples of widely varying quality against a standard numerical scale of quality. None are found to be ideal for this application, so Sect. 7.5 introduces the concept of a quality ruler, which is specifically designed for calibrated quality assessments. The properties that ideally should be possessed by an attribute varying in a quality ruler are enumerated in Sect. 7.6. Finally, the use of quality rulers to assess individual attributes of quality, rather than overall quality, is explained in Sect. 7.7. Practical implementations of quality rulers are described in detail in the next chapter.

7.2 Paired Comparison and Rank Order Methods

Although paired comparison is the best method available for determining small quality differences, it conversely is relatively unsuitable for quantifying larger sample differences because of its limited dynamic range, which does not significantly exceed 1.5 JNDs or so, as discussed in Sects. 2.4 and 3.4. In principle, though, if a long series of closely spaced reference samples were available, this approach could be used to cover a wide range of quality. The problem encountered is that a potentially large number of comparisons could be required to deduce the test sample quality accurately. Because the investigator cannot predict ahead of time which reference samples will be perceived as being closest to each test sample by each observer, in a static design (in which the same pairs are shown to all observers for all scenes), most paired comparisons are useful only directionally because of saturation.

Paired comparison experiments involving hardcopy stimuli usually require an administrator to be present throughout the test to provide the stimuli one pair at a time and to record the results. With softcopy display, automation of the entire process is practical. Furthermore, computer control can enable an adaptive approach to presenting paired comparisons, in which the stimuli presented are adjusted based on previous responses of an observer, as described in the next chapter.

A related test method is rank ordering. It is slightly more demanding than paired comparison in terms of experimental setup because a larger working space, uniformly lit and at an appropriate viewing distance must be provided so that more samples can be considered simultaneously. Such a test can typically be self-administered; the observer can simply leave a stack of ordered hardcopy images behind, or can write down their own rankings when the task is completed. The rank order task is more complex for the observer than is paired comparison, particularly as the number of samples increases. In principle, a proper rank ordering is equivalent to a full series of paired comparisons. In practice, observers often compare samples in a somewhat haphazard fashion and then deem the rank ordering to be complete when they tire of the task. The data obtained is consequently often inferior to that from a full set of paired comparisons, but the rank ordering can take much less time. If the time saved can be used to assess additional samples, the results of which can be pooled with those of the earlier ranking, a better overall result can often be obtained from rank ordering than paired comparison in a fixed period of time.

Various methods exist for converting paired comparison and rank order data for a set of samples to approximate interval scales (Engeldrum, 2000), usually based on Thurstone's law of comparative judgments (Thurstone, 1927). These methods may suffer from a variety of difficulties associated with saturation and missing data, and no analysis can correct for deficiencies in the spacing of the stimuli. For example, if there are any large gaps in quality between two nearest samples, so that the higher quality sample is rarely confused with the lower quality sample, the size of the gap cannot be accurately quantified and the resulting approximate interval scale is compromised. In cases where reference samples are included in the experiment, if they are sufficiently closely spaced and fully span the quality of the test samples (to avoid saturation), this situation can be avoided, but at the cost of an increased number of assessments or a decreased number of test stimuli.

7.3 Categorical Sort Method

The categorical sort method is simple to set up and explain, and is readily self-administered. The individual assessments are rapid, with four per minute being a typical rate. An observer is asked to sort individual samples into a finite number of categories (usually five to nine in number), typically labeled with adjectives such as good, fair, poor, etc. Although categorical sorting is a very simple task for the observer and requires minimal effort by the researcher, the method suffers from several deficiencies, including those listed on the following page.

1. The method is subject to strong range effects. Observers seem to want to use most of the categories, perhaps so that they feel they are making some distinctions among the samples, but usually they do not use the end categories much or at all, possibly in case a sample appears that is of much higher or lower quality than seen previously. A skeptic might claim that regardless of the adjectives used and the properties of the stimuli presented, the final distribution of ratings depends mostly on the number of categories. This is less of an exaggeration than might be imagined.

2. The individual assessments are coarsely quantized because only a small number of categories are provided, and often even fewer are actually used. Quantization and perhaps the ambiguity of the application of adjectives leads to large uncertainties in the individual assessments, so many data must be pooled to obtain a reasonable signal-to-noise ratio.

3. The conversion of the ratings to an approximate interval scale is somewhat cumbersome and involves assumptions that are not easily tested (Engeldrum, 2000). The adjectives are assigned numerical values, usually consecutive integers, to produce numerical ratings. These usually are transformed using Torgerson's law of categorical judgment (Torgerson, 1958), which is analogous to Thurstone's law of comparative judgments (Thurstone, 1927).

Range effects (Point #1) and interval scale generation (Point #3) are of much less concern when reference samples are included in the experiment, allowing the ratings to be mapped to a standardized scale of quality, but little can be done about quantization and other noise (Point #2) besides averaging many data. The adjectival mappings obtained are of some utility, as discussed in Sect. 6.4.

7.4 Magnitude and Difference Estimation Methods

The experimental method nearly always used to generate approximate ratio scales is magnitude estimation. Typically, an observer is asked to compare each test sample with a fixed reference sample that has been assigned a numerical value. The observer is instructed to provide a rating twice as high as the reference value if the test sample quality is twice as good as that of the reference, half as high if half as much, and so on. In some variations, the observer is permitted to assign their own numerical value of the reference sample so that numbers used are as natural to them as possible. If each observer is permitted to assign their own scale value to the reference image, then each observer's ratings should be multiplied by a factor that causes their reference rating to be mapped to a constant value, so that the ratings from different observers can be compared or averaged. When averaging ratings on an approximate ratio scale, geometric means are generally used rather than arithmetic means to reflect the nature of the scale. It may be more convenient to take the logarithm of observer ratings to make an approximate interval scale so that arithmetic means may be taken instead.

The magnitude estimation technique largely avoids the deficiencies listed for the categorical sort method. There are no adjectives to be interpreted, nor any quantized categories, and the ratings provided by the observer should approximate a ratio scale without requiring the use of complicated transformations. Range effects should be largely eliminated because the scale is not bounded except at zero, although the possibility remains that each observer may have a range of numbers with which he or she is comfortable working, thereby influencing the ratings given. Any such minor range effect should be compensated when the ratings are transformed to a standard scale of image quality.

One minor drawback of magnitude estimation is that assigning ratio values is difficult for many observers because they think more naturally in terms of intervals than ratios. For example, suppose that the reference sample is defined to have a scale value of 100, and the first test sample shown is assessed by the observer to be one-half as high in quality, and so is given a 50. Now if the second test sample seems to have a quality halfway in between the reference sample and the first test sample, an observer's tendency might be to give the sample a rating of 75, whereas in an ideal ratio scale the rating should be ≈70 (the geometric mean of 50 and 100). The impact of this interpolation problem is that the rating scales obtained may less closely correspond to an ideal ratio scale, but this effect should be corrected when the rating scale is transformed to a standard scale of image quality.

It is a bit surprising that a method like magnitude estimation, but based on differences rather than ratios, is not used more frequently. This may, in part, be because of the difficulty of wording the instructions. Consider the following statements, from the instructions for a magnitude estimation experiment.

> The reference sample is assigned a value of 100. If the test sample is twice as high in quality, then give it a rating of 200; if instead it is half as high in quality, give it a rating of 50, and so on.

These statements are very simple and seem quite clear, even though it is actually very difficult to envision what a factor of two in quality might look like. Consider now an attempt to write analogous instructions for a difference estimation experiment.

> The reference sample is assigned a value of 100. If the test sample is substantially higher in quality, then give it a rating of 150; if instead it is substantially lower in quality, give it a rating of 50, and so on.

Two problems have arisen in this case: (1) some kind of adjectival descriptor (in this example, "substantial") must be used to indicate how quickly the scale values should change with quality; and (2) an interval scale cannot be bounded at zero, so negative numbers may arise and provide a source of discomfort or confusion. Being able to use a phrase like "twice as high in quality" in the magnitude estimation experiment neatly avoids both of these problems.

One approach to solving the two problems in the difference estimation instructions would be to provide two reference samples rather than one, and to assign them values so that negative ratings would be unlikely. Instructions for such an experiment might read as follows.

> The reference sample on the left is assigned a value of 200, and that on the right a value of 100, as shown in their labels. If the test sample has quality halfway between that of the two reference samples, give it a rating of 150. If the test sample is half as much better than the 200 reference as the 200 reference is compared to the 100 reference, give it a rating of 250, and so on.

If the two reference samples approximately span the range of quality of the test samples, most ratings will be between 100 and 200, and negative ratings would be unlikely. These instructions are not a great deal more complicated than those for magnitude estimation, and have the advantage of allowing the observer to provide ratings on an approximate interval scale, avoiding the need for geometric interpolation.

If the approximate interval scale generated in a difference estimation were assumed to be an ideal interval scale, and if the two reference samples had known standard quality scale values, the experimental ratings could be transformed to the standard scale of quality without the observer rating included reference samples. This would be highly advantageous because more test samples could be assessed in a given time and because the regression between experimental ratings and standard scale values would be replaced by a direct solution of simple equations. For example, if the standard scale and

Calibrated Psychometrics Using Quality Rulers

experimental rating scales were both ideal interval scales, they would be related by a linear transform. Because the experimental ratings and scale values would be known for the two reference samples, it would be trivial to solve for the slope and intercept of the linear transform.

A similar approach could be taken even if the standard scale were an ideal ratio scale, because its logarithm could be taken to produce an ideal interval scale, the above analysis could be performed, and the deduced values could be exponentiated. Although, in practice, the assumption that the experimental scale is an ideal interval or ratio scale would be risky, and certainly would not be recommended, discussion of this approach is conceptually useful because it leads naturally to the idea of a quality ruler, as described in the next section.

7.5 The Quality Ruler Concept

It is evident from the discussions above that none of the more common psychometric methods are ideally suited for evaluation of samples of widely varying quality in a fashion yielding results that are calibrated against a standard scale of quality. Problems with these methods include:

1. an excessive number of required assessments (paired comparison, unless the sample presentation is adaptive);

2. an involved procedure for generating approximate interval scales (paired comparison, rank order, and categorical sort);

3. a poor signal-to-noise ratio because of quantization and task ambiguity (categorical sort);

4. a need for assessments of a number of included reference samples and subsequent regression of known standard scale values against experimental ratings (paired comparison, rank order, categorical sort, and magnitude estimation); and

5. an underlying assumption that is questionable (difference estimation with two reference samples).

The method showing the greatest promise is difference estimation with two reference samples, in part because no post-experimental analysis is required to convert a rating to a standard scale value. A table of possible ratings and their equivalent standard scale values can be prepared before any assessments are made, highlighting the efficiency of the method. The only drawback of the method is the assumption that the experimental ratings form an ideal interval scale, which is unlikely to be true unless the test samples span a narrow range of quality and the quality levels of the two reference samples are positioned near the extremes of the test sample quality range. If these conditions were met, most assessments would represent interpolations over a narrow quality range. Over a small enough quality range, the relationship between any reasonably well-behaved rating scales should approximate a line segment, so the evaluations should closely approach an ideal interval scale. This situation minimizes potential systematic error in translation to the standard scale of quality, and so enables assessments of high accuracy. Because most assessments are interpolations over a limited range, good reproducibility (precision) would also be expected in such a case.

If the quality range spanned by the test samples were larger, many assessments would become extrapolations, and both accuracy and precision would suffer. If the reference samples were chosen to be far apart in quality, to span most of the quality range of the test samples, the number of extrapolations would be reduced, but the interpolations would then become less accurate and precise. The obvious solution to this problem is to provide a series of reference samples that are closely spaced in quality, but in aggregate span a wide range of quality. This concept is the basis for the quality ruler method. The principle challenge in such an approach is to enable the observer to efficiently select the reference samples from a series that are closest in quality to a given test sample, while maintaining matched viewing conditions between the test sample and the references. This problem is tractable, and details of hardcopy and softcopy solutions are described in the next chapter.

So far, we have defined a quality ruler to have the two properties listed below.

1. A ruler is composed of a series of reference samples of known standard scale values that are closely spaced in quality but in total span a wide range of quality.

2. The series of reference samples is presented in a fashion that facilitates rapid identification of those reference samples closest in quality to a given test sample and enables matched viewing between the test sample and those reference samples.

A final property of a typical implementation of a quality ruler is as follows.

3. The series of images comprising an individual quality ruler depict a single scene and vary in only a single attribute.

Calibrated Psychometrics Using Quality Rulers 95

The merits of univariate standards were discussed in the previous chapter; in brief, they are: (1) freedom from inversions and ambiguities; and (2) simplicity of simulation and objective verification. Because the images in a series depict only a single scene and vary in only one attribute, the observer can quickly and reliably evaluate the quality levels of the reference series, enabling rapid and robust assessments. Typically a set of quality rulers, each depicting a different scene, would be maintained so that: (1) one or more rulers depicting scenes similar to that of the test image would be likely to be available; and (2) multiple assessments could be made of a single test image, using different rulers, to improve the signal-to-noise ratio through averaging.

7.6 Attributes Varied in Quality Rulers

The choice of the attribute to be varied in a univariate quality ruler is important. An ideal attribute would:

1. have low variability of observer sensitivity and scene susceptibility;

2. be capable of strongly influencing quality so that a wide quality range could be spanned;

3. be produced at widely varying levels by practical imaging systems, so that the ruler samples would not appear contrived;

4. be easily and robustly characterized by objective measurements; and

5. be readily simulated through digital image processing.

One attribute that meets these criteria is sharpness. We have used sharpness-based quality rulers to conduct the majority of the research reported later in this volume. Sharpness varies widely in consumer images (addressing Point #3), and some of the more extreme cases of blur, often arising from camera lens misfocus or camera motion during exposure, can be rated as not worth keeping for quality, even though they have nothing wrong with them except unsharpness (Point #2). Sharpness can be readily characterized by routine modulation transfer function (MTF) measurements (Point #4), using any of a variety of targets such as slanted edges (ISO 12233, 1998). Techniques of spatial filtering in digital image processing are well known and simple to apply (Point #5), and, conveniently, their impact on MTF is readily predicted by linear systems theory (which will be discussed in Ch. 22). This leaves only the first point, regarding variability of observer sensitivity and scene susceptibility.

Figure 7.1 shows independent ratings of quality by 11 expert observers for 13 levels of sharpness. The ratings were made against a primary multivariate standard depicting a wide variety of images representative of consumer and professional photography. The values shown are the average ratings for eleven scenes. The x-axis shows an objective metric correlating with perceived sharpness. The y-axis is 50% JNDs of quality relative to the mean quality of the sharpest level, with negative values corresponding to quality loss. Note both the relatively small variation among observers and the approximate linearity of quality with objective metric. Although the number of observers is small, and the expertise of the observers high, the results are consistent with what we have observed with larger groups of inexperienced observers.

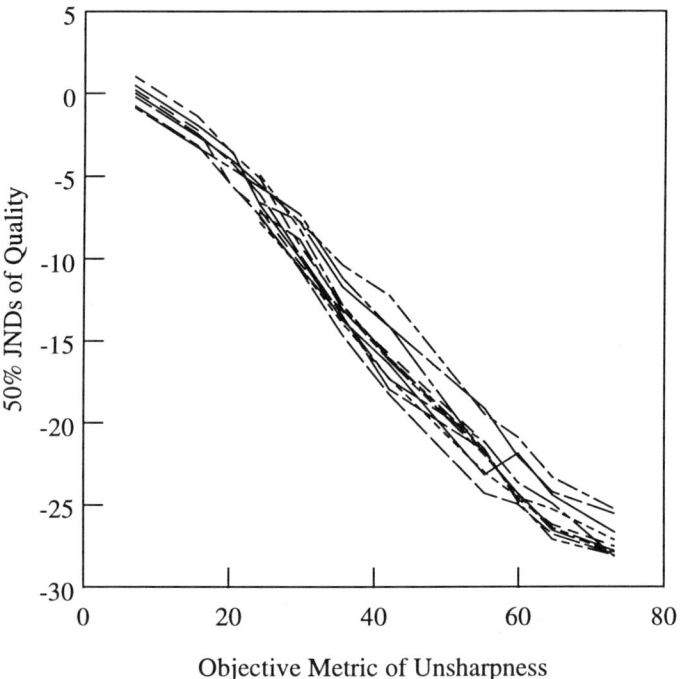

Fig. 7.1 Ratings of quality of samples varying in sharpness against a multivariate primary standard by eleven expert observers, averaged over eleven scenes. The consistency between the observers is very good; this low variability makes sharpness a good candidate for the attribute varied in a quality ruler.

Calibrated Psychometrics Using Quality Rulers

Figure 7.2 has the same axes and is derived from the same assessments, but they have been averaged over the 11 observers, rather than the 11 scenes, so that the variability of scene susceptibility can be evaluated. The difference in net quality change from the lowest to highest quality positions is only about 25% higher in the most susceptible scene than in the least susceptible scene, despite an effort to include scenes with widely varying characteristics. As expected, the more susceptible scenes usually have considerable fine detail that is quickly lost with progressive blur, whereas the less susceptible scenes have less high frequency content. Although the variability of scene susceptibility to sharpness is relatively small, it is still desirable when translating quality ruler ratings to standard scale values that scene-specific scale value calibrations be used. The errors resulting

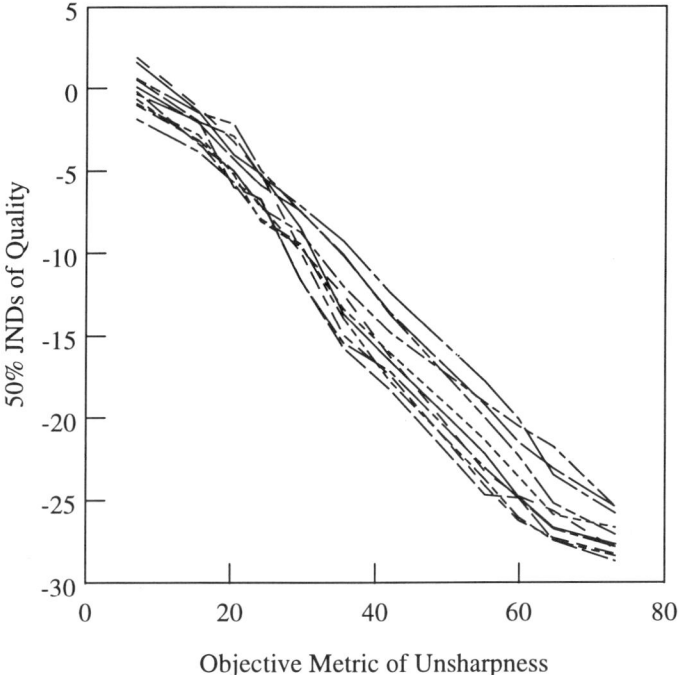

Fig. 7.2 Ratings from previous figure, but averaged over the 11 observers, rather than the 11 scenes, so that variability of scene susceptibility is evident. The consistency among scenes is quite good, and like the low variability between observers, this property makes sharpness a suitable choice for variation within a quality ruler.

from using a single average calibration could be minimized by avoiding the use of scenes particularly endowed with, or devoid of, fine detail.

One minor disadvantage of varying sharpness or any other image structure attribute in a quality ruler is that the associated scale values are only valid at a single viewing distance (as can be appreciated by recognizing that images that differ noticeably in sharpness at a close viewing distance may be essentially indistinguishable at longer viewing distances). Consequently, viewing distance must be constrained when making assessments using such a quality ruler; however, this is not a significant difficulty in practice.

7.7 Using Quality Rulers to Assess Individual Attributes

In addition to evaluation of overall quality, quality rulers may also be used to assess the levels of their varied attributes in test samples. For example, a quality ruler varying in MTF may be used to evaluate the sharpness of a test image, as well as to estimate a corresponding objective metric value if the metric values of the ruler samples are known. When assessing the level of an attribute in a test sample against a ruler varying in that attribute, the observer is doing some form of appearance matching. Aspects of the test sample appearance that do not vary in the ruler are essentially neglected. For example, if the ruler varies in sharpness, but the test sample has poor color balance, the observer will match the appearance of high-frequency detail, while largely ignoring the difference in color balance. In our experience, the success of observers in performing this type of appearance matching is remarkably good, so the level of an attribute can be assessed in this manner even if a number of other attributes significantly influence the overall quality of the sample.

The component of a vector in N-dimensional space that lies along one of the N unit coordinate axes is computed by a process called projection of the vector onto the axis (which is accomplished by taking the dot product of the vector with the unit coordinate axis vector). By analogy, we refer to the assessment of one image quality attribute in a multivariate sample, by appearance matching with a quality ruler varying in that attribute, as a projection process. Projection is very useful in analyzing multivariate samples, as discussed further in Ch. 16.

It is important to recognize that, when assessing overall image quality, appearance matching is exactly the wrong thing to do! If the test sample has poor color balance, the assessed quality should reflect it. As might be expected, observers rarely fall into the trap of appearance matching when the attributes varying in the test samples and the ruler are rather different in nature, e.g., if the test samples differ in noisiness but the ruler varies in sharpness. In contrast, if the test samples were to co-vary in noisiness and sharpness, observers might primarily match sharpness and tend to pay less attention to noisiness variations.

We have found that this undesirable condition can be almost completely avoided through emphasis of the nature of the comparison to be made in the instructions to the observer. We ask the observer to imagine that the image displayed captures one of their treasured memories, and that the observer must choose between possessing either the test image or the ruler image being considered at that moment. Focusing on the question "Which image would I rather own?" seems to place observers in the proper frame of mind for considering all aspects of the image quality. We have run several experiments in which observers rated a set of test samples for one attribute, and then in a separate session were asked to rate the same samples against the same ruler for overall quality. The results have been completely consistent with those expected based on the above considerations, indicating that the mode of assessment can be reliably controlled through the nature of the instructions.

One advantageous feature of quality ruler experiments, or indeed any experiment calibrated to a primary standard, is that once the test samples have been evaluated, they may be treated as secondary standards. Particularly if such samples are univariate in nature and consist of series of levels in each of several scenes, they may be used to construct new quality rulers, which can be useful in several ways. First, as rulers varying in a number of attributes are accumulated, it becomes possible to assess more attributes of multivariate samples via projection. Second, if there is a concern regarding possible appearance matching of test samples with a primary quality ruler, a secondary ruler varying in a different attribute may be used instead.

In the following chapter, practical implementations of quality rulers will be described and measures of their performance will be presented.

7.8 Summary

A quality ruler is a device for making precise and rapid visual assessments of quality that are calibrated against a standard numerical scale of quality. A quality ruler is a series of calibrated samples that are closely spaced in quality but in aggregate span a wide range of quality. The ruler samples are displayed in a manner that facilitates rapid identification of the images that are closest in quality to a given test sample and enables matched viewing between the test and ruler samples. The images in an individual quality ruler depict a single scene and vary in only a single attribute. Usually, sets of quality rulers, each depicting a separate scene, are maintained, so that arbitrary test images may be evaluated against a ruler depicting a similar scene, or to permit averaging of assessments of a test image against multiple rulers.

The attribute varying in a quality ruler ideally should: (1) be capable of strongly influencing quality; (2) possess low observer and scene variability; (3) be produced at widely varying levels by practical imaging systems; (4) be correlated strongly with routinely available objective measurements; and (5) be amenable to variation via digital image simulation. Our most common choice of a quality ruler attribute is sharpness, which meets each of these criteria. When image structure attributes such as sharpness are varied in a quality ruler, the viewing distance must be constrained at the value for which the calibration is determined.

A desirable result of a quality ruler experiment is that the samples assessed constitute a set of secondary standards, which may, in turn, be assembled as new quality rulers. In addition to being used to rate overall quality, rulers may be used to reliably assess the attribute varying in the ruler. This technique, referred to as projection, is often helpful in separating the impact of individual attributes when multiple attributes simultaneously influence the quality of an image.

8

Practical Implementation of Quality Rulers

with Robert E. Cookingham
Eastman Kodak Company
Rochester, New York

8.1 Introduction

We have extensively used two embodiments of the quality ruler, one involving hardcopy images and the other softcopy display. These two implementations differ not only physically, but also in the nature of the assessment task. This chapter, which explains both approaches in detail, is organized as follows. Sections 8.2 and 8.3 describe the hardcopy and softcopy quality rulers, respectively. Instructions to the observers are discussed in Sect. 8.4. Finally, Sect. 8.5 reviews various aspects of quality ruler performance.

8.2 Hardcopy Quality Ruler

A photograph of the hardcopy quality ruler in use is shown in Fig. 8.1. A series of 4 × 6 inch reference prints depicting a given scene are positioned in a print holder having a numerical scale along the bottom edge. The print holders are fabricated from two plastic sheets that are laminated together. Milled slots permit the prints to be inserted between the sheets and viewed through apertures cut in the top sheet. Because rulers for images with portrait (vertical) orientation are taller, the test sample holder, just above the ruler, slides vertically between two positions to allow room for either type of ruler, while still permitting the test

Fig. 8.1 Hardcopy quality ruler, which slides so that any ruler image can be shifted below the test sample holder for critical comparison.

image to be positioned directly adjacent to the ruler in both cases. The ruler assembly rides in a low-friction track attached to a drafting table. The observer can easily slide the ruler either by pushing on an end or by holding one of two small tabs protruding from the face of the ruler; only eight ounces of force are required to shift the ruler. The observer slides the ruler so that the two reference samples most similar in quality to the test sample are symmetrically below it. In this triangular configuration, the viewing distance, illumination level, and viewing geometry of the three images are matched to tight tolerances.

The drafting table surface and all attached ruler parts are approximately 20% reflectance gray tones. The observer adjusts the chair height so that his or her forehead rests comfortably against the padded headrest. The light source is a bank of 5000 K daylight balanced fluorescent tubes with collimators. A black cloth stretched from the light source to the headrest blocks stray light, and the lighting and viewing geometry is such that the images exhibit no specular reflections. Observers wear dark lab coats to avoid reflections from light-colored clothing, and cotton gloves are worn to prevent fingerprints being left on the glossy surfaces of the test samples. The viewing geometry is $\approx 45°/0°$, i.e., light is incident on the image at a 45° angle, and viewing is perpendicular to the image. The test sample holder is angled about 5° to match the viewing angle of the ruler samples. The illumination level is ≈ 750 lux and the viewing distance is 16 inches. The latter is slightly longer than the average hand-held viewing distance of a small print, but matches the distance for which reading glasses and bifocals are typically set. If significantly shorter constrained viewing distances are used, many older observers will have difficulty focusing on the images.

The reference images are typically separated by about three JNDs, which is a good compromise in being large enough that sample differences are readily apparent, but narrow enough that interpolations are accurate and precise. To give some idea of the sample variations within a hardcopy quality ruler, a series of six images that decrease in sharpness and are spaced by approximately three 50% JNDs of quality are shown in Figs. 8.2–8.7. The numerical scale beneath the ruler images shows equally spaced positive integers starting with one, with multiples of three in larger type and directly underneath an image. The observers are asked to provide integer ratings, so their assessments are quantized in approximately one JND increments. In our experience, allowing two intermediate ratings in between reference images (e.g., intermediate ratings 13 and 14 between image values of 12 and 15) leads to an unbiased usage of the possible ratings, whereas when three interpolated ratings are available (e.g., 13, 14, and 15 between image values of 12 and 16), the one-quarter and three-quarters interpolations (13 and 15 in this example) are decidedly underutilized, as if the observers frequently dropped the least significant bit in the rating.

Fig. 8.2 Sharpest image in a series of six, in which adjacent samples differ by approximately three 50% JNDs of quality.

Practical Implementation of Quality Rulers

Fig. 8.3 Second sharpest image in a series of six, in which adjacent samples differ by approximately three 50% JNDs of quality.

Fig. 8.4 Third sharpest image in a series of six, in which adjacent samples differ by approximately three 50% JNDs of quality.

Practical Implementation of Quality Rulers

Fig. 8.5 Fourth sharpest image in a series of six, in which adjacent samples differ by approximately three 50% JNDs of quality.

Fig. 8.6 Fifth sharpest image in a series of six, in which adjacent samples differ by approximately three 50% JNDs of quality.

Practical Implementation of Quality Rulers

Fig. 8.7 Least sharp image in a series of six, in which adjacent samples differ by approximately three 50% JNDs of quality.

Figure 8.8 shows a series of six final image modulation transfer functions (MTFs) that could be suitable for use in a hardcopy quality ruler. These MTF shapes are based on the diffraction-limited lens MTF formula, and are reasonably similar to those resulting from different amounts of lens defocus or from blur caused by camera or subject motion, so the resulting images resemble unsharp consumer images. The spacing between these positions, at a viewing distance of 16 inches, is approximately three 50% JNDs of quality in an average scene. The images in Figs. 8.2–8.7 were prepared from continuous tone images with MTFs similar to those of Fig. 8.8, but the MTFs of the images in this book have been further modified by the halftone reproduction process, and so do not match the MTFs shown in Fig. 8.8.

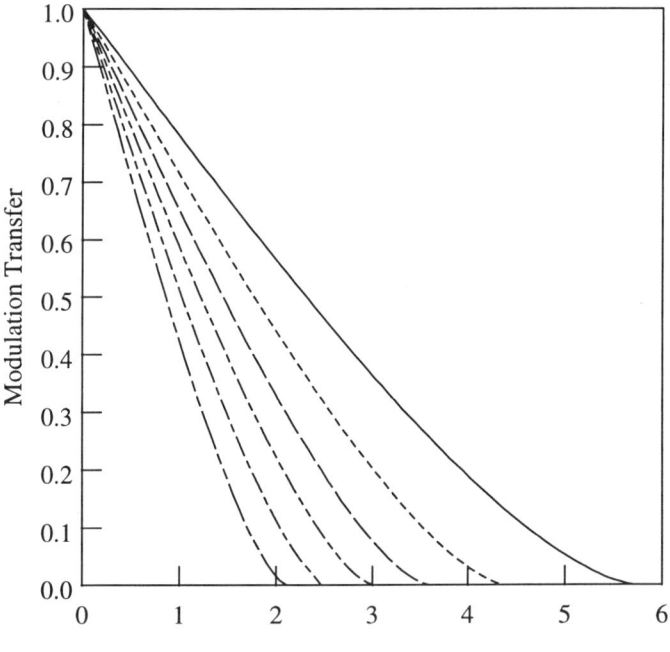

Spatial Frequency in Final Image Plane (cycles/mm)

Fig. 8.8 Final image MTFs of possible quality ruler positions. The spacing between the levels, at a viewing distance of 16 inches, is approximately three 50% JNDs of quality in an average scene.

8.3 Softcopy Quality Ruler

Most of our image structure research is done using the hardcopy quality ruler because very high-resolution output is needed to properly simulate many spatial artifacts. The majority of our color and tone reproduction is performed using the softcopy quality ruler, which has several desirable features, as listed below.

1. Observer responses are recorded electronically, so that an administrator is needed only during the initial instruction period.

2. Cycle time is reduced because digital images can be viewed immediately after they are computed, and ratings are available in electronic form as soon as an observer completes an evaluation.

3. Adaptation studies are facilitated because the image is luminous and therefore ambient lighting may readily be varied nearly independently of the average image luminance.

4. With simple monitoring and calibration procedures, very tight color and density tolerances can be maintained.

5. The high dynamic range and color gamut volume permit a wide range of stimuli to be produced.

A photograph of the softcopy quality ruler arrangement is shown in Fig. 8.9. Two matched, high-resolution, cross-calibrated monitors are placed next to each other and angled so that the observer views them perpendicular to their faceplates. Pre-processed reference and test images are stored on a high capacity storage drive, which, in combination with the high performance graphics cards, allows either of the 8 × 12 inch softcopy images to be replaced with a different image in just two seconds. A headrest is positioned to maintain a viewing distance of 0.9 m. At this distance, the angular separation of the two images is small enough that each can be comfortably viewed by a change in gaze angle, without the necessity for head movement. The structure of the monitor screen is difficult to discern at this viewing distance, which is about 3000× the raster pitch. The horizontal and vertical subtended image angles are ≈10% smaller than in the hardcopy ruler. The viewing room is darkened to avoid flare, but the neutral wall behind the monitors (which subtends much of the observer's visual field of view) is reasonably uniformly illuminated by multiple shielded fluorescent tubes to control adaptation. To emulate hardcopy viewing, the luminance of this wall is adjusted to 25% of that of monitor white.

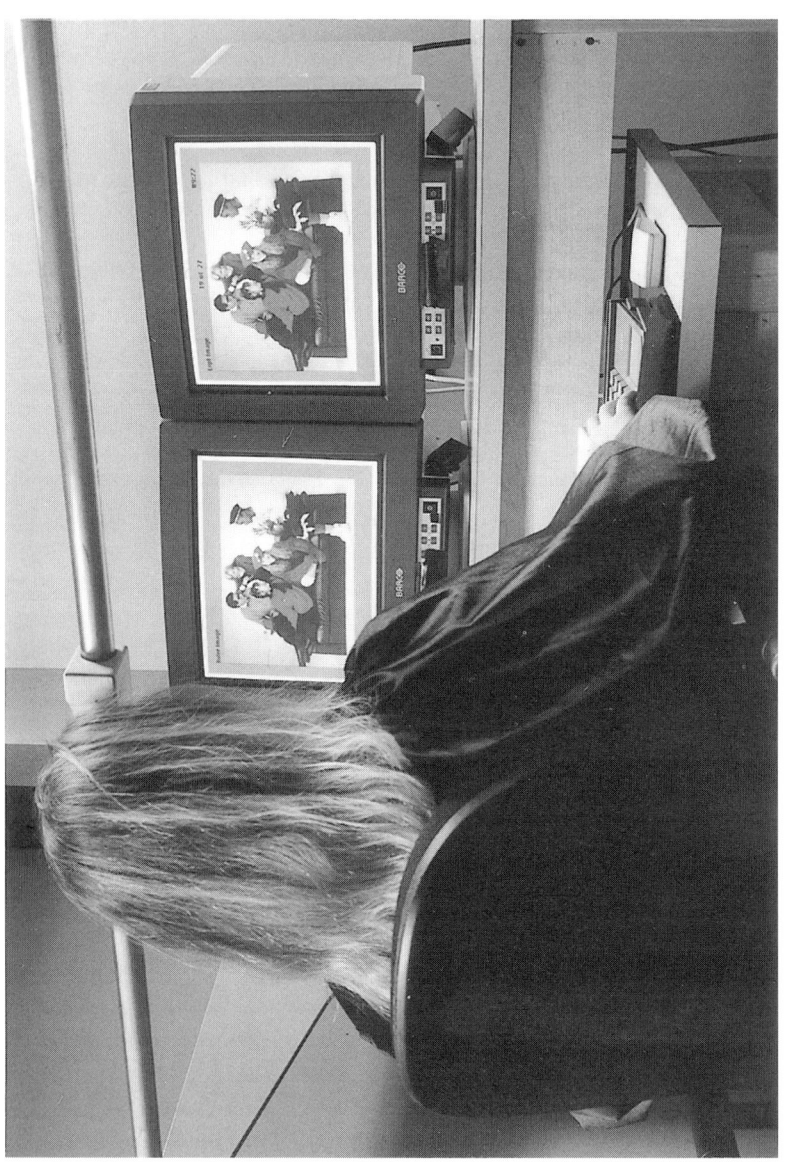

Fig. 8.9 Softcopy quality ruler, in which paired comparisons are automatically presented based upon a binary search routine.

Image presentation is controlled by automated software. After obtaining the name of the observer, and presenting some trial images for practice, the software randomly selects the following: (1) a scene; (2) a test image depicting some attribute level in that scene; (3) a ruler image depicting the same scene; and (4) the monitor (left or right) to which the test image will be displayed. The observer uses a dimly illuminated numeric keypad with specially labeled keys to indicate which image is of higher perceived quality. The index and middle fingers can be rested on the keys in such a way that one finger can be depressed when the left image is better and the other when the right image is better. Another key allows a test sample to be re-evaluated if the observer believes he or she has made an error.

After the first paired comparison assessment, a binary search algorithm is used to select the next ruler sample to be presented. If the ruler sample was chosen as being of higher quality, a lower quality ruler image is displayed, and vice versa. In each iteration, the new ruler image is the one closest to being halfway between the highest quality ruler image that has lost a paired comparison and the lowest quality image that has won. Until a first ruler sample has won (lost) a paired comparison with the test sample, the highest (lowest) quality ruler sample is used as a proxy. This process is repeated until the quality of the test sample has been determined within the desired tolerances. This is an example of an adaptive method that overcomes the necessity for excessive numbers of paired comparisons to determine the quality of test samples spanning a wide range of quality. For $2^N + 1$ or fewer equally spaced ruler samples, a maximum of $N + 1$ paired comparisons are required to define the test sample rating to within plus or minus one-half interval as long as its quality falls within the range spanned by the ruler. Of course, if an incorrect choice is made in a paired comparison because of overlapping perceptual distributions (Sect. 2.2), as becomes more likely as the binary search produces closer quality matches, the rating may be in error by more than half an interval. We use about thirty ruler samples spaced by one 50% JND of quality for an average scene, so a maximum of six paired comparisons are needed to define the test sample rating to within ±0.5 JNDs. Even if the randomly chosen starting ruler image is far different in quality from the test image, little time is wasted because the quality difference will be so large in the earlier paired comparisons that the observer can make a choice almost instantly.

8.4 Instructions to the Observer

A sample set of instructions for a hardcopy ruler experiment is included in Appendix 2. These instructions were used in a study of misregistration, which is discussed in detail in Ch. 14. Misregistration is a spatial shift between color records of an image, as can be seen in extreme form in Sunday comics, where the colors frequently do not match up with the outlines drawn by the cartoonist. Very small amounts of misregistration cause images to look simply a bit unsharp, but larger amounts produce colored ghost images. The instructions in Appendix 2 explicitly describe the origin and appearance of misregistration. The administrator shows the observer preview samples that demonstrate different degrees and types of misregistration. Some researchers believe that observers should not be told anything about the nature of the samples being assessed to avoid influencing them in any way. They view the observer as initially existing in some pristine and fundamentally more valid state that must not be compromised. What they fail to recognize is that, in an almost quantum mechanical fashion, the stimuli that they present to the observer are likely to alter that observer's assessment criteria, as demonstrated by the range effect described in connection with Fig. 6.1.

If nothing else, during an experiment an observer will obtain practice in recognizing the attributes that are varied in the test. Suppose that the samples under study have different degrees of a digital artifact. If the observer has not been significantly exposed to digital imaging, they may be unfamiliar with that artifact. If they are told nothing, they may not even notice the artifact until partway through the test, when a sufficiently egregious sample is presented. But once they have seen the artifact, they are likely to remain aware of it during the remainder of the test. This, in fact, is just what would happen if they started using a digital imaging system that exhibited the artifact; after they noticed the defect in one of their personal images, it would not soon be forgotten. In the case where the observer became aware of the artifact partway through a psychometric test, the data obtained would be of little use except perhaps for predicting transient behavior. By educating the observer about the attribute in question, consistent data may be obtained, which can then be used to predict the final stable response of the observer to the artifact.

Is there a danger of over-sensitizing the observer to the attribute, so that their results become unrepresentative, if the attribute being studied is identified to the observer? Although this might be a concern with other techniques, it should not be a problem in a quality ruler experiment. The observer is also told that the ruler varies in a single attribute, which is likewise identified to them. Therefore, any sensitization that occurs should have approximately equal effect on perception of the test and ruler sample attributes, and so the effects should roughly cancel out, preventing inflated results.

8.5 Performance Characteristics of Quality Rulers

The average time for an assessment of a test image with both hardcopy and softcopy quality rulers is ≈30 seconds, with 85% of observers having mean times within 30% of this value. A one-hour session allows sufficient time for instructions and evaluation of 70–80 test samples except in the case of a small percentage of observers. If more test samples must be rated, it is recommended that they be evaluated in multiple sessions to prevent observer fatigue. In such cases, the sample presentation should be blocked or randomized to prevent bias.

To test for systematic error in quality ruler experiments, we have performed a number of experiments involving assessment of test samples that were, unknown to the observer, identical to ruler samples. An average bias of ≈1 JND has been observed in the hardcopy quality ruler when the test samples match the highest quality ruler sample, but not elsewhere along the ruler. Observers may subconsciously expect that test samples will be inferior to the highest quality ruler position and this may influence their ratings. Notably, in the softcopy quality ruler, this bias is not observed, presumably because the observer is not normally aware of their current position on the ruler, as only one ruler image is shown at a time in the paired comparison format, and the first image shown is randomly chosen. The hardcopy ruler bias is easily avoided by slightly degrading the test samples relative to the best ruler sample. For example, if using a sharpness ruler, the test samples can be slightly blurred to prevent the quality of the test samples from too closely approaching that of the sharpest ruler sample. In the study of attributes that are preferential in nature, some test sample positions may be significantly preferred over the ruler position, so a somewhat larger test sample degradation may be needed. For example, if degree of colorfulness is studied with a sharpness ruler, no matter what level of colorfulness is selected for the ruler, some observers would prefer the colorfulness of certain test sample positions to that of the ruler, and so would rate the test sample higher in quality than a ruler sample having equal sharpness. Applying a modest level of blur to the test samples should prevent the quality of the test samples from exceeding that of the best ruler sample, so that the preference can be evaluated without bias or extrapolation.

The root-mean-square (RMS) uncertainty in a single quality ruler assessment (one observer rating one test sample) is consistently found to be ≈2.5 50% JNDs of quality, regardless of whether expert or untrained observers evaluate the samples and whether the samples vary in one or many attributes. From this value, estimates of precision of the mean may be made for pooled data from different numbers of observers and scenes. For example, if 6 assessments are averaged, whether they are from different observers, scenes, or a combination thereof, the standard error of the mean would be approximately $2.5/6^{1/2} \approx 1.0$ JNDs. We have also calculated single assessment RMS errors for carefully performed experiments using other techniques and obtained substantially higher values, e.g., ≈ 4.3 JNDs for magnitude estimation and ≈ 7.8 JNDs for a categorical sort with 6 choices. As might be expected, this trend correlates negatively with the number of reference samples employed in each technique (many in the quality ruler, one in magnitude estimation, and zero in categorical sort). Although the lesser assessment time of the simpler techniques partly compensates for their higher noise, the total assessment time to obtain data that could be averaged to yield a desired precision is still significantly lower for the quality ruler, particularly when the extra time required to assess reference samples of known standard scale value is included for the other techniques (without which they are uncalibrated). The categorical sort value is slightly larger than the width in JNDs of the adjective categories used, whereas if quantization were the only source of noise, the RMS error should be only $\approx 29\%$ of the category width (Bendat and Piersol, 1980), so other sources of error, such as the ambiguity of interpretation of adjectives, appear to contribute to the assessment uncertainty. The reasons why even the quality ruler does not approach the theoretical minimum of one JND of uncertainty are not fully understood, but probably have to do with the wide range of potential quality of the test samples, which requires that a number of comparisons between test samples and different ruler samples be made before a rating can be assigned.

In most research applications, we use digital image simulation techniques to generate test samples and ruler samples at the same time, in a fashion that they are closely matched in all respects except for the attribute variations being studied. It might be expected that assessing an arbitrary test sample against a quality ruler that depicted a different scene would yield data with higher noise and perhaps systematic error. We have compared the results for the same samples rated against hardcopy quality rulers with matched and unmatched scenes and the mean values and variability are not statistically significantly different within the quality range spanned by the ruler. This is shown in Fig. 8.10, where matched scene ratings are plotted as a solid line and unmatched scene ratings from three different rulers are plotted as discrete symbols.

The performance of the quality ruler has now been reported in terms of the objective measures of assessment time, accuracy, and precision, but some aspects of the performance are difficult to describe quantitatively, and so a few last comments of a more anecdotal nature will be made. Observers are enthusiastic about the quality ruler technique partly because (paraphrasing their comments) they feel that they are providing data that they could closely reproduce on another day. Neither categorical sort nor magnitude estimation experiments leave the majority of observers feeling this way, because they are troubled by the vagueness of the meaning of the adjective categories, or the difficulty of envisioning what an image twice as good as another would look like. The quality ruler task is very well defined, and combined with the explicit directions, seems to give observers a clear idea of what they are doing and enhances their sense of accomplishment. Observers particularly like using the hardcopy ruler, which from their perspective may have the following advantages: (1) the task is carried out in a normally lit room with an administrator present; (2) all ruler images are visible simultaneously and the physical position of the ruler provides visual and tactile feedback regarding the state of the evaluation; and (3) the rating task is under the complete control of

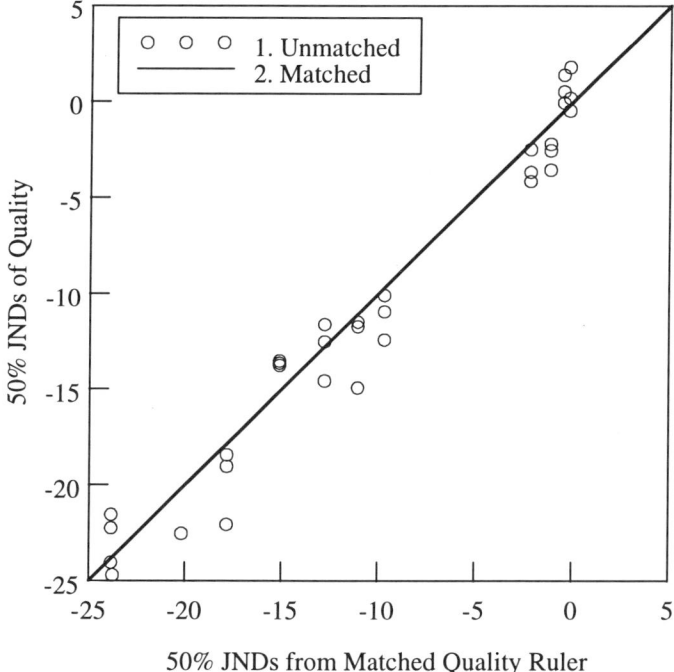

Fig. 8.10 Comparison of assessments made against quality rulers depicting the same scene as in the test image (solid line) and those depicting different scenes (circles). Neither the means nor variances of the two sets of data are statistically significantly different, implying that the ruler scene need not match the test image scene.

the observer and can more easily be envisioned in its entirety. Most of our observers are willing to participate repeatedly in quality ruler tests, despite the time and effort required.

From an analyst's point of view, the data obtained from quality ruler experiments seems exceptionally clean and well behaved, even beyond that expected based on the low noise level. Perhaps systematic error, which can be difficult to estimate, is more serious in other techniques than is generally appreciated. Whatever the actual causes, the development of the quality ruler method has substantially improved our ability to obtain quantitative assessments of image quality.

8.6 Summary

The hardcopy quality ruler that we have used extensively involves a sliding ruler with slots for reflection prints, allowing the test sample and particular ruler prints to be compared while adjacent to each other, in the same lighting and at the same viewing distance, without the observer moving (excluding using his or her hands to slide the ruler). The softcopy quality ruler displays the test image on one monitor and a ruler image on a second, matched monitor. The observer indicates which of the images is higher in quality, and controlling software selects a new ruler image to be displayed depending on the answer given, using a binary search algorithm, so as to converge on the ruler position having quality equal to that of the test image. The hardcopy implementation is preferred for image structure studies in which fine spatial structure must be accurately reproduced. The softcopy ruler is preferred for color and tone reproduction studies because of stable color calibration, high color gamut volume, automated data collection, and faster cycle time.

The uncertainty in a single quality ruler assessment has been found to be considerably lower than that of other common psychometric techniques, and lower noise levels can be achieved in a given period of evaluation time compared to other methods. In a one-hour session, the maximum recommended time, nearly all observers can receive instructions and complete evaluation of 70–80 test samples. Although in research applications it is common to generate test samples that match the quality rulers in scene content through digital image simulation, quality rulers have been found to work approximately equally well in evaluating images with different scene content.

9

A General Equation to Fit Quality Loss Functions

with Robert E. Cookingham
Eastman Kodak Company
Rochester, New York

9.1 Introduction

In Chapter 4, a quality loss function was defined to be the relationship between the quality change in JNDs arising from an attribute and the value of an objective metric correlated with that attribute. In that chapter, the quality loss function of a contrast metric was determined, thereby quantifying the loss of quality caused when the preferential attribute of contrast deviated by different amounts from its optimum value for a particular scene and observer. The quality loss function of the contrast metric was fit with a quadratic function primarily for mathematical convenience, and although the psychometric data was fit well (Fig. 4.4), that data spanned a fairly narrow range of quality (ca. 9 JNDs) and so represented a relatively undemanding case. Researchers typically fit each new set of psychometric ratings versus objective metric values with whatever mathematical function seems to be suggested by the data. In some cases, less reliable data have been fit with complex functions to account for saturation and range effects, which were largely artifacts of the psychometric methods employed rather than perceptual phenomena of primary interest. The wide range of properties of the ratings being analyzed has contributed to the proliferation of specialized treatments.

In this chapter, a generally useful equation for fitting quality loss functions is presented. This equation is called the integrated hyperbolic increment function for reasons that will become evident; its acronym, IHIF, will be used repeatedly in the remainder of this volume. We have successfully fit essentially all the psychometric data we have collected, which has involved many attributes and objective metrics, with this single, three-parameter function. In part, the general applicability of this function results from expressing the perceptual data in terms of JNDs.

Use of a single function for relating quality changes to objective metrics is advantageous in several respects, which are listed below.

1. A degree of standardization is achieved, which facilitates interpretation of fit parameters, communication of results, re-use of code in software, etc.

2. Observer and scene variability are readily characterized by changes in fit parameters rather than changes in functional form, simplifying implementation.

3. More robust extrapolation may be provided by building the desired asymptotic dependencies into the general equation.

4. Analytical derivatives can be calculated once and re-used in subsequent nonlinear regression analyses (analytical derivatives are not required in such analyses, but they can facilitate convergence and calculation of regression tolerances).

5. Familiarity with the assumptions made in the derivation of the equation may suggest modifications that can be made to a trial objective metric to better fit psychometric data.

This chapter is organized as follows. Section 9.2 describes a simple model of the dependence of JND increments on the values of their associated objective metrics. In Sect. 9.3, integration of this model according to Eq. 5.8, which was used previously to calibrate psychometric scales, leads to a general equation for fitting quality loss functions. Finally, the dependence of the shape of this function on its three fit parameters is explored in Sect. 9.4.

9.2 Dependence of JND Increment on Objective Metric

Equation 5.8 was derived in the context of conversion of an arbitrary rating scale to an ideal ratio or interval scale; however, the arbitrary rating scale s can equally well be replaced by an objective metric Ω. If the parameter Δt_j in Eq. 5.8 is set equal to unity, the resulting scale is an ideal interval scale in which one unit is equal to one JND, which is thus a scale of quality in JND units. These substitutions yield:

$$Q(\Omega) = Q_r - \int_{\Omega_r}^{\Omega} \frac{d\Omega'}{\Delta\Omega_J (\Omega')} \tag{9.1}$$

where Ω' is the variable of integration, Ω_r is a reference position, $\Delta\Omega_J$ is a JND increment, Q is quality in JNDs, and Q_r is the quality at the reference position. Dividing the infinitesimal objective metric change $d\Omega'$ by the JND increment $\Delta\Omega_J$, which has units of objective metric change per JND, yields infinitesimal changes in JNDs, which are summed in the integration to yield cumulative JNDs of change from the reference position. The minus sign reflects the convention, which will be followed hereafter, that a positive change in objective metric would lead to a negative change in quality (except below threshold, where it would have no effect). Thus, objective metrics will be measures of impairment, and a negative number of JNDs will correspond to a quality loss. The reference position Ω_r is usually chosen to be at threshold for artifacts, so that if $Q_r = 0$, zero JNDs corresponds to the perceived absence of the artifact. In matters of preference, the reference position is usually chosen to correspond to the optimal position, so that if $Q_r = 0$, the highest quality attainable corresponds to zero JNDs.

To make further progress we must assume a mathematical form for the JND increment as a function of objective metric value. It is not obvious that this will be a fruitful activity. There seem to be few assumptions that can be made that are not arbitrary in nature and will not lead to an overly restrictive function, which may then fail to fit psychometric data well. Nonetheless, experience has shown that a plausible set of assumptions can indeed lead to a generally useful equation that has robust extrapolation behavior and easily interpreted fit parameters. Consider first an artifactual attribute, which, when evident, reduces quality. Suppose that an objective metric exists that positively correlates with this attribute, so that, when the artifact is evident, higher values of the metric correspond with greater quality reductions. Two assumptions shall be made:

1. below some threshold level, the attribute is perceptually undetectable and therefore does not affect quality; and

2. well above threshold, the JND increment of the objective metric approaches a constant value.

The threshold assumption implies that as the objective metric decreases and approaches the threshold value, the JND increment must become infinite. This ensures that no change in quality can occur below threshold, because the denominator in the integrand of Eq. 9.1 is infinite. An objective metric meeting the constant JND increment assumption is convenient because of simplified

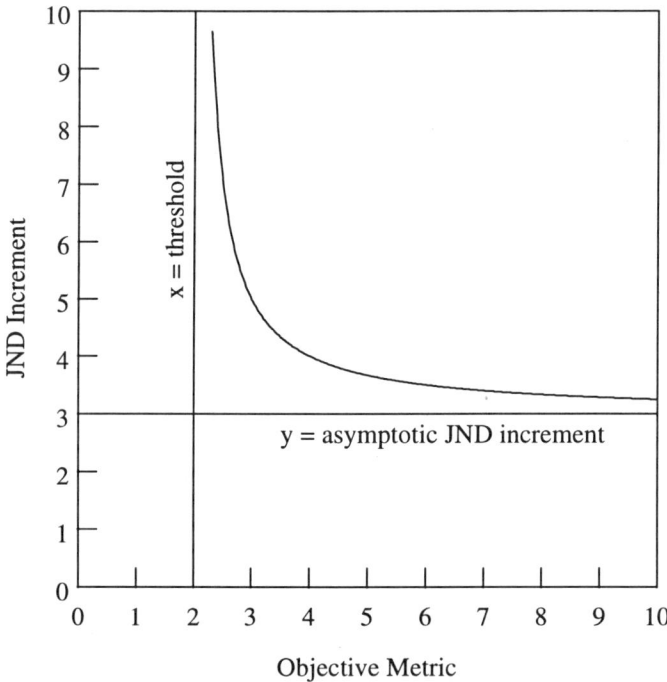

Fig. 9.1 Assumed hyperbolic dependence of JND increment on objective metric value, as in Eq. 9.2. At threshold (x = 2 in this plot), further reduction in objective metric does not improve quality, so the JND increment diverges; well above threshold, it asymptotically approaches a constant value (y = 3).

interpretation and improved reliability of extrapolations of metric values based on straight lines. Although a trial objective metric may not initially meet this criterion, it is usually possible to transform the metric so that it does. For example, well above threshold, one JND of perceived noisiness corresponds roughly to a constant percentage change in RMS granularity (Zwick and Brothers, 1975). Although RMS granularity would not meet the constant JND increment assumption, the logarithm thereof would do so. We have seen a number of cases where logarithmic transforms of trial metrics approximately meet the constant JND increment criterion. Similar results were obtained in many early psychophysical studies, as embodied in Weber's Law, which states that discriminable differences correspond to constant ratios (percentages) of change in physical measurables (Bartleson and Grum, 1984).

Although there are many functions meeting the threshold and constant JND increment assumptions, a particularly simple one is given in Eq. 9.2:

$$\Delta \Omega_J = \Delta \Omega_\infty + \frac{R_r}{\Omega - \Omega_r} \qquad (\Omega > \Omega_r) \qquad (9.2)$$

where Ω_r is the threshold of detection (or the optimum of preference), $\Delta \Omega_\infty$ is the asymptotic JND increment well above threshold, and R_r is a curvature parameter that determines how rapidly above threshold the asymptotic behavior is reached. Figure 9.1 shows a plot of the hyperbola of Eq. 9.2 for $R_r = 2$, $\Omega_r = 2$ and $\Delta \Omega_\infty = 3$. The JND increment function is bounded by lines having the equations $x = \Omega_r$ and $y = \Delta \Omega_\infty$.

As discussed further in Ch. 20, Eq. 9.2 may also be applied to attributes of preference if the threshold of detectability is reinterpreted as the threshold of distinction from the preferred position, and if the objective metric reflects a necessarily positive distance from that position.

9.3 The Integrated Hyperbolic Increment Function (IHIF)

Substituting Eq. 9.2 into Eq. 9.1 with the change of variables $u = 1/(\Omega'-\Omega_r)$, implying $du = -u^2 \cdot d\Omega'$, yields the definite integral:

$$Q(\Omega) = Q_r - \int_{1/(\Omega-\Omega_r)}^{\infty} \frac{du}{u^2 \cdot (\Delta \Omega_\infty + R_r \cdot u)} \qquad (9.3)$$

which is available in standard tables of integrals. Performing the definite integration yields the following result.

$$Q(\Omega) = Q_r + \frac{R_r}{\Delta\Omega_\infty^2} \cdot \ln\left(1 + \frac{\Delta\Omega_\infty \cdot (\Omega - \Omega_r)}{R_r}\right) - \frac{\Omega - \Omega_r}{\Delta\Omega_\infty} \quad (9.4)$$

Equation 9.4 is the IHIF, which is the primary result of this chapter. Being based on Eq. 9.2, this equation is valid for $\Omega > \Omega_r$. Below the reference position, in the subthreshold regime, $Q = Q_r$. As shown in Appendix 3, at the reference position, these two functions ($Q = Q_r$ and Eq. 9.4) are equal and have equal first derivatives with respect to the objective metric, so the transition is a smooth one.

Appendix 3 also gives an extended form of Eq. 9.4 that is valid for all objective metric values, and provides the analytical derivatives of the extended equation with respect to each of the three fit parameters, $\Delta\Omega_\infty$, R_r, and Ω_r (as well as Q_r, which is usually assigned by convention rather than being used as a fit parameter). The extended equation and partial derivatives are useful in nonlinear regressions to determine the three fit parameters from psychometric data. It is also proven in Appendix 3 that the parameter R_r is the radius of curvature of the quality loss function at the reference position, providing a simple geometric interpretation of this fit parameter. The radius of curvature of a function at some point is the radius of a circle that matches the shape of the function at that point. A small radius of curvature corresponds to a tight bend in a function, whereas a large radius of curvature approximates a straight line.

A slightly simpler form of the IHIF results if only the change in quality arising from an attribute, relative to the reference position, is considered. This quality difference, $Q - Q_r$, which is denoted by ΔQ, is zero when an artifact is subthreshold or a preferential attribute is at the optimal position, and is negative otherwise. Using this quantity in Eq. 9.4, the IHIF assumes the following form.

$$\Delta Q(\Omega) = \frac{R_r}{\Delta\Omega_\infty^2} \cdot \ln\left(1 + \frac{\Delta\Omega_\infty \cdot (\Omega - \Omega_r)}{R_r}\right) - \frac{\Omega - \Omega_r}{\Delta\Omega_\infty} \quad (9.5)$$

9.4 Effect of Fit Parameters on IHIF Shape

To better visualize Eq. 9.5, it is helpful to investigate how the IHIF shape varies as a function of each of its three fit parameters, $\Delta\Omega_\infty$, R_r, and Ω_r. The fourth

A General Equation to Fit Quality Loss Functions

parameter in Eq. 9.4, Q_r, is the reference quality and usually is a matter of convention, not a fit parameter; it simply shifts the quality loss functions along the y-axis.

Figure 9.2 shows what will become a very familiar plot format: JNDs of quality (relative to a reference position) on the y-axis, and objective metric values on the x-axis. The IHIF is shown for a series of values of the reference objective metric value Ω_r, with increasing values of Ω_r plotting further right. The parameters R_r and $\Delta\Omega_\infty$ have been held constant in this series. Changes in the Ω_r parameter simply shift the curves along the x-axis, as expected from Eq. 9.5, where Ω and Ω_r always appear together in the difference term $\Omega - \Omega_r$.

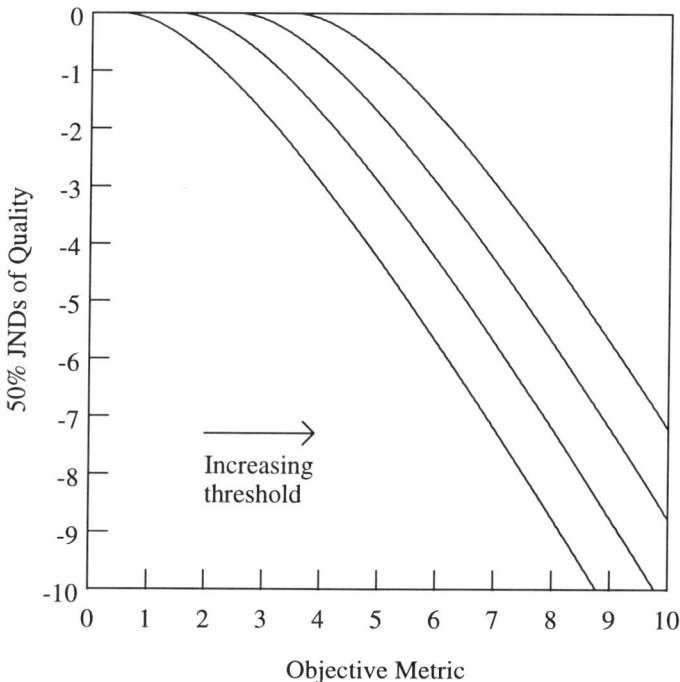

Fig. 9.2 Dependence of the shape of the integrated hyperbolic increment function, Eq. 9.5, on threshold fit parameter Ω_r. As threshold increases, the quality loss function is shifted to the right.

Figure 9.3 shows a series in the asymptotic JND increment $\Delta\Omega_\infty$ with other parameters fixed. This fit parameter controls the slope reached well above threshold, where the curve becomes linear. Curves having increasing values of $\Delta\Omega_\infty$ plot further to the right and have weaker (less negative) slopes. The limiting slope is given by $-1/\Delta\Omega_\infty$.

Finally, Fig. 9.4 shows a series in R_r, the radius of curvature at threshold. All curves have the same threshold and the same asymptotic slope; curves with increasing R_r plot further right. R_r controls the width of the transition region

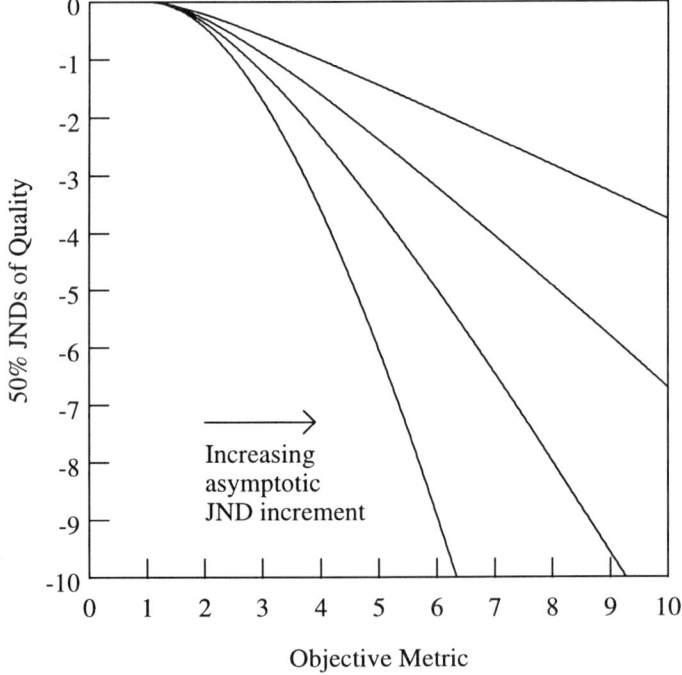

Fig. 9.3 Dependence of IHIF shape on the asymptotic JND increment fit parameter $\Delta\Omega_\infty$. Increases in this parameter, which is the objective metric difference needed to change quality by one JND when well above threshold, lead to less negative limiting slopes.

A General Equation to Fit Quality Loss Functions

from the subthreshold regime, where the curves are flat (y = 0), to the asymptotic region, where the curves become linear with negative slope. Increasing R_r produces a more gradual transition, as expected from a greater radius of curvature.

The next chapter, which treats variations in observer sensitivity and scene susceptibility, includes several examples of the practical application of the IHIF for quantifying variability. Fits based on the IHIF will also be found in many of the figures in Part II of this book, which concerns the design of objective metrics.

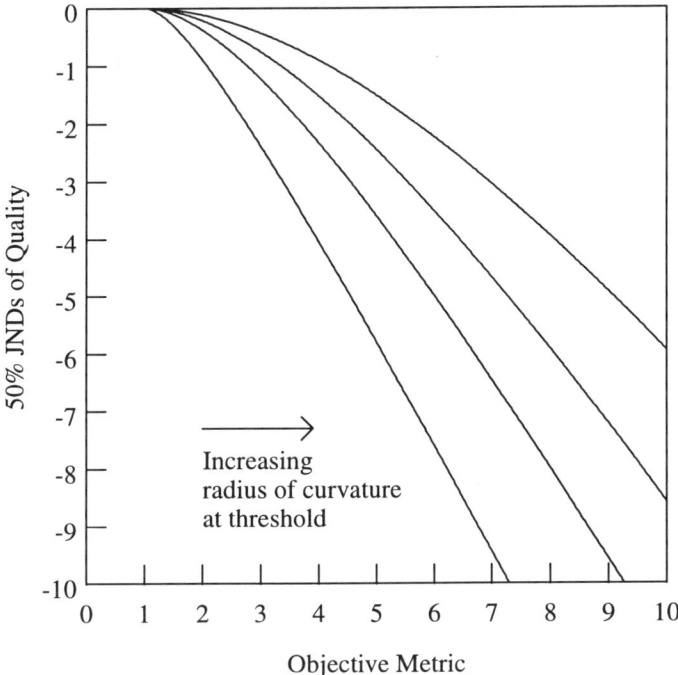

Fig. 9.4 Dependence of IHIF on the threshold radius of curvature fit parameter R_r. Increases in this parameter lead to slower transition from subthreshold (flat) to asymptotic (linear, negatively sloped) behavior.

9.5 Summary

In this chapter, the integrated hyperbolic increment function (IHIF), a general equation for fitting quality loss functions, has been developed. Advantages of using a standard equation to describe quality loss functions include: (1) simplified interpretation and communication of results; (2) effective quantification of variability; (3) robust extrapolation; (4) re-use of nonlinear regression routines; and (5) insight regarding properties of objective metrics.

In the development of the IHIF, it was assumed that below a certain threshold the attribute is undetectable (in the case of an artifact) or is equivalent to the optimum position (in a case of preference). At this reference position, the JND increment diverges, because additional changes in the metric value produce no change in quality. Far above threshold, the JND increment was further assumed to approach a constant value. This condition is desirable because it makes the objective metric more easily interpretable, and improves the robustness of extrapolations, because they are based on straight lines. We often find that linear metrics must be logarithmically transformed to meet this criterion, as might be expected based on Weber's Law.

The IHIF has three fit parameters: (1) the asymptotic JND increment (i.e., the constant value achieved well above threshold); (2) the objective metric value at threshold; and (3) the radius of curvature at threshold. The first parameter controls the ultimate slope of the curve; the second shifts the curve along the objective metric axis; and the third controls the rapidity of the transition from the subthreshold (flat) part of the curve to the asymptotic region. A fourth parameter, the reference quality, is usually assigned by convention rather than being used as a fit parameter.

We have successfully used this equation to describe quality loss functions in many varied cases, as well as characterizing the variability arising from different observer sensitivities and scene susceptibilities through the use of different values of some subset of the three fit parameters. The IHIF is usually handled well by nonlinear regression techniques; analytical derivatives are given in Appendix 3 in support of such applications.

10
Scene and Observer Variability

10.1 Introduction

This chapter describes how variation in scene susceptibility and observer sensitivity may conveniently be characterized mathematically using IHIF fit parameters and how such information may be useful in different types of analyses. Selection of scenes and observers for psychometric studies, which was briefly considered in Sect. 6.2, is addressed in greater detail here.

In this chapter, data from a study of misregistration will be used to illustrate various points; the reader will recall that this artifact arises from misalignment of color records in an image, as seen in extreme form in Sunday comics, where the printed colors do not match up with the drawn outlines.

This chapter is organized as follows. Section 10.2 shows representative variations in scene susceptibility and observer sensitivity. Sect. 10.3 discusses factors influencing the selection of scenes and observers in calibrated psychometric studies. Use of the IHIF to characterize variability of scene susceptibility and observer sensitivity is described in Sect. 10.4. Finally, the utility of variability information is outlined in Sect. 10.5.

10.2 Scene Susceptibility and Observer Sensitivity

A typical quality loss function determined in a quality ruler experiment is shown in Fig. 10.1. The x-axis is an objective metric of misregistration, based on a visually weighted angular subtense of color record misalignment (see Ch. 14). The y-axis is 50% JNDs of quality loss arising from misregistration, with zero JNDs corresponding to a subthreshold effect. The data shown by circles are mean values over 21 observers and 12 scenes from a series of levels in which the green record was shifted by varying amounts. The IHIF fit is plotted as a solid line, and dashed lines show the 95% confidence interval of the regression. The width of the confidence interval may reflect the following factors:

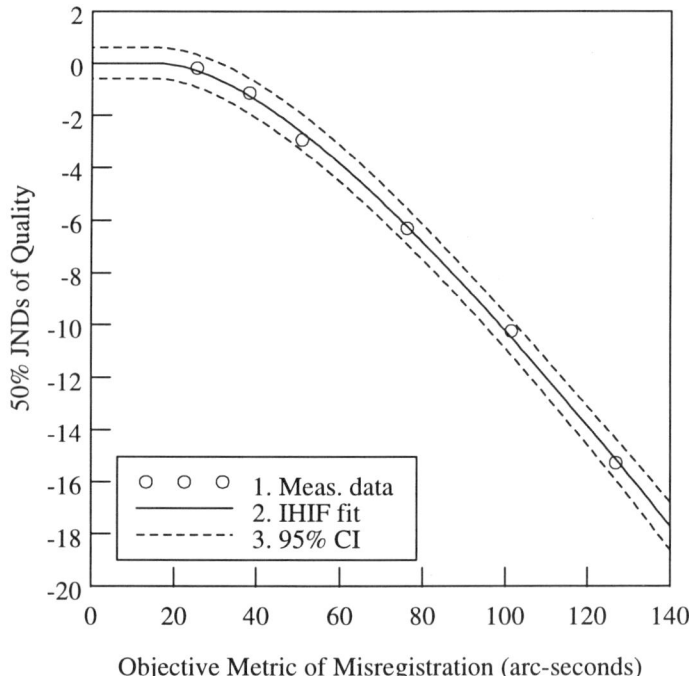

Fig. 10.1 Quality loss function and IHIF regression fit for green record misregistration. The 95% confidence interval of the regression primarily reflects the uncertainty in individual assessments; these results are typical of those from quality ruler experiments.

Scene and Observer Variability

1. assessment uncertainty;

2. lack of predictiveness of the objective metric; and

3. failure of the IHIF to fit the shape of the quality loss function.

The random, rather than systematic, deviation of the data from the IHIF fit indicates that the IHIF shape is appropriate and that the third factor above should not be a significant contributor. In this simple case, involving only one dimension of sample variation (magnitude of green record shift), the objective metric should be well correlated with the quality loss, and so the second factor would be expected to be negligible. This might not have been the case if the samples had varied in the color record shifted as well as the amount of the shift (as described in Ch. 14); an objective metric might then fail to predict the relative importance of which record was shifted compared to the magnitude of the shift. Given these considerations, we expect that the regression uncertainty largely reflects the noise of the individual assessments.

An observer's sensitivity to a given attribute is treated as a constant characteristic of the observer; thus, variability of observer sensitivity refers to variation among observers, not within an individual observer. Because all observers contribute an equal number of assessments to each mean datum in Fig. 10.1, variability of observer sensitivity should not affect the regression confidence interval. If the group of observers selected were more sensitive to misregistration than the population of observers as a whole, a bias in the mean values would result, but it would not increase the width of the regression confidence interval. Similarly, variability of scene susceptibility does not influence the regression uncertainty. Any variation between repeated assessments by an individual observer of a particular scene is reflected in the width of the perceptual distribution of Sect. 2.2, rather than being part of the variability of scene susceptibility and observer sensitivity.

Figures 10.2 and 10.3 provide a simple pictorial demonstration of variations in scene susceptibility. These two images of different scenes have the same objective amount of the digital artifact of streaking (see Sect. 13.2). In Fig. 10.2 the streaking is evident as horizontal stripes of higher and lower density and of differing width, which are particularly noticeable in the slowly varying sky. In Fig. 10.3 the same pattern of streaking has been digitally added to a different scene. Because of the abundance of high-frequency signal, which visually masks the streaking, and the extensive darker areas in the image, where image structure is less visible (Sect. 15.5), the streaking is nearly undetectable in this image. The scene of Fig. 10.2, with its slowly varying sky, is much more susceptible to streaking than is the scene of Fig. 10.3, which hides the artifact very effectively.

Fig. 10.2 The digital artifact of streaking, which is most evident as horizontal stripes of variable density and width in the areas of sky.

Scene and Observer Variability 133

Fig. 10.3 The same pattern of streaking as in Fig. 10.2, but in a much less susceptible scene.

Fig. 10.4 Artifact of contouring, causing hazy (dark gray) shadows and abrupt density jumps (e.g., in the legs of the rightmost person).

Fig. 10.5 A level of streaking perceived by the average observer to cause quality loss equaling that arising from contouring in Fig. 10.4.

Figures 10.4 and 10.5 attempt to demonstrate variability of observer sensitivity, which is difficult to do with a single reader serving as the sole observer. Figure 10.4 exhibits the artifact of contouring, which is discussed in detail in Chs. 15 and 17. In brief, it is a digital artifact resulting from overly coarse quantization of signal, which causes an area that should vary smoothly to instead exhibit distinct jumps in density between one flat area and another. In Fig. 10.4, the contouring is mostly evident as shadows that are dark gray, rather than black, and have no evident detail, but it also can be seen elsewhere, e.g., as discrete density steps in the legs of the rightmost person. Figure 10.5 shows a new level of streaking, which is mostly apparent in the sky in the upper right corner of the image. The average observer assesses the quality loss arising from streaking in Fig. 10.5 to be approximately equal to the quality loss caused by contouring in Fig. 10.4. If the reader perceives the quality of Fig. 10.5 to be noticeably poorer (better) than that of Fig. 10.4, then the reader is relatively more (less) sensitive to streaking compared to contouring, than is the average observer. Many readers will have a fairly strong opinion which image is of higher quality, demonstrating that observers vary in their sensitivity to different attributes.

In Sect. 7.6, the properties of ideal attributes upon which to base a quality ruler were considered. One such property was low variability of quality assessments of samples differing only in that attribute. To demonstrate this property for sharpness, ratings for different scenes, which had been averaged over multiple observers, were compared to characterize scene susceptibility variations. Similarly, ratings from different observers, averaged over multiple scenes, were compared to evaluate observer sensitivity variability. A similar approach will be adopted here, using misregistration data, which exhibits behavior more typical of the attributes we have studied than does sharpness.

To depict variability in scene susceptibility, the means of data pooled over the observers is compared in Fig. 10.6. The x-axis shows the grand mean JNDs of quality loss for each of the six green record shifts; these are the same values as on the y-axis of Fig. 10.1. The y-axis in Fig. 10.6 shows the mean values for five selected scenes of twelve, from data pooled over the twenty-one observers. A perfectly average scene would have data falling along a 45° diagonal line (not shown to avoid clutter); a more susceptible scene would fall below such a line, and a less susceptible one above it. Within quantization error, the five scenes shown approximately represent the extrema and quartile positions, i.e., the 0, 25, 50, 75, and 100% cumulative percentile positions of susceptibility.

The level of variation shown in Fig. 10.6 is typical of attributes we have studied. The set of quality loss functions occupies a region of space shaped like a tilted cone. There is close agreement near threshold, where the quality loss is small, but a greater range of responses occurs at larger degradations. The degree of

Scene and Observer Variability

variation is very roughly a constant fraction of the level of degradation. For example, the range of data in Fig. 10.6 is very small near threshold (zero JNDs), about 4 JNDs at −6 JNDs, and about 7 JNDs at −15 JNDs. Thus, to a crude approximation, the range of data is about half as large as the level of degradation. A similar result would be obtained if the standard deviation of the data at each level of degradation were used instead of the range of data. This behavior implies that the coefficient of variation, which is the ratio of the standard deviation to the mean, should be fairly independent of the level of degradation; it consequently provides a convenient measure of variability. Most of the attributes we have studied have coefficients of variation of scene susceptibility in the range 10–40%, with a median value being near 15%.

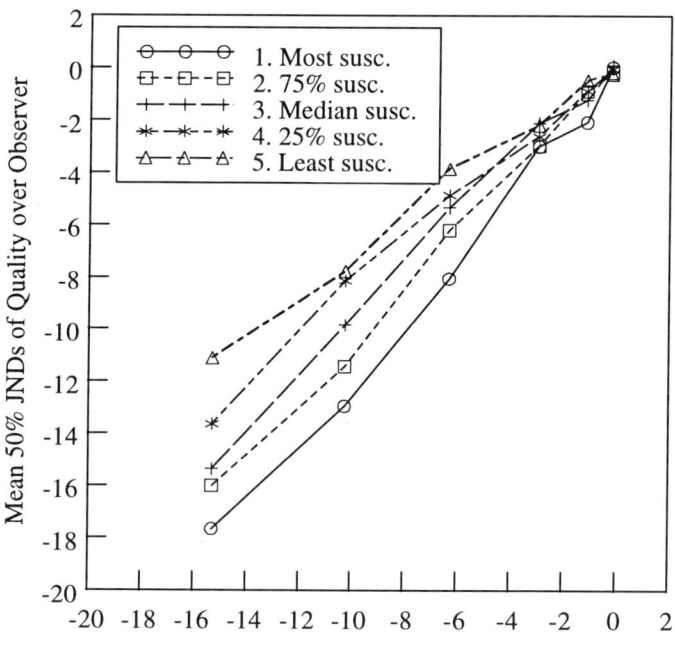

Fig. 10.6 Mean misregistration quality loss data over observer plotted by scene, against mean response over all observers and scenes. An average scene would plot as a 45° diagonal line; the systematic variations exhibited, which are typical in magnitude for artifacts we have studied, arise from variability in scene susceptibility.

Figure 10.7 shows a similar plot for observer sensitivity. The x-axis data is the same as in Fig. 10.6, whereas the y-axis shows data pooled across the 12 scenes to show observer sensitivity variation. In analogy with Fig. 10.6, the 5 of 21 observers shown correspond approximately to the extrema and quartile positions. The level of observer sensitivity variability depicted here is about half as large as we have observed on average. Most of the attributes we have studied exhibit coefficients of variation in the range of 10–40%, with a median value being near 25%. This value is distinctly greater than the 15% median variability in scene susceptibility.

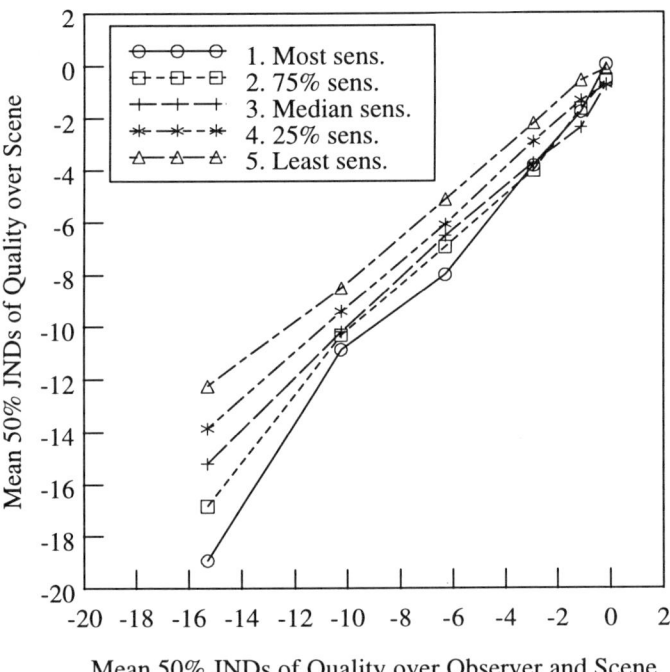

Fig. 10.7 Complement of Fig. 10.6 in which data is averaged over scene and plotted by observer. An average observer would plot as a 45° diagonal line; the range of behavior shown, which is about half as great as is typical, arises from variability in observer sensitivity.

10.3 Selection of Scenes and Observers

The importance of selection of representative sets of scenes and observers in experiments quantifying the JND increments of standard scales of image quality was discussed briefly in Sect. 6.2. Somewhat different factors are of importance when assessing the impact of particular attributes against an already calibrated standard scale. Observers more or less sensitive to image quality differences will provide similar ratings using a quality ruler, if their results are averaged over a number of attributes. Suppose one observer is very sensitive to image quality attributes, and a second observer is relatively insensitive to the same. The first observer, rating a sample containing a moderate amount of an artifact, perceives a substantial quality loss arising from the artifact. When lining up the ruler to match the quality, this observer perceives large differences in quality between the adjacent ruler images. The observer's high sensitivity to image quality attributes in general affects both the perception of the ruler and the test sample; these two effects approximately cancel when the test sample is rated against the ruler. The second observer perceives little quality loss in the test sample, but also perceives only small differences between the ruler samples, and so the placement of the test sample against the ruler is similar to that of the first observer. Essentially, the overall sensitivities of these observers have been mapped to that of the average participant in the paired comparison experiment used to calibrate the standard scale of image quality.

So far, we have been discussing overall sensitivity to image quality, which we have seen is normalized when ratings are made against a calibrated scale. What is not normalized, and what causes the residual variability exemplified by Fig. 10.7, is the distribution of ratios of sensitivity to the test attribute compared to that of the ruler attribute. An observer who is very sensitive to image quality attributes overall may still be relatively more or less sensitive to individual attributes. It is this inter-attribute relative sensitivity variability that is reflected in Fig. 10.7, which depicts the ratio of sensitivity of misregistration compared to the attribute varying in the quality ruler, which in this case is sharpness. Similarly, Fig. 10.6 depicts the relative scene susceptibility variability of misregistration compared to sharpness.

An important consequence of calibration to a standard scale of quality is that the requirement for representative sets of observers is greatly relaxed. Because overall sensitivity to image quality differences is compensated, the characteristics of the observer group have little effect on the quality loss functions obtained. In artifact studies, we have only rarely seen statistically significant differences between assessments by expert and novice observers, or between different demographic groups. Similarly, we have seen a number of comparisons of data from internal (company employee) and external (customer) studies of artifacts and the agreement usually has been very good. Internal studies are often advantageous in being faster, less expensive, and more confidential than external studies.

Figure 10.8 compares the misregistration data from expert and novice observers (classified based on training and photographic activity). The differences between the fits are small compared to the regression confidence interval widths. This degree of agreement between experts and novices is somewhat but not greatly superior to what is typically found in quality ruler experiments. In the rare instances when the equivalence of two regression fits has not been obvious upon visual inspection, we have determined statistical significance based on an extra sum-of-squares F-test comparing the case of fitting the two data sets with a

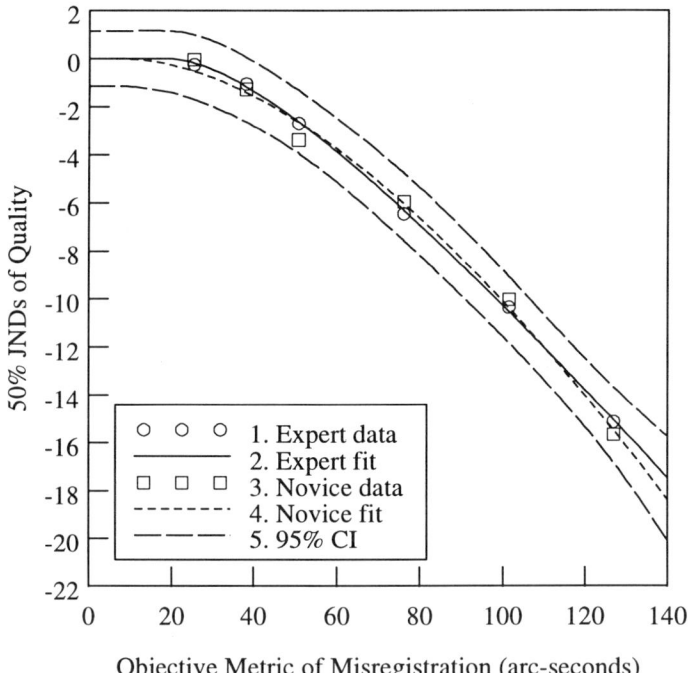

Fig. 10.8 Comparison of quality loss functions for expert observers and novices. This level of agreement is slightly, but not substantially, better than what is typically observed in quality ruler experiments.

single curve versus using two distinct regression curves (Draper and Smith, 1981).

Although overall sensitivity to image quality differences is normalized by calibration of the quality loss function against a standard scale, the preference distribution is unaffected by such a calibration, because it depends only on rankings. This implies that in studies of preferential attributes, selection of representative observers is important even when an experiment is calibrated to a standard scale. Although we have infrequently found differences between preferences of groups segregated based on expertise or demographics, it is still prudent to test a sufficiently diverse group of observers to determine whether distinct preference groups exist, and to quantify the different behavior of any such groups that are found.

Somewhat different factors influence scene selection. It is important that the scenes selected span the characteristics that might influence the impact of that attribute under study. For example, as demonstrated by Figs. 10.2 and 10.3, streaking is most visible is slowly varying (nearly uniform) areas, and is significantly less noticeable in scenes with substantial high-frequency content that can mask the artifact. Consequently, in a study of streaking, scenes possessing and scenes lacking uniform areas should be included to characterize the resulting variability. It is not critical that such a scene set be perfectly representative with regard to its characteristics, i.e., it need not contain the same proportion of scenes with and without uniform areas as does the population of customer images as a whole. Instead, it is adequate to characterize the distribution of the test scenes and contrast that with the population at large.

In some cases a substantial difference between a scene set and the population of images as a whole may be desirable to improve the signal in a psychometric experiment. For example, suppose that 25% of customer images contained fairly uniform areas, and that the susceptibility to streaking in the remaining 75% of images was much lower. An experiment using a representative scene set would expend 75% of its assessments on scenes generating relatively little signal, which could make it difficult to test for subtle failures of the objective metric. A better strategy might be to devote more assessments to scenes generating strong signal and scale back the results based on the actual distribution in customer images. As long as the frequencies of influential factors are known for both the study's scene set and the population of images pertinent to the intended application, the relevant susceptibility distribution can be reconstructed. An example of this approach is described in Sect. 21.2, where the effect of head size on the impact of skin-tone reproduction of quality is quantified.

10.4 Fitting Variability Data

There are one or two subtleties associated with using the IHIF to fit quality loss functions associated with subsets of observers and/or scenes. Consider Fig. 10.9, which compares fits to data averaged over observer, for the scenes most (top 25%) and least (bottom 25%) susceptible to misregistration. The regression curves cross near threshold, which is unlikely to reflect the actual behavior of the scene groups, because scenes that are more susceptible should probably show equal or greater quality loss at all levels of misregistration. The data subsets are only one-quarter as large as the total sample, so the confidence interval of the mean assessments will be twice as large as that of the grand mean

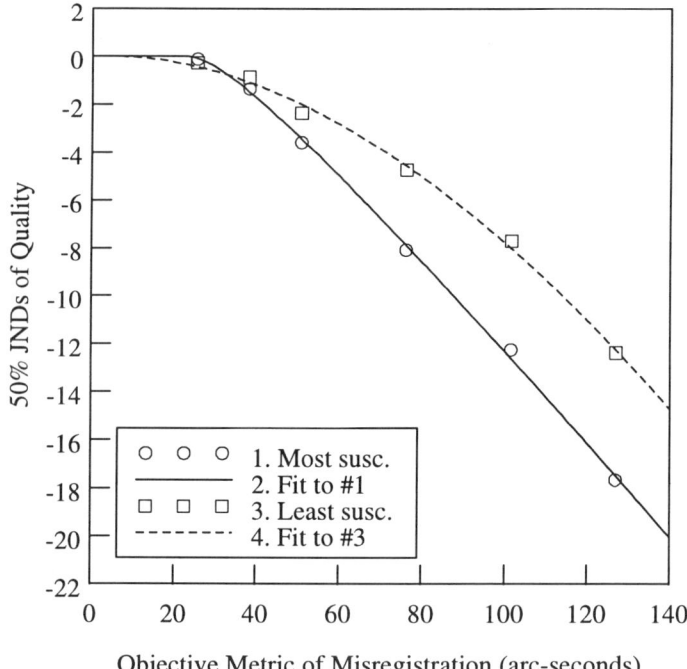

Fig. 10.9 Curve crossing inconsistency between IHIF fits of the 25% most and least susceptible scenes, assessed by the average observer. It is not surprising that these subsets, representing only one-quarter of the full data set, are likely to yield less robust regression results.

data. Consequently, some lack of robustness in the regression fits is hardly surprising.

A method we commonly use to improve regression robustness is to fix one of the three IHIF fit parameters at its value from the grand mean regression, which constitutes the best determined case because of its maximum sample size. Typically, the best parameter to constrain can be chosen by visual inspection of the subset fits from the three alternatives. In the case of misregistration, as with many artifacts studied, fixing the threshold radius of curvature worked best. Figure 10.10 shows the same data as in Fig. 10.9 but with the subset regressions so constrained. As desired, the crossing of the IHIF curves has been eliminated, without a significant loss in the degree of fit of the subset regressions.

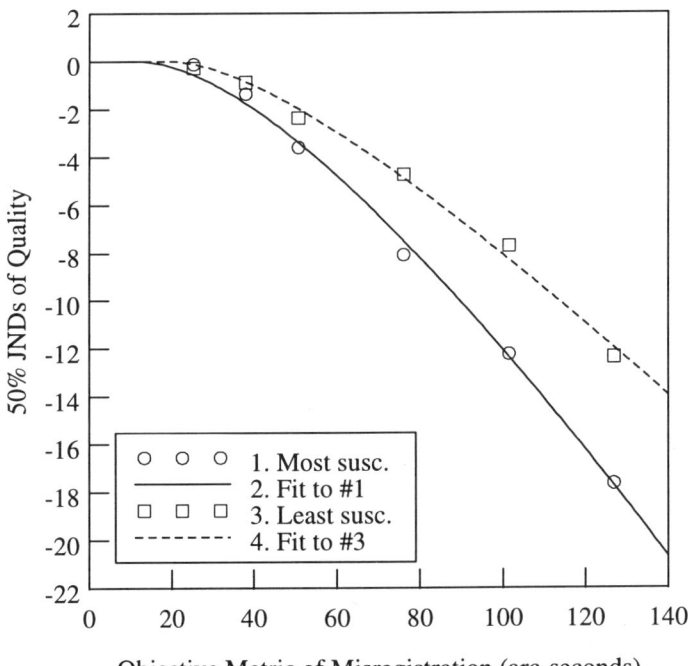

Fig. 10.10 Same data as in Fig. 10.9 but with threshold radius of curvature fixed at the value from the regression of the full data set. The fits are nearly as good as those of the previous figure, but the crossing of the curves has been eliminated, leading to improved consistency.

Essentially, the degree of fit of individual subset regressions has been traded for consistency among the subset regressions, which is a sensible exchange as long as the degree of fit of each of the individual regressions remains commensurate with the uncertainty in the data being fit. Sometimes, fixing the value of one parameter may be too restrictive, in which case restricting its value within a bounded interval may work better.

Figure 10.11 depicts a second type of potential inconsistency between independently fit subset regression curves. Curve #1 reproduces the IHIF fit to the grand means of the fully pooled data set from Fig. 10.1. The scale is changed from previous figures to emphasize the region near threshold. Curve #2 is the

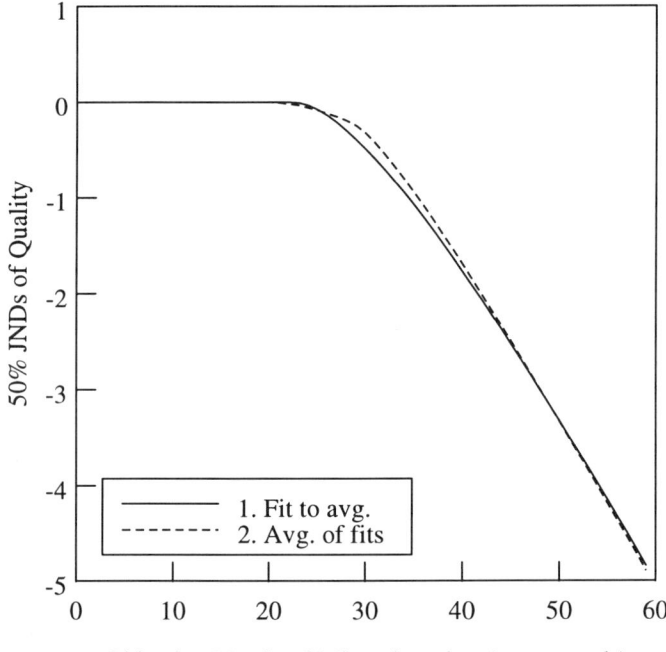

Fig. 10.11 Comparison of the IHIF fit to the full data set with the average of the fits of 50% more and less sensitive observers. These curves should match closely because the union of the non-overlapping subsets is equal to the full data set, but a very small discrepancy is present near threshold (note the change in the quality axis scale).

Scene and Observer Variability 145

average of the IHIF fits to data for 50% more and less sensitive observers (averaged over scene), with all three fit parameters being allowed to vary. Because the union of these two subsets, each comprising one-half the data, is identical to the full data set, the average of the quality loss data from the two subsets is equal to the grand mean quality loss data. Consequently, the average of the two subset IHIF fits (Curve #2) should, ideally, equal the grand mean IHIF fit (Curve #1). The two curves are indeed very close, although there is a slight discrepancy near threshold.

There is no guarantee that fitting subset data and averaging the regression curves (averaging fits) will give a comparable result to pooling the subset data first and then fitting the grand mean (fitting an average). Given the nonlinearity of the quality loss functions, changing the order of operations might be expected to introduce discrepancies. As in the case of undesired curve crossing just discussed, such inconsistencies can be reduced to an inconsequential level by fixing one of the fit parameters at its value in the grand mean regression. When the threshold radius of curvature in the subset regressions is fixed at the grand mean value, the average of the subset curves becomes so similar to the grand mean curve that it cannot be distinguished on the scale of Fig. 10.11, and so is not shown.

In summary, inconsistencies of both types discussed above (curve crossings and averages of fits versus fits of averages) are usually very minor effects. The examples shown in this section depicted larger discrepancies than are typical, so that the effects could be seen clearly in the figures. Both types of inconsistencies can be rendered negligible by placing restrictions on subset regressions, such as fixing one fit parameter at its value in the grand mean regression.

10.5 Usefulness of Variability Data

Once consistent sets of regressions have been obtained for a number of combinations of observer sensitivities and scene susceptibilities, they may be used to accomplish two principal tasks:

1. performing Monte Carlo calculations to predict quality distributions; and

2. setting quality specifications customized for a particular application.

The first task listed above, Monte Carlo modeling, involves generation of the distribution of image quality that would be produced by an imaging system as a result of variability in scene susceptibility, observer sensitivity, and many other factors. Further discussion of Monte Carlo modeling is deferred to Ch. 28, which is devoted entirely to this topic. The second task, setting customized specifications, involves identifying allowed limits for an objective metric based on a particular combination of observer sensitivity and scene susceptibility. Although this task is also discussed in detail in the latter portion of this book, a brief example, emphasizing choice of scene and observer characteristics, is described here.

Setting custom specifications involves selection of the appropriate observer sensitivity and scene susceptibility for determining imaging system requirements. For example, suppose that a noise specification were being set for a digital still camera marketed primarily for professional portrait photography. Portraits often have fairly extensive, slowly varying areas such as background sweeps, in which noise is particularly evident. Furthermore, noise in human faces may produce the appearance of unattractive blemishes or mottling, and so is minimally tolerated. Consequently, use of a regression from a subset containing more susceptible scenes might be appropriate in setting a camera noise specification. The purchase of a portrait imaging system might depend more on a professional photographer's assessment of its image quality than the impressions of consumers purchasing the final images. Because the sensitivity of professional photographers to noise might be quite high, IHIF fit parameters from a subset containing more sensitive observers, as well as more susceptible scenes, might be used. Once the IHIF fit parameters were chosen, an allowed quality loss arising from noise would be chosen, and the value of objective metric (e.g., RMS granularity) producing that quality loss would be found by numerical methods. The specification might then be stated in terms of the highest allowed RMS granularity at a particular camera exposure.

The responses of more sensitive observers in a calibrated psychometric experiment are not necessarily the same as those of observers with higher absolute sensitivity to image quality variations, because the overall sensitivity of the observers has been normalized, in a sense. For example, in a quality ruler experiment, as discussed above, observers identified as being more sensitive are those with a relatively higher sensitivity to the attribute being studied (in this case, noise) compared to their sensitivity to the attribute varied in the ruler (e.g., sharpness). Restated, more sensitive observers are those with higher sensitivity ratios, rather than higher absolute sensitivities. In principle, one might conduct paired comparison experiments to determine JND increments for particular groups such as professional photographers, so that no adjustment for overall sensitivity would be required. In practice, the use of high sensitivity ratio observers as a proxy for observers of high absolute sensitivity has proven to be satisfactory.

10.6 Summary

Variability of quality loss arising from scene susceptibility and observer sensitivity increases approximately proportionally with quality loss, so it is convenient to express the variability in terms of a coefficient of variation (standard deviation divided by the mean, expressed as a percentage). We have found variability of observer sensitivity spanning a range of ≈15–40%, and averaging ≈25%; and of scene susceptibility over a range of ≈10–40%, averaging ≈15%.

Overall observer sensitivity to image quality attributes is compensated when assessments are made against a standard scale of quality. Consequently, when studying artifactual attributes, it is not critical that a representative observer set be used. In cases of preferential attributes, the preference distribution may be a function of observer characteristics such as expertise or demographics, so greater caution is required in observer selection.

When studying the effect of an attribute on quality, it is desirable to select a set of scenes that spans the characteristics that might influence the impact of that attribute. The relative frequency of occurrence of different scene types need not be the same in the test set as in customer images as a whole, as long as the frequencies for both groups are known. It may be advantageous to bias the scene set towards scenes that generate better signal and subsequently scale back the results for relevancy.

One way to characterize the variability associated with observer sensitivity and scene susceptibility is to classify scenes and observers into small numbers of groups, and to form subsets of assessments based on combinations of the groups, e.g., more sensitive observers and more susceptible scenes. The data from the different subsets can be separately fit with IHIF regressions. These sets of results are useful in Monte Carlo calculations of quality distributions and in setting customized specifications for particular applications. In some cases it may be desirable to trade a small degree of fit of one or more subsets to increase the consistency between the different regressions. Frequently, fixing one IHIF fit parameter at the value obtained in the grand mean regression noticeably increases consistency and robustness of the subset fits.

11

Predicting Overall Quality from Image Attributes

11.1 Introduction

There are so many attributes that influence image quality that it is clearly impractical to perform factorial experiments that investigate the dependence of quality on combinations of attribute levels. Suppose one were interested in the effect of two attributes on image quality. To span the quality range of practical interest, with acceptable resolution, might require that 8 variations of each attribute be investigated. If 10 scenes were included in the study, and a full factorial design were followed, at 80 quality ruler assessments per one-hour session, it would take $10 \cdot 8^2/80 = 8$ hours of evaluation time per observer to rate every sample once. A similar design in which 3 attributes were varied would require 64 hours per observer, and so on.

Although more efficient statistical designs than a full factorial experiment could be employed, because experimental size would grow exponentially with the number of attributes under study, the resources required would be prohibitive for most sets of attributes that were sufficiently comprehensive to be of practical interest. Consequently, development of a method for predicting the overall quality associated with combinations of different attribute effects is essential.

Prediction of multivariate quality from contributing attributes is the topic of the present chapter, which is organized as follows. Section 11.2 discusses the nature of interactions between attributes and describes how to identify such interactions with simple experiments. Section 11.3 explains the assumptions behind the multivariate formalism, which has proven to be very successful in predicting overall quality from individual attributes that do not interact. The equations used in the multivariate formalism are presented in Sect. 11.4. Finally, in Sect. 11.5, the agreement between experimental measurements and calculations is demonstrated, and some of the implications of the multivariate formalism are analyzed.

11.2 Attribute Interactions

The problem being addressed in this chapter is to predict the overall quality of an image affected by a set of attributes, given the effect on quality that each of the attributes would have had by itself. Conceptually, this problem may be broken down into two steps:

1. predicting the effects of any interactions in which the presence of one attribute influences the perception of another attribute; and

2. predicting the overall quality from the perception of the individual attributes in the multivariate environment.

The first step is potentially complex because interactions between attributes may be expected to be specific to each pair of attributes (or even to higher-order interactions). Consequently, it could be difficult to develop an underlying theory explaining all interactions, although advanced models of the human visual system can predict certain aspects of some attribute interactions.

In contrast, the second step might be amenable to a general treatment because once the effects of the attributes, adjusted for any interactions in the first step, were expressed in strictly comparable units, the identity of the source of the quality changes might no longer matter. This section will discuss attribute interactions, and the next section will address the second step, which is accomplished using the multivariate formalism.

Suppose that two artifacts are under study, and that samples may be produced through digital image simulation containing any combination of these artifacts. To test for interactions between the two artifactual attributes, three sample series should be produced:

Series #1: variations in the magnitude of Artifact #1, with Artifact #2 absent;

Series #2. variations in the magnitude of Artifact #2, with Artifact #1 absent; and

Series #3. variations in which both Artifacts #1 and #2 are present and assume different combinations of magnitudes from Series #1 and #2, respectively.

To determine whether there is an interaction between the two attributes, samples from Series #3 are assessed for Attributes #1 and #2 by projection against Series #1 and Series #2, respectively, which are used like quality rulers. A consistent discrepancy between the matched objective values provides evidence for an attribute interaction.

For example, if Attribute #1 were contouring, and Attribute #2 noisiness, observers could be asked to identify the sample in Series #1 with contouring most similar to that of a sample from Series #3. If, on average, samples from Series #3 were matched up with the samples from Series #1 having the same objective level of contouring, it would be concluded that the presence of noisiness does not affect the perception of contouring. In contrast, if the samples from Series #3 were systematically matched with samples from Series #1 having lower values of the objective metric of contouring, it could be concluded that noisiness reduces perceived contouring. In such a case, the interaction can be addressed by constructing a model for the impact of noise on contouring in either objective or subjective space. For example, an objective metric of contouring might be modified by subtracting a term depending on RMS granularity. Examples of such interaction models involving modification of objective metrics are presented in Ch. 17.

It is important to recognize that perceptual independence does not imply that the attributes may not be affected in a correlated fashion by some objective change in the image. For example, blurring an image decreases both its perceived sharpness and noisiness; however, this fact does not imply that sharpness and noisiness interact perceptually. Rather than either attribute affecting the perception of the other, both are influenced by the same physical change. Objective metrics of image sharpness and noisiness should properly reflect the impact that the blurring operation would have upon each of the perceived attributes. A perceptual interaction refers only to a case in which the same objective amount of one attribute is perceived differently depending on the level of a second attribute.

In practice, there are relatively few such interactions that are of real perceptual significance within the range of levels of attributes that are of interest in the design of imaging systems, particularly in the case of artifactual attributes.

Although there may be changes in the appearance of images as a result of interactions between attributes, frequently such changes do not affect quality; i.e., the images may look different, but not particularly better or worse. As discussed in detail in Ch. 17, if a set of perceptual attributes are defined with care, and their objective metrics formulated with orthogonality in mind, it is usually possible to make them perceptually independent. Simple comparisons of the type described above may be helpful in clarifying questionable cases.

In summary, the interaction of perceptual attributes in practice is less pervasive than might be expected, and simple, robust comparisons may be made to assess whether the assumption of perceptual independence between two attributes is violated. If an interaction is significant, a model of the interaction may be constructed based on experiments involving combinations of attributes.

11.3 Multivariate Formalism Assumptions

Unlike the problem of interactions, in which attribute-specific solutions are required, the effect of combinations of independent attributes is amenable to a general description, which we shall refer to as the multivariate formalism. The use of the term formalism reflects that the problem is not simply solved by writing an equation that relates a specified unknown quantity to specified known quantities. Although it is clear that overall quality is the quantity sought, and that the known variables are quality changes arising from individual attributes in isolation, none of these quantities are well-defined physical quantities such as mass or time, and so the fashion in which they should be measured is not obvious. Restated, writing an equation relating overall quality to contributing attributes is only the second half the task; before this can even be attempted, a measurement space must be found in which it is actually possible to write such a generally valid equation. Various spaces have been tried by different researchers; e.g., Prosser, Allnatt, and Lewis (1964) related impairment harmonically to the complement of a 1–5 scale of quality; de Ridder (1992) defined impairments as fractions of a maximum quality loss; and Bartleson (1982) used 1–9 categorical scales of sharpness, graininess, and overall quality, which he assumed constituted interval scales. The equations developed in these spaces have fit only single experiments, and have not been used successfully to predict the results of other experiments (Engeldrum, 1999).

In the multivariate formalism, it is assumed that the proper measurement space for quantifying overall quality and individual attributes is JNDs of overall image quality. The distinction was made in Sect. 6.3 between JNDs of overall quality arising from an attribute, and the JNDs of the attribute itself. Recall that the JND increment of an attribute itself relates to the ability to distinguish different degrees of that attribute, whereas the JND increment of quality arising from the attribute relates to the extent to which the attribute differences matter. The multivariate formalism requires that the impact of each attribute be expressed in identical units. JNDs of quality arising from an attribute meet this criterion because all assessments are made against a common standard, and involve the same task. JNDs of attributes themselves do not necessarily meet the identical units criterion, because the relationship of detectability and significance is not necessarily fixed among different attributes.

In the multivariate formalism, it is assumed that a universal relationship exists between a list of JNDs of quality changes arising from a set of independent attributes, in isolation, and the overall quality change in JNDs when all attributes are present in the same sample. The four requirements listed below are placed upon this relationship.

1. A list of only one quality difference must map to itself, i.e., if only one attribute affects quality, the overall quality difference must equal the quality difference arising from that attribute. This first identity requirement is a mere formality, but is stated for completeness.

2. Adding an element equal to zero to the quality difference list does not change the overall quality difference. This second identity requirement seems trivial, but it is easily violated inadvertently, e.g., by using an average of the elements of the list in the relationship.

3. When attribute quality differences are small in magnitude, they approximately sum. This additivity requirement is intuitively plausible; if each of three attributes in isolation degrades quality by one JND, it seems reasonable to expect that the overall quality loss corresponds to approximately three JNDs, because such subtle changes are unlikely to affect one another very much.

4. When one or more attribute quality differences are large in magnitude, modest changes in other attribute quality differences have little impact on overall quality. This criterion is called the suppression requirement because the presence of a serious degradation suppresses the impact of minor degradations on overall quality. Conversely, and more importantly, if an imaging system has several minor flaws and one major flaw, fixing the minor flaws will not yield much improvement because the major flaw largely determines the quality. In common parlance, the suppression requirement reflects the notion that the worst problem dominates.

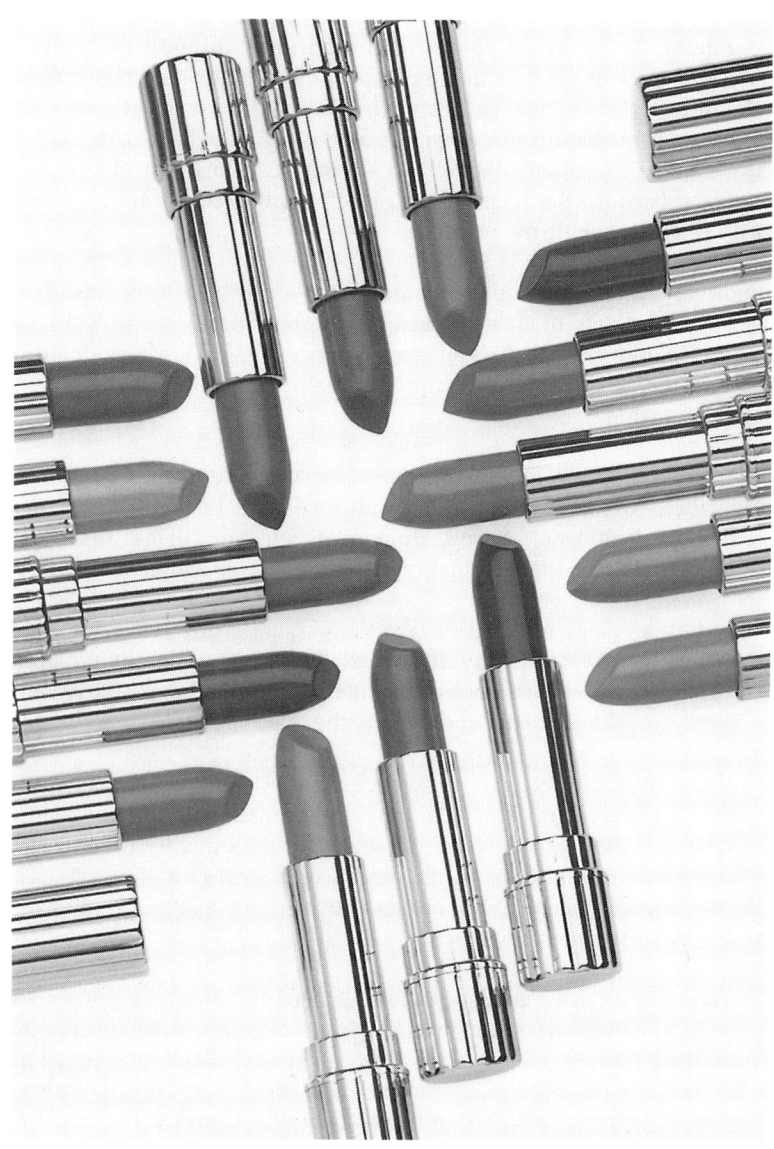

Fig. 11.1 Starting image for multivariate demonstration, having no intentionally introduced artifacts.

Predicting Overall Quality from Image Attributes

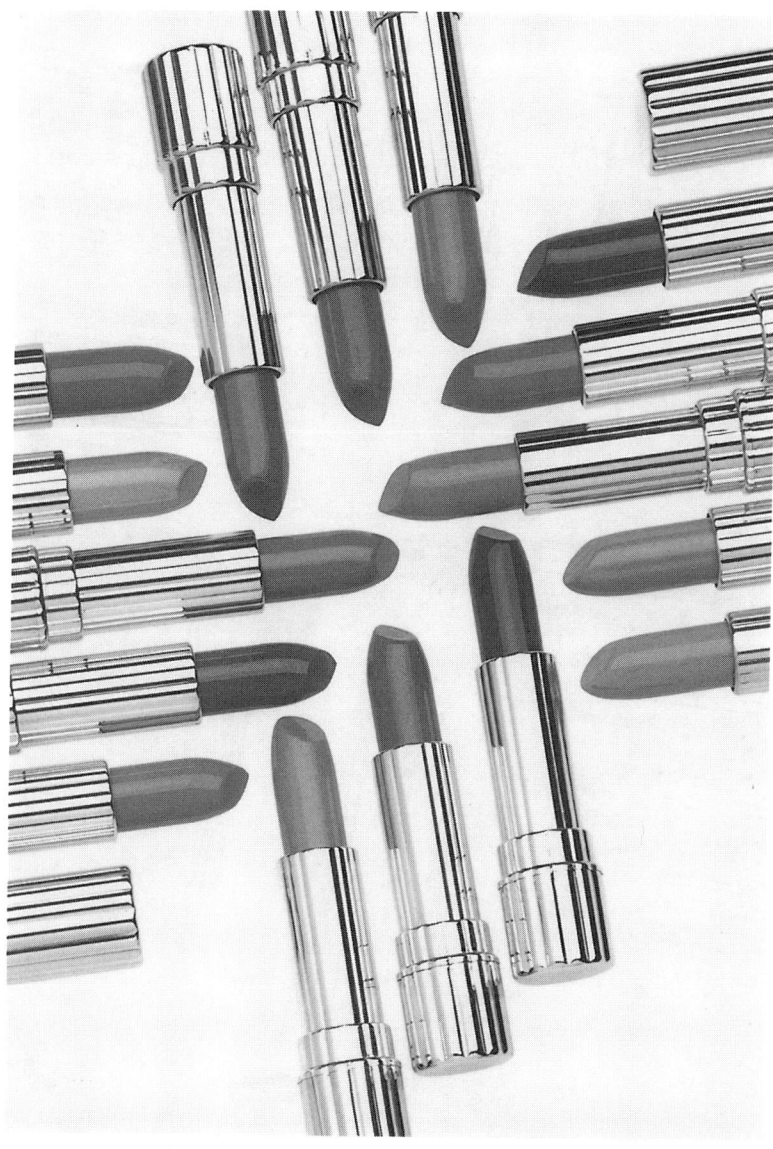

Fig. 11.2 Image slightly degraded by contouring, evident as areas of constant density separated by discrete jumps in the lipstick material.

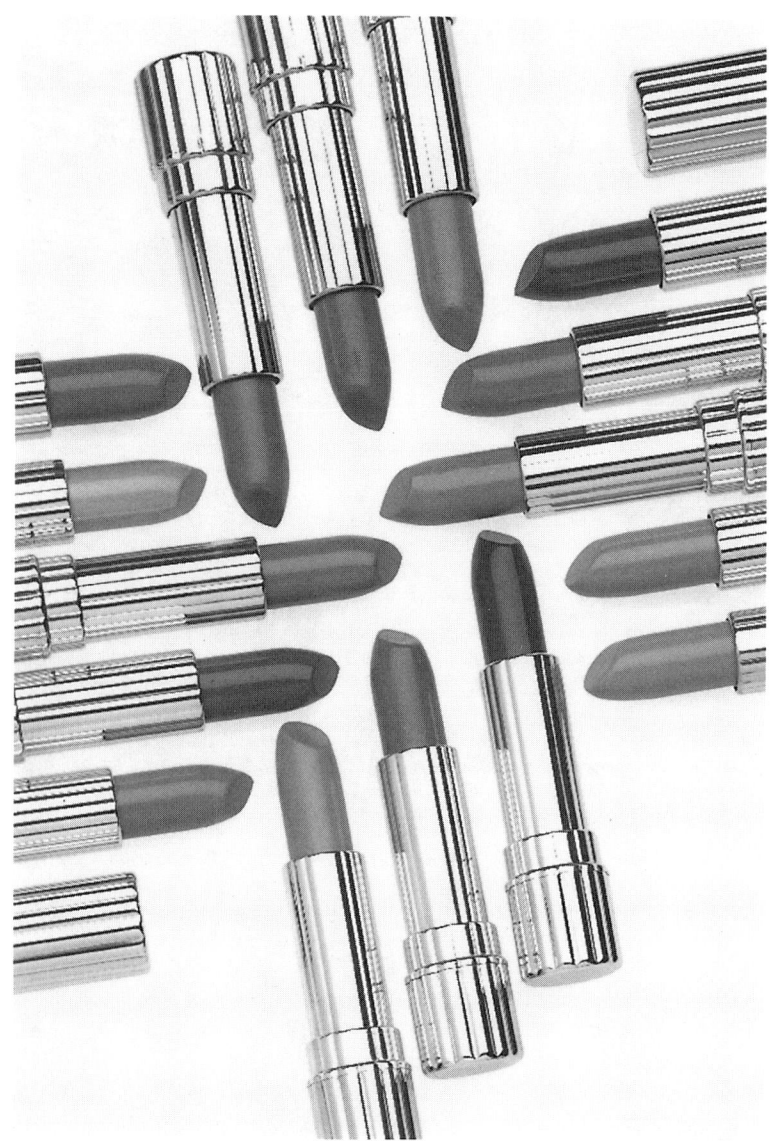

Fig. 11.3 Image slightly degraded by aliasing, evident in the waviness or jaggedness of lines of the lipstick tubes.

Predicting Overall Quality from Image Attributes

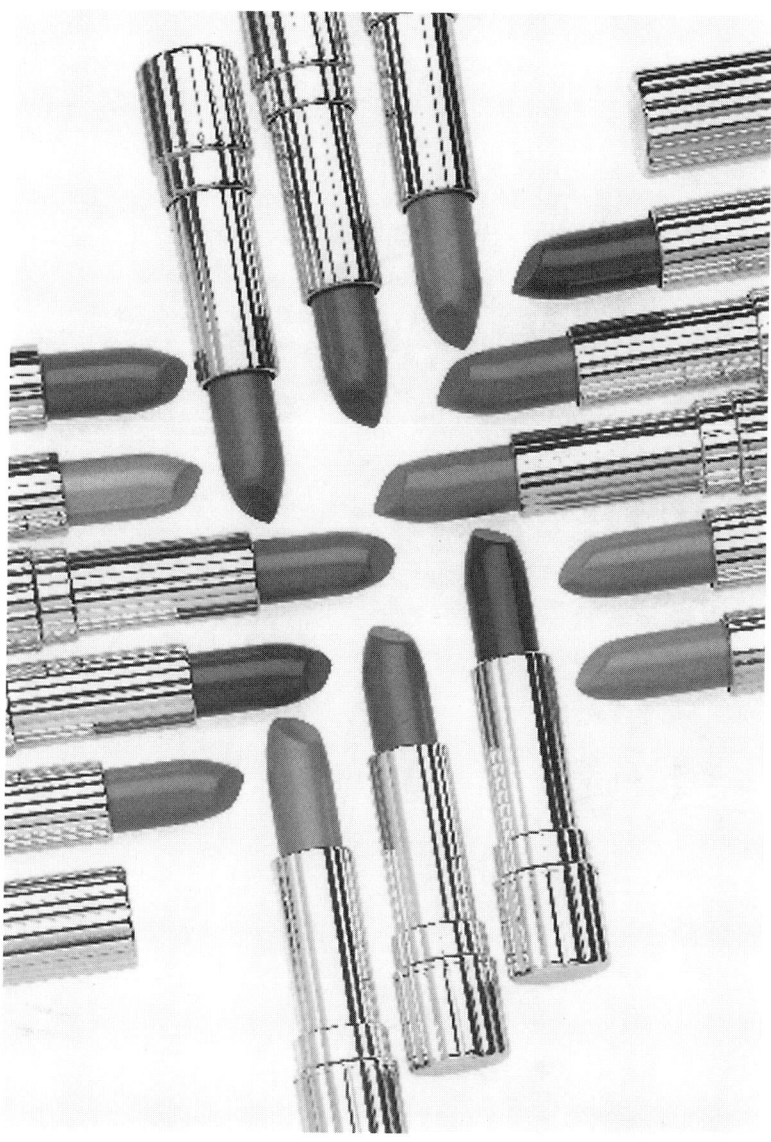

Fig. 11.4 Image strongly degraded by aliasing.

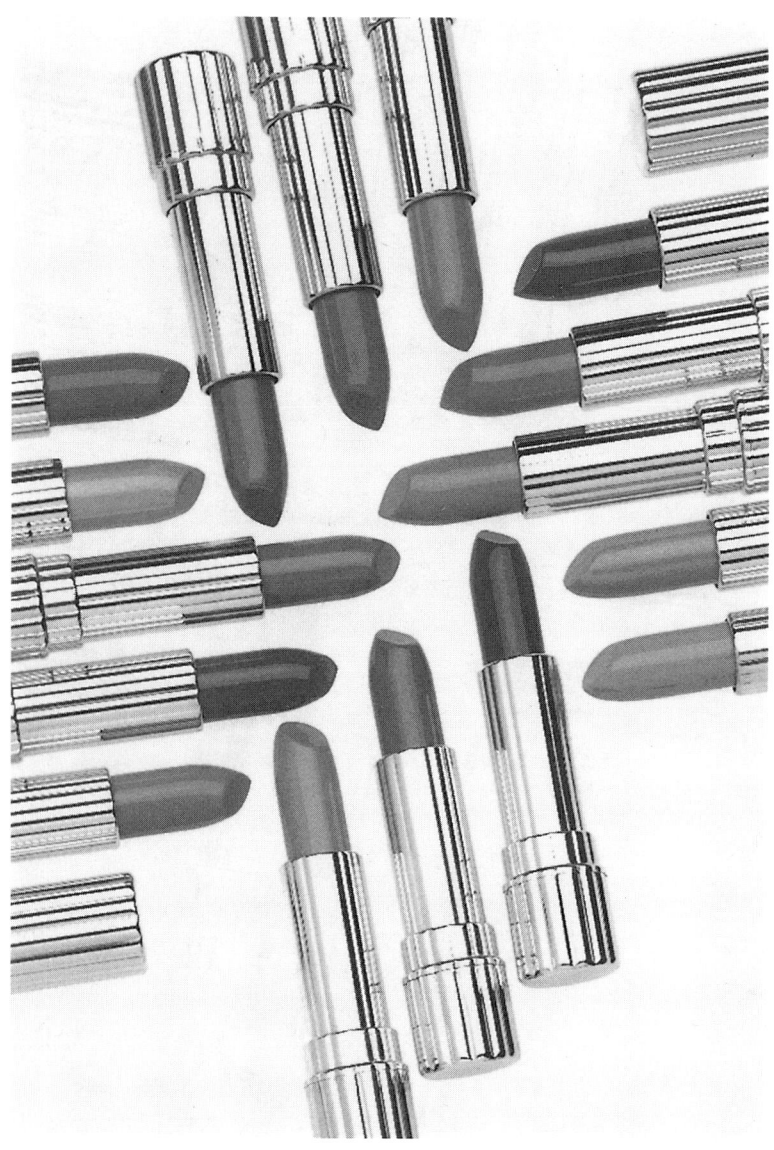

Fig. 11.5 Image slightly degraded by both contouring (as in Fig. 11.2) and aliasing (as in Fig. 11.3), the effects of which approximately add.

Predicting Overall Quality from Image Attributes

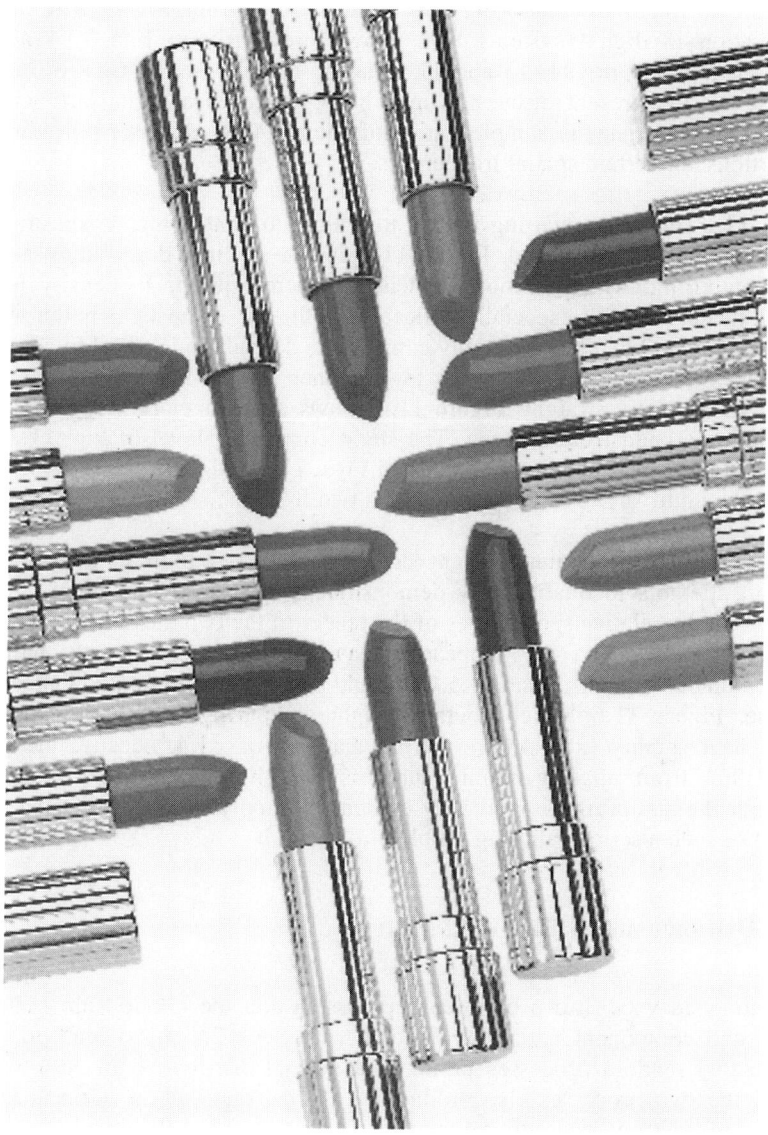

Fig. 11.6 Image strongly degraded by aliasing (as in Fig. 11.4), which suppresses the slight contouring (at the same level as in Fig. 11.2).

Figures 11.1–11.6 attempt to pictorially demonstrate additivity and suppression using the attributes of contouring and aliasing. Contouring, arising from signal quantization in digital systems, was shown previously in Fig. 10.4 and is discussed in detail in Chs. 15 and 17. Aliasing is addressed in Ch. 18, but for purposes of the present discussion may be regarded as a digital artifact that occurs when a signal is sampled at insufficiently high resolution, leading to "distortions" of certain spatial frequencies.

Figure 11.1 shows the starting image, to which no contouring or aliasing has been intentionally introduced. Figure 11.2 depicts a slight degradation arising from contouring, which is most evident as discrete jumps in density in the lipstick material, and is especially noticeable in the darker lipsticks in the sets of three. Figure 11.3 exhibits a roughly comparable degradation caused by aliasing, which is most obvious as wavy or jagged lines on the lipstick tubes, which should be perfectly straight. Figure 11.4 shows a much more severe level of degradation arising from aliasing. The slight contouring level of Fig. 11.2 and the slight and severe levels of aliasing in Figs. 11.3 and 11.4, respectively, are used in the multivariate samples in the next two figures.

Figure 11.5, which contains slight degradations from both contouring and aliasing, attempts to qualitatively demonstrate additivity. Although the effect depends on the relative sensitivities of the reader to the two attributes, the intent is that Figs. 11.2 and 11.3 appear to have quality approximately halfway between that of Figs. 11.1 and 11.5, as should occur if small quality changes are additive. Figure 11.6, which contains slight contouring and severe aliasing, should have quality little different from that of Fig. 11.4, because the large degradation from aliasing should suppress the slight effect of contouring. Although the contouring is still very evident, it simply does not matter in the context of such a serious aliasing problem.

11.4 Distance and Minkowski Metrics

An analogy may be drawn between suppression and the relationship between overall and component variability. Recall from Sect. 2.4 that if each of two uncorrelated random processes produces a normal distribution of results, the sum of the two processes also produces a normal distribution of results, the variance of which is equal to the sums of the variances of the two component processes. Consequently, the standard deviation of the overall process (the sum of the components) is the square root of the sum of the squares of the component standard deviations (i.e., their root-mean-square (RMS) sum). Suppose that two processes have uncorrelated normal distributions with standard deviations of 1.0

and 3.0 units; the standard deviation of the sum of these processes would be $(1.0^2 + 3.0^2)^{1/2} \approx 3.16$ units, which is only ≈5% larger than if the first process, with the smaller standard deviation, had not contributed at all. In this case, the variability of the second process dominates, and modest changes in the variability of the first process would have little impact on the overall variability.

The relationship between the standard deviations of two component processes and their sum is like that between the lengths of the sides of a right triangle. The component standard deviations correspond to the two legs of a right triangle, and the composite standard deviation corresponds to its hypotenuse. More generally, for N components, the overall standard deviation is the N-dimensional length associated with N component standard deviations, each lying along one of N coordinate axes. The N standard deviations are then the magnitudes of the projections of the overall vector onto each of the coordinate axes (i.e., the dot products of the overall vector with the N orthogonal, unit coordinate vectors).

There are many modeling applications in which an optimum compromise position between multiple factors is to be found. One approach to such problems is to evaluate a variety of potential solutions in terms of a cost (demerit) function, and to select the solution having the least cost. Such cost functions often are assumed to be N-dimensional distance metrics. Typically, the deviations of each contributor from its optimal value are weighted and then combined using an RMS sum. The weighting step places all contributions into equivalent units, and the RMS sum emphasizes the larger weighted deviations.

An RMS sum is a special case of a Minkowski metric, which is based on taking the nth root of the sum of nth powers. In our application, this takes the form:

$$\Delta Q_m = -\left(\sum_i (-\Delta Q_i)^{n_m} \right)^{1/n_m} \tag{11.1}$$

where ΔQ_i is the quality change arising from the ith attribute, ΔQ_m is the overall quality change, and n_m is the power of the metric (which need not be an integer). The negative signs ensure that only positive numbers are raised to a power. When $n_m = 2$ the Minkowski metric is an RMS sum; higher powers lead to greater degrees of suppression. For example, if standard deviations of 1.0 and 3.0 are added as the cube root of sums of cubes ($n_m = 3$) instead of RMS sums, the overall standard deviation would be ≈3.04, instead of ≈3.16, so that the less variable component would have even less effect than in the RMS case.

Minkowski metrics generally meet the two identity requirements of the previous section. They also meet the suppression requirement as long as $n_m > 1$; however, they violate the additivity requirement unless $n_m = 1$, which is simply linear addition of the components. For example, regardless of the magnitude of the component contributions, the Minkowski sum of N equal components is $N^{1/n_m} \times$ as large as any individual component, instead of N times as large, as stipulated by the additivity requirement at lower degradation levels. For example, if three attributes each produce one JND of degradation, a Minkowski sum of power two would be $-3.0^{1/2} \approx -1.7$ JNDs, well short of the -3.0 JNDs that would result if the effects were perfectly additive.

A Minkowski metric cannot rigorously meet both the additivity and suppression requirements, because the former implies $n_m = 1$ and the latter $n_m > 1$; however, if the definition of a Minkowski metric is extended by allowing the power to be a function rather than a fixed constant, both requirements can be met. Specifically, the power must have a value near unity when all attribute quality degradations are small, and must increase as one or more of the attribute quality degradations becomes large. We have had excellent success predicting overall quality of multivariate samples using a variable-power Minkowski metric having these properties, with the power being given by:

$$n_m = 1 + c_1 \cdot \tanh\left(\frac{(-\Delta Q)_{max}}{c_2}\right) = 1 + 2 \cdot \tanh\left(\frac{(-\Delta Q)_{max}}{16.9}\right) \quad (11.2)$$

where $(-\Delta Q)_{max}$ is the most severe component degradation. Because the hyperbolic tangent of a positive argument ranges from zero to one, the power n_m varies from one to three, and so the component quality changes add in a fashion varying continuously from summing linearly for small changes, to adding as the cube root of a sum of cubes at large changes. When the maximum degradation is $-16.9 \cdot \tanh^{-1}(1/2) \approx -9.3$ JNDs, the degradations add as an RMS sum.

11.5 Predictions and Measurements of Multivariate Quality

Figure 11.7 substantiates the ability of the multivariate formalism, embodied in Eqs. 11.1 and 11.2, to explain the results of diverse multivariate experiments. The first experiment, with data plotted as circles, involved assessment of samples varying in sharpness and noisiness. The second experiment, plotted as squares, used samples varying in these two attributes plus a third attribute, tonal clipping, caused by camera misexposure. As discussed in Ch. 15, tonal clipping is the truncation of information in light and/or dark areas of an image, leading to featureless dull whites and/or hazy blacks. The third experiment, plotted as plus symbols, continues the series, employing samples varying in the previous three attributes as well as a fourth attribute, streaking (see Fig. 10.2). The fourth study, plotted as asterisks, involved two attributes, contrast (see Figs. 4.1 and 4.2) and color balance. The four experiments were each carried out independently and by different investigators, using three psychometric methods (categorical sort, hardcopy quality ruler, and softcopy quality ruler). The attributes varied span both image structure and color/tone reproduction, and include both artifactual and preferential types.

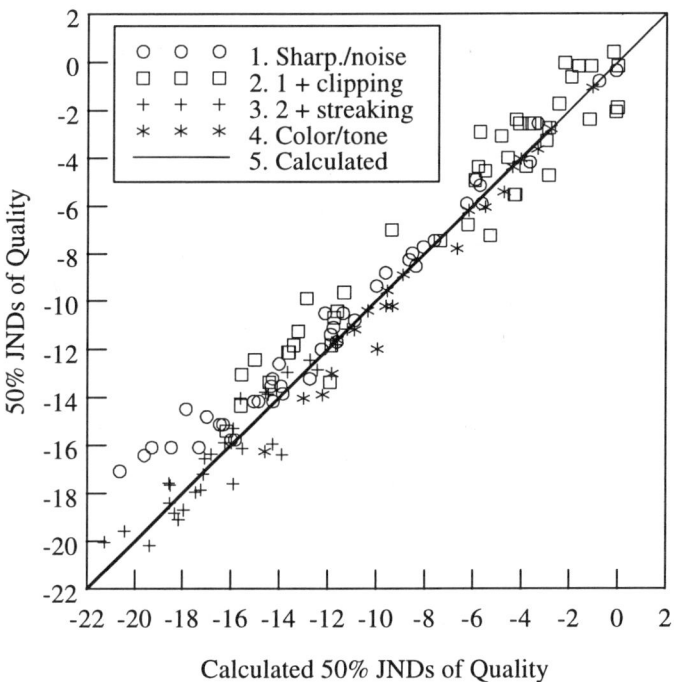

Fig. 11.7 Calculated versus measured overall quality of multivariate samples from four independent studies (see text for details). The agreement is excellent over the full range of quality tested, strongly substantiating the general validity of the multivariate formalism.

To derive the form of the Minkowski power function, Eqs. 11.1 and 11.2 were assumed to apply, and initially the constants c_1 and c_2 in Eq. 11.2 were varied as fit parameters. The two fit parameters were sufficiently correlated (one increasing as the other decreased) that there was no significant advantage to varying both as fit parameters, as long as $c_1 \geq 2$ (at smaller values, poorer fits resulted, whereas at larger values, equally good fits were obtained). The constant c_1 was somewhat arbitrarily chosen to be exactly two to minimize the range of possible Minkowski powers, given by $1 \leq n_m \leq 1 + c_1$, while still fitting the data well. With this choice, the simultaneous best fit to the data of the four experiments was obtained with $c_2 = 16.9$ JNDs, as shown in Fig. 11.7. The calculated values for each of the four experiments fall on the 45° solid line. All four sets of data agree closely with the multivariate formalism computations over the quality range of more than twenty JNDs spanned in the experiments. This is a remarkable result, particularly given the use of essentially only a single fit parameter. Both the diversity of types and numbers of attributes varied, and the variety of psychometric and display methods employed, argue for the general validity of the multivariate formalism.

To develop further intuition regarding the implications of the multivariate formalism, it is helpful to plot its predictions for two attributes in two different ways. Figure 11.8 shows iso-quality contours in the two-dimensional space of the individual attributes. The x- and y-axes are quality changes arising from the two attributes in isolation, expressed in 50% JNDs of quality. The contours connect points having the same overall quality; these contours are at −1, −3, ..., −15 50% JNDs. The contours are easily identified because, for example, the −15 JND contour must pass through (0, −15) and (−15, 0) by the two identity requirements. At small degradations (lower left corner), the attribute changes nearly sum, and so the contours are almost straight lines at a 45° angle, in accordance with the additivity requirement. The contours become nearly horizontal or vertical when the quality change from one attribute is substantially greater than that of the other attribute, reflecting suppression. For example, starting on the x-axis at the −13 JND contour, where the quality change from the first attribute is −13 JNDs, and then increasing the degradation from the second attribute as shown by the long arrow, requires ≈ −8.6 JNDs of change in the second attribute to shift the overall quality by just −2 JNDs, to the −15 JND contour. In contrast, from the position on the −13 JND contour where the contributions of the two attributes are equal, only ≈ −3.4 JNDs of shift in the second attribute (short arrow) is required to change the overall quality by the same amount. When the attribute effects are approximately balanced, changes in either attribute will significantly affect overall quality.

The contours in Fig. 11.8 look approximately circular away from the origin. This is expected because the variable Minkowski power is constrained to vary from one to three, and a fixed intermediate value of two would yield circular contours. The differences are sufficient, however, that use of any fixed power metric (not just $n_m = 2$) yields predictions that differ systematically from measurements, especially as the number of attributes increases. The small cusps in the curves along the diagonal x = y, which will also be seen in the next figure, are of no perceptual or practical significance. They result from using only the maximum degradation in Eq. 11.2 for simplicity. Including other attributes in the Minkowski power with low weightings reduces these cusps but does not improve the agreement with measured data significantly, so the additional complexity is deemed to be unwarranted.

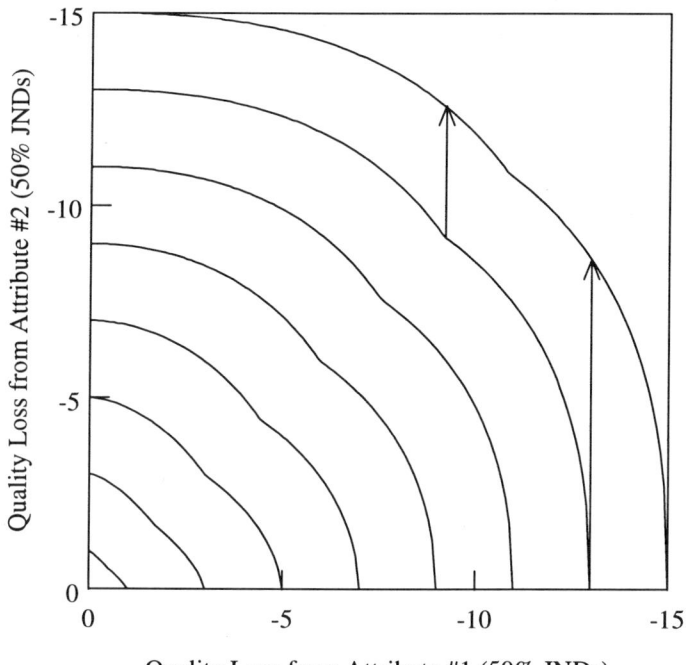

Fig. 11.8 Predicted iso-quality contours for two attributes based on the multivariate formalism. At smaller quality losses the contours are nearly linear (additivity), whereas at larger quality losses a significantly worse attribute will dominate, curving the contours (suppression).

Frequently, image quality attributes may be affected in opposing ways by a process or a variation in system design parameters. For example, as mentioned earlier, performing a blurring operation increases unsharpness and decreases noisiness (if it is not already subthreshold). If the sharpness of an image were very high but the noise were objectionable, a blurring operation might improve overall quality by better balancing the attributes. Similarly, an unsharp image with little noise might benefit from a sharpening operation. Figure 11.9 shows how overall quality might vary when the balance of two attributes is changed. The quality losses arising from the two attributes are constrained to sum to a constant amount, so that if one improves by a certain amount, the other becomes worse by the same amount. The x-axis shows the ratio of the first attribute to the sum of the attributes; this fraction varies from zero to one. The y-axis shows

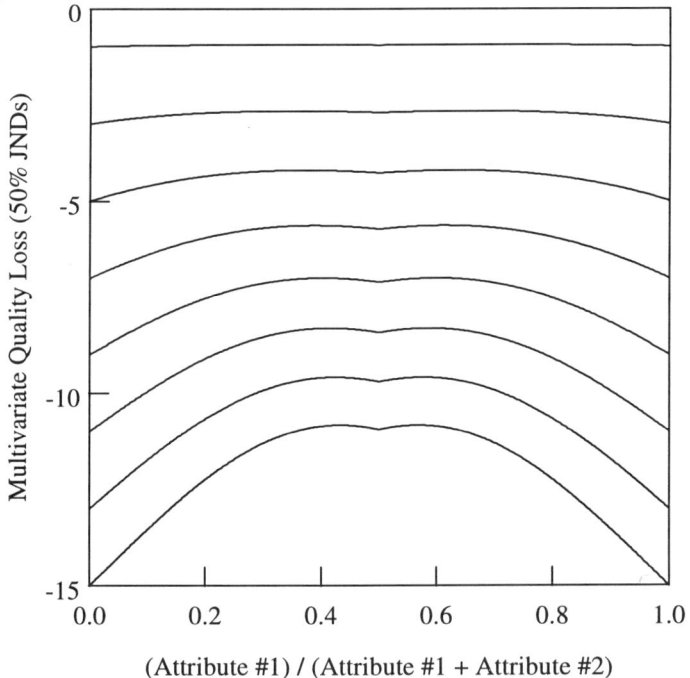

Fig. 11.9 Predicted overall quality loss for two attributes that sum to a constant amount. The best overall quality occurs when the two component losses are approximately equal, balancing their contributions so that neither attribute dominates.

overall quality. Each curve depicts the relationship for a different sum of attributes, having values of −1, −3, ..., −15 JNDs. At low amounts of degradation, the balance between the attributes has little effect on quality because the effects are nearly additive. In contrast, at higher quality losses, the balance significantly influences overall quality, with the best quality occurring when the magnitudes of the two attributes are similar (i.e., near $x = 0.5$). It is a general property of well-designed imaging systems that no single attribute consistently dominates overall quality, but rather that a balance is maintained.

This chapter completes Part I of this book, which has focused on perceptual and psychometric issues relating to the definition, measurement, and mathematical description of image quality and its component attributes. Part II of this volume will describe the development of objective correlates of image quality attributes.

11.6 Summary

It is not usually practical to perform factorial experiments that fully map out the dependence of overall quality on different attributes of importance in a practical imaging system, because the amount of experimentation grows exponentially with the number of attributes involved. The multivariate formalism greatly simplifies the task of predicting overall image quality from a knowledge of individual attribute levels, provided that the attributes are approximately perceptually independent. Where an interaction between attributes is suspected, simple diagnostic tests can be run to evaluate the significance of the interaction. If present, such interactions may be modeled as described in Ch. 17, after which the adjusted results may be used in the multivariate formalism. In practice, significant interactions between carefully defined attributes are quite uncommon, especially among artifactual attributes.

The multivariate formalism is based on the assumption that, if the impact on overall quality of a number of attributes is known, when the attributes are in isolation, then the total impact of their presence together in an image can be predicted from a universal combination rule. The effects of the isolated attributes must be expressed in identical units based on analogous assessment tasks; a natural choice for this purpose is JNDs of overall quality. Several requirements are specified regarding the combination rule, including that small quality losses be approximately additive, and that serious degradations dominate minor ones, in the sense that fixing a small problem in the presence of a large one does not significantly improve overall quality.

The variable-power Minkowski metric of Eqs. 11.1 and 11.2 meets the stated requirements and explains the results of four independent experiments involving: (1) two, three, and four simultaneously varying attributes; (2) both artifactual and preferential attributes, spanning sharpness, noise, digital artifacts, tone reproduction, and color reproduction; and (3) images displayed both in hardcopy and softcopy modes. These results substantiate the accuracy with which predictions of overall quality may be made using the multivariate formalism. Consequently, the systematic study of a number of image quality attributes may be undertaken with a reasonable expectation that the results from the different investigations may ultimately be integrated into a unified model for predicting overall image quality.

Part II

Design of Objective Metrics

Part II starts with a discussion of the desirable characteristics of objective metrics that correlate with perceived attributes of image quality (Ch. 12). Methods of analysis for testing and refining such metrics based on psychometric data are reviewed (Ch. 13), and a detailed practical example is presented (Ch. 14). The treatment of artifacts that vary across an image, such as those that depend on density, is subsequently considered (Ch. 15). Methods for analyzing experiments with several varying attributes, based on the multivariate formalism, are described (Ch. 16), followed by examples of extensions of objective metrics to account for perceptual interactions (Ch. 17). Artifacts that present a continuum of appearances (Ch. 18), and those having a strong scene dependence (Ch. 19), are addressed next. Finally, methods for accommodating preference are demonstrated using color and tone reproduction results (Chs. 20–21).

12

Overview of Objective Metric Properties

12.1 Introduction

Part II of this book is concerned with the design of objective metrics. The definition of an objective metric is as follows.

> An objective metric is a single number that may be determined through objective means and is correlated with a perceived attribute of quality in an image, accounting for its viewing conditions and the properties of the human visual system.

The primary reason for defining and using objective metrics is that they often permit replacement of resource-intensive perceptual experiments with efficient measurements and/or modeling. Although creation of a reliable objective metric requires an initial research investment, the benefits can be considerable if an attribute influences the quality of many different imaging systems. If objective metrics correlating with different attributes are measured or calculated, they may be converted to quality changes in JNDs using IHIF regressions. These quality changes arising from individual attributes may then be combined to predict overall quality using the multivariate formalism, providing a powerful analytical capability.

This chapter, which provides an introduction to objective metrics and other objective quantities, is organized as follows. Section 12.2 lists some of the applications in which objective quantities are useful. Methods for determining the value of a defined objective metric are considered in Sect. 12.3. Section 12.4 classifies objective quantities into four types and describes their properties. Finally, an example of a benchmark metric is presented in Sect. 12.5.

12.2 Usefulness of Objective Metrics

The intrinsic value of objective metrics arises from the combination of their objective nature and their correlation with attributes of image quality. We have found objective metrics to be useful in the following applications:

1. setting product specifications based on perceptually relevant criteria;
2. predicting the quality expected from novel imaging systems to evaluate their viability;
3. supporting advertising claims;
4. providing product information to customers;
5. monitoring manufacturing processes for quality assurance purposes;
6. quantifying technological progress of imaging components and systems;
7. benchmarking competitive products;
8. substantiating improvements to obtain patent protection; and
9. establishing industry standards related to image quality.

The first of these applications is emphasized in Part III of this volume and warrants additional explanation. Given the constant desire to reduce product development cycle time, it is a substantial advantage to be able to set specifications in a matter of hours or days, rather than weeks or months. In many cases, time does not permit empirical experimentation to be pursued at all, in which case a decision is sometimes made by one or a few people casually examining a small number of samples. This is a far less robust approach than basing the decision on an objective means backed by rigorous psychometric experimentation involving many sample levels, scenes, and observers. A typical approach to developing such a specification is to define the image quality requirements in perceptual terms of JNDs of quality, for particular observer sensitivity and scene susceptibility classifications, and then to map these requirements to equivalent objective metric terms for convenience. For example, it might be required that the quality loss arising from noise in an imaging system should not exceed one JND for the average observer and scene. From IHIF regression results, this requirement could be translated to a maximum permissible value of an objective metric of noisiness, so that the system specification could be stated in terms of objective quantities.

12.3 Determination of Objective Metric Values

Objective metrics usually are determined by one of three methods:

1. direct experimental measurements;

2. system modeling; or

3. visual assessment against calibrated standards.

In the first method, experimental measurement, a target is propagated through an imaging system and the output image is measured. Differences between the initial properties of the target and the final image properties are used to characterize the system. For example, color and tone reproduction may be characterized by photographing a target consisting of colored patches of known spectral properties, allowing it to pass through the entire imaging system, and measuring the spectra of patches in the output image. Another example is the imaging of a uniform field to determine the system noise power spectrum (NPS), which is useful in characterizing several different types of noise and non-uniformity (see Ch. 13). A third example is the propagation of a slanted edge, sine wave, or square wave target through an imaging system to determine its modulation transfer function (MTF).

The second method, system modeling, involves the prediction of the output characteristics of an imaging system based on properties of its components. These properties may be experimentally measured (Ch. 24) or they may be calculated using component models based on engineering parameters (Ch. 23). For example, an objective metric of unsharpness might be computed from the MTF of an imaging system. If the MTF of each of the system's components were known, and if the system were linear, the system MTF could be predicted from linear systems theory (Ch. 22), permitting computation of the objective metric. Sequences of components, called subsystems, can be measured as black boxes, reducing the number of measurements needed, but also reducing the flexibility of modeling that can be undertaken using the measurements.

In the third method, visual assessment, either targets or pictorial images that have passed through the imaging system are assessed against a set of physical standards that are calibrated in terms of the objective metric of interest. The process of projection, involving appearance matching with a univariate quality ruler (see Sect. 7.7), is one example of this method. Another example of this approach is the assessment of print grain index (PGI; Kodak Publication E-58, 1994), an objective metric of noise, using the Grain Ruler (PIMA IT2.37, 2000). The grain ruler consists of uniform neutral mid-tone patches containing specified amounts of noise having an appearance similar to that of color film grain. The patches are perceptually uniformly spaced and span a sufficient range of noise to encompass nearly all practical pictorial imaging situations. An observer holds the ruler up against the test images that have passed through the imaging system, identifies the patch having the most closely matching noise level, and reads the associated print grain index value off the ruler.

It might seem unnecessary to determine an objective metric by visual assessment, because if observers were already evaluating an image, they could simply rate its overall quality directly, so that the objective metric would not be needed. This would be true if the only information sought were the overall quality of the exact system producing the images evaluated. The reason that a visual assessment of one attribute may be useful is that it supports predictions of the quality of systems that differ in some regard from the system tested. For example, if each of the attributes of an imaging system were known, and if a potential design change would affect only one of those attributes, in a fashion that were known, the overall quality of the modified system could be predicted using the multivariate formalism. The greatest flexibility results when the properties of each component affecting each attribute of importance are individually known, so that new combinations and variations upon the imaging system design may be investigated by modeling.

12.4 Other Types of Objective Quantities

In this volume, four types of objective quantities are distinguished, as summarized in Table 12.1. The objective metric, defined in Sect. 12.1, is a single number that correlates with an attribute of image quality. Objective metrics reflect properties of the final image produced by a system, as it is viewed by an observer. Objective metrics are often calculated from objective measures such as MTF and NPS, which are functions of one or more variables such as spatial frequency, exposure, image density, etc. As described in Ch. 22, objective measures often can be used to characterize system components individually, the properties of which are combined to predict the behavior of the full imaging system. Engineering parameters are usually single numbers that describe a property of one component in an imaging system. Examples of engineering parameters include the number of megapixels in a digital camera and the resolution (dots per inch) of a digital printer. Although an engineering parameter may be correlated with an attribute of image quality under restricted conditions, because engineering parameters characterize only one property of one component, they do not account for the effects of other components, image viewing conditions, and characteristics of the human visual system on final image quality. Therefore, they do not constitute objective metrics and cannot generally be used to predict image quality directly. They may, however, be used as input to parametric models that estimate component objective measures, which, in turn, may be used to calculate objective metrics (see Ch. 23).

Objective Quantity	See also	Examples	Definition	Utility
objective metric	Part II	print grain index	objectively determined single number correlating with a quality attribute in a viewed image	prediction of image quality
objective measure	Ch. 22	MTF, NPS, sensitometry	function of at least one variable, characterizing component or system behavior	basis for predicting one or more objective metrics
engineering parameter	Ch. 23	megapixels, dots per inch	single number describing property of a component in an imaging system	estimation of objective measures via parametric modeling
benchmark metric	Ch. 12	enlargeability factor	single number derived from one or more objective metrics and characterizing "quality" of single component	comparing component performance, between models or over time

Table 12.1 Properties of the four types of objective quantities.

To compare objective metrics, objective measures, and engineering parameters, consider the example of noise in a digital imaging system. As discussed in Sects. 22.7 and 23.3, many factors influence the final image noise, but one contributor that exemplifies an engineering parameter of relevance is the dark noise of the sensor in the digital still camera. If all other properties of the imaging system were held constant, increases in dark noise would be likely to lead to increases in perceived noisiness in the final image, although they might not lead to any change at all. This latter result might occur if other types of noise in the sensor dominated dark noise, so that increases in dark noise had no noticeable effect, or if the final image noise were subthreshold even after dark noise increased. An example of an objective measure for characterizing the noise in the digital system would be its NPS. The NPS of the system components might depend on various factors, e.g., the digital camera NPS probably would be a function of the camera exposure and would be different in the three color channels. Another objective measure, RMS noise, which is analogous to granularity, might be employed for simplicity instead of the frequency-dependent NPS. An example of an appropriate objective metric of viewed image noisiness would be the print grain index mentioned in the previous section. It is based on final image RMS granularities but accounts for additional factors such as the impact of different color records, the importance of different image densities, and viewing distance.

The fourth and final type of objective quantity is a benchmark metric. A benchmark metric is a single number derived from one or more objective metrics. The intent of a benchmark metric is to characterize some aspect or aspects of the impact of a single component on quality, within the context of a reference imaging system. All properties of the reference system are fixed except those properties of the component in question that affect the attributes of interest, so that comparisons between benchmark metrics reflect the performance differences between various examples of a component. To benchmark the noise of different models of digital still cameras, one might determine the print grain index of images of a particular size, created through a standard digital image processing pathway, written with a specified printer onto a certain medium, and viewed at a chosen distance. A single-attribute benchmark metric may be regarded as a special case of an objective metric that is specifically constrained to facilitate the comparison of imaging system components, and so might not be worthy of a separate classification. But benchmark metrics may also depend on multiple objective metrics and reflect several attributes of image quality, as exemplified by the enlargeability factor described in the next section.

Although objective metrics, objective measures, and engineering parameters are each discussed in greater detail later in this book, the benchmark metric concept will not be needed subsequently. Nonetheless, benchmark metrics are useful for monitoring the evolution of component technology over time, and for comparing and characterizing the performance of different products of a particular type. Consequently, before closing this chapter, an illustrative example of a more complex benchmark metric is presented.

12.5 Example of a Benchmark Metric

Although most objective metrics pertain to a single attribute of image quality, the example considered in this section, a benchmark metric called enlargeability factor, involves two attributes, sharpness and noisiness. A practical question of interest to workers optically printing or digitally manipulating and rendering images is the degree of magnification possible while maintaining some minimum acceptable quality. For example, if an 8 × 10-inch print were desired from a film negative, and the composition of the resulting image could be improved by cropping (using a region of the image rather than the entire image), it would be helpful to know the maximum amount of cropping that would still yield acceptable quality. Enlargeability factor characterizes the capability of a camera film to be magnified and can provide guidelines pertinent to a question of this type. At the end of this section, the way in which enlargeability factor might be applied to digital systems is described briefly, but the present discussion will be limited to film capture because of its greater simplicity.

The film affects two final image attributes, sharpness and noisiness, to an extent that is strongly influenced by printing magnification. Printing magnification is the ratio of the size of a feature in the final print to that in the film. For example, if an uncropped 4R (4 × 6-inch) print were made from a 35-mm format negative (slightly smaller than 1 × 1.5 inches), the printing magnification would be a bit above 4×. Usually, as magnification increases from greater cropping, sharpness decreases and noisiness increases (provided the noise is above threshold), and so overall quality decreases. Other film properties may influence image quality attributes but not in a fashion that significantly depends on magnification, and so they may be neglected in designing a benchmark metric of enlargeability. The enlargeability factor metric is based upon a standard system involving optical enlargement at varying magnification to a specified print size. The properties of each component affecting sharpness and noisiness, other than those of the camera film, are fixed at representative positions. System modeling of sharpness, noisiness, and overall quality (based on the multivariate formalism) is performed for a series of printing magnifications, and the highest magnification maintaining quality above a specified level is identified as the enlargeability factor. A convenient aspect of this benchmark metric is that it is readily interpreted in terms of a physical parameter, the printing magnification, and so is easily understood.

An example of the results of an enlargeability factor calculation for three films of widely varying speed is shown in Fig. 12.1. The x-axis is printing magnification and the y-axis JNDs of overall quality. A threshold of −15 JNDs was adopted, leading to enlargeability factor values of approximately 5, 15, and 25 for films with ISO speeds of 1600, 400, and 25, respectively. This trend is in the direction expected, because higher speed films are generally higher in grain and usually lower in MTF than analogous lower speed materials.

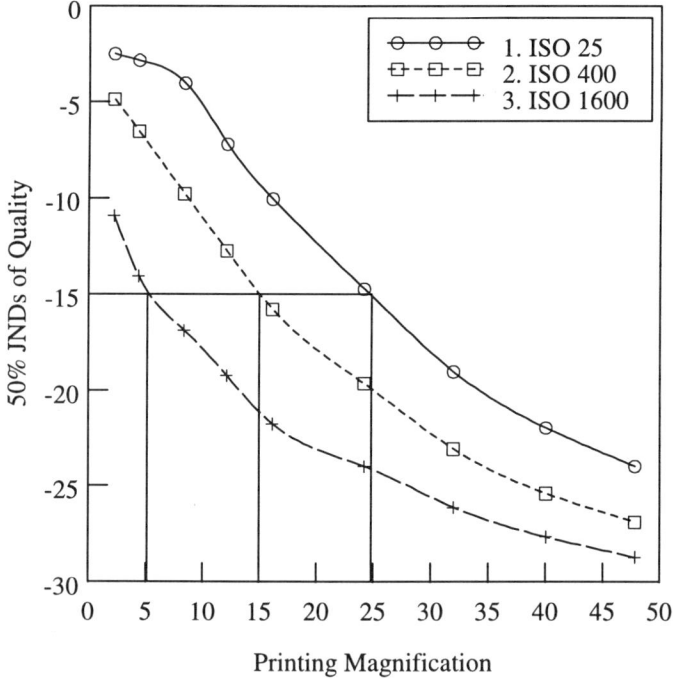

Fig. 12.1 Derivation of the benchmark metric of enlargeability factor at three ISO film speeds. This factor is the printing magnification at which the quality loss arising from unsharpness and noisiness equals a specified level (here, −15 JNDs) in a standardized system.

The standard printing magnification of a photographic system is an engineering parameter; e.g., as mentioned above, production of standard 4R prints from 35-mm format negatives corresponds to a printing magnification of slightly over 4×. In contrast, enlargeability factor is a benchmark metric, even though it is expressed in terms of a printing magnification, because it is based on objective metrics that reflect the quality attributes of a viewed image produced by a standardized system. This system has every property except printing magnification and camera film MTF and NPS fixed at standard values, so that differences in enlargeability factor reflect the impact on quality that film MTF and NPS variations would have in the context of the standardized system.

An analogue of enlargeability factor could be created for digital still cameras, but additional factors would have to be accounted for in the benchmark metric, which might be called the electronic zoom factor instead. In addition to sharpness and noisiness, several other attributes affected by the camera properties are functions of the magnification factor between the camera sensor and the final image. These include the effects of aliasing and reconstruction error, which are discussed in Ch. 18, and also artifacts caused by color filter array interpolation (see Sect. 24.3).

Calculations of the type used in computing enlargeability factor are useful in products where consumers manipulate their own images on a monitor to print customized hardcopy output. If the user zooms in on the main subject, cropping out undesired background, the composition may be improved, but the sharpness and noisiness may suffer. It may be difficult for an untrained user to evaluate this quality loss based on the appearance of image on the softcopy display, so it can be helpful to issue a warning when the degree of electronic zoom exceeds a critical level.

The remainder of Part II provides examples of design, verification, and generalization of objective metrics correlating with a number of different image quality attributes and exhibiting a wide range of properties.

12.6 Summary

An objective metric is a single number that may be determined through objective means and is correlated with a perceived attribute of quality in an image, accounting for its viewing conditions and the properties of the human visual system. Development and verification of objective metrics is one of the primary reasons for performing psychometric experiments. Once defined, objective metrics allow replacement of further perceptual assessments by predictive analysis, which is usually faster, less expensive, and more robust. If objective metrics correlating with different attributes are measured or calculated, they may be converted to quality changes in JNDs using IHIF regressions. These quality changes arising from individual attributes may then combined to predict overall quality using the multivariate formalism, providing a powerful analytical capability. Alternatively, a product specification may initially be expressed in perceptually relevant terms (e.g., allowable quality change in JNDs arising from some attribute), and by mapping back through an appropriate IHIF relationship, may be restated in purely objective terms.

Occasionally, it may be desirable to estimate objective metric values through visual assessment, as in the projection technique, which may be performed using a univariate quality ruler. Much more frequently, objective metrics are calculated from objective measures, such as MTF and NPS, which may be functions of various factors such as camera exposure, image density, etc. Objective measures of an imaging system may be determined by propagating test targets through the system and measuring the output. It is advantageous if objective measures may also be determined for individual components of imaging systems, and from them the full system properties calculated. This permits the quality of a large number of systems (consisting of different combinations of components) to be predicted from a modest number of component measurements. In some cases, it may be possible to estimate an objective measure of a component from a model based on various engineering parameters, such as sensor pixels or printer resolution, allowing even greater flexibility in modeling.

When benchmarking the performance of one component in an imaging system, it is often helpful to define a standard system in which all other system properties are fixed, so that the resulting objective metric values reflect the component's impact on quality. These benchmark metric values may be used to compare the quality of different products, or may be tracked over time to measure advancement of the component technology.

13

Testing Objective Metrics Using Psychometric Data

with Paul J. Kane
Eastman Kodak Company
Rochester, New York

13.1 Introduction

This chapter addresses some of the generic aspects of testing and revising objective metrics based upon comparison of predictions with quality loss functions determined in calibrated psychometric experiments. The concepts reviewed in this chapter are applied frequently elsewhere in Part II of this book. The examples shown in this chapter are drawn from studies of streaking and banding, two digital artifacts that are distinct types of noise differing from the more common isotropic noise exemplified by film granularity. A comparison of these three kinds of noise is provided in Sect. 13.2. Section 13.3 describes the use of series of levels varying only in the magnitude of the attribute under study, to establish a reference quality loss function along the primary dimension of the experimental design. The utility of comparing data from levels varying in other attribute dimensions with the reference quality loss function is demonstrated in Sect. 13.4. Finally, in Sect. 13.5, an example of the identification of limitations of objective metrics is presented.

Fig. 13.1 Isotropic noise, as from film grain, producing a fine speckled appearance that is most obvious in the sky in this image.

Fig. 13.2 The artifact of banding, evident in the sky as a regular pattern of density oscillations parallel to the long side of the image.

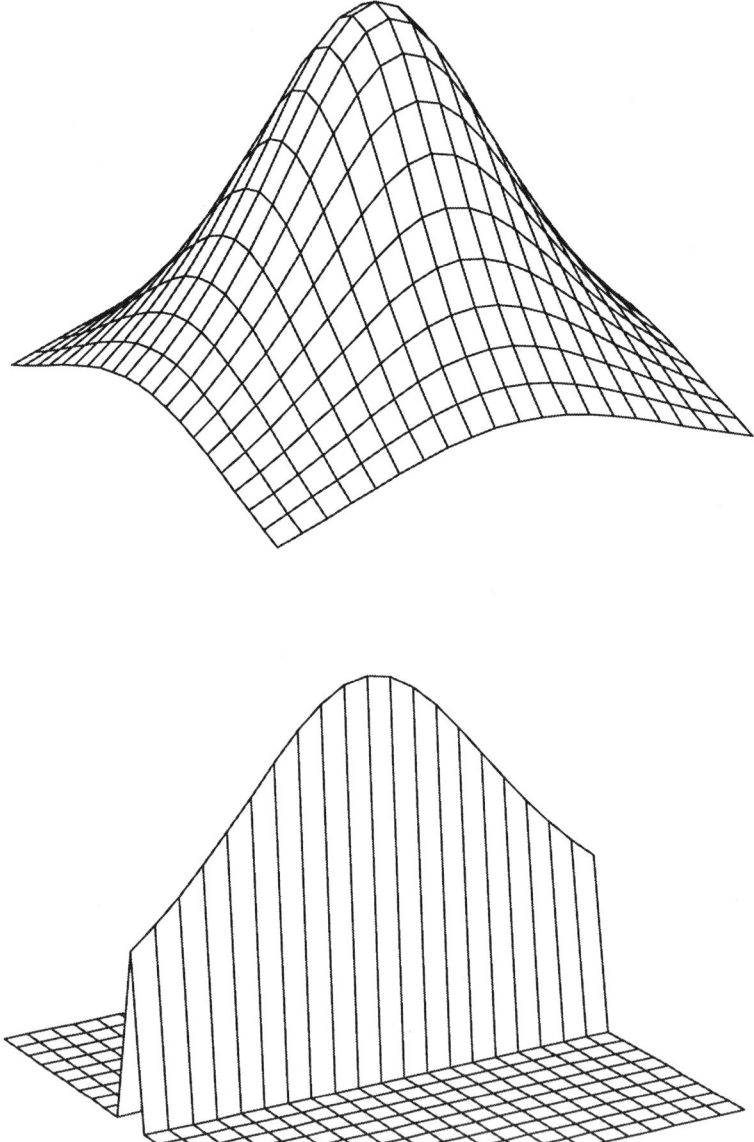

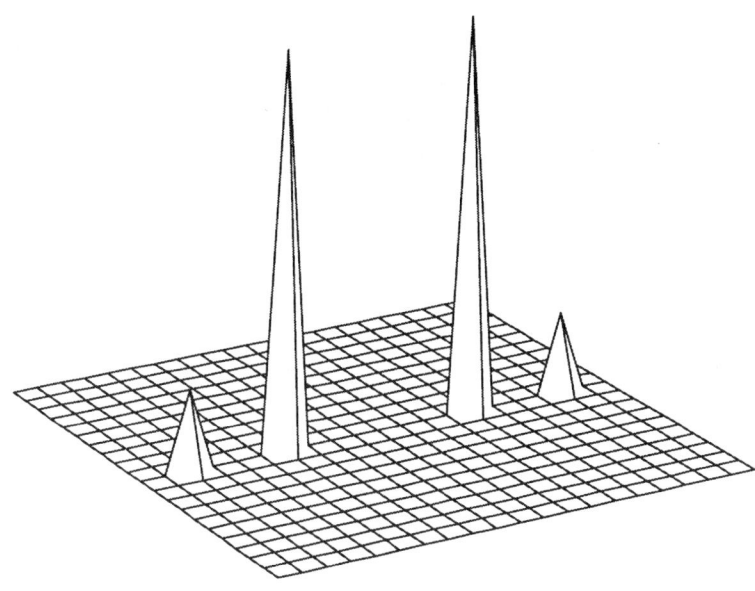

Fig. 13.3 Two-dimensional NPS for isotropic noise (opposite page, top), streaking (opposite page, bottom), and banding (above). The x- and y-axes are spatial frequency in cycles per mm, with the origin located in the center of the area shown. The z-axis is noise power (spectral density) in mm^2. Isotropic noise is random in both spatial directions, and so is broadband in nature and rotationally symmetric. A common source of such noise is film grain. Streaking is a type of noise that is random in one direction but deterministic in the other. Its NPS is zero except along the axis of random variation. Streaking is usually produced by random mismatch between marking elements in linear arrays of digital writing devices. Banding is a type of noise that is periodic in one direction and deterministic in the other direction. Its NPS is zero except at the frequency (and overtones) of, and along the axis of, the periodic variation. Banding is usually a result of cyclical variations in a property of a digital writing device, e.g., polygonal mirror facet reflectance.

13.2 Comparisons of Isotropic Noise, Streaking, and Banding

Pictorial examples of streaking have already been shown, as in Fig. 10.2. Figures 13.1 and 13.2 show examples of isotropic noise and banding, respectively, in the same scene as depicted in Fig. 10.2. All three types of noise are most obvious in slowly varying areas of an image, such as the clear sky in this scene. Each of these types of noise is described in greater detail below, but in brief, their visual characteristics are as follows. Isotropic noise, such as that caused by film grain, produces a speckled appearance because the density fluctuations are random in both spatial dimensions. Streaking is random in one dimension but deterministic in the other, producing a pattern of parallel "stripes" of different densities and widths. Banding is periodic in one direction, and deterministic in the other, so it produces an effect like that of streaking except that the stripes are of uniform width and form a regular oscillatory pattern.

Isotropic noise is produced by a number of sources or processes that have no unique orientation in two-dimensional space. Film granularity is usually very nearly isotropic, as are most components of electronic sensor noise of importance in digital still cameras. One reason that detectors often produce approximately isotropic noise is that the exposure itself is microscopically non-uniform, because photons are randomly, not uniformly, separated in space and time, leading to shot (Poisson) noise. Additional contributors of isotropic noise to film granularity include the non-uniform distributions of silver halide grain positions and sensitivities in a film coating. In sensors, thermal events generate dark current that is usually isotropic, reflecting the distribution of structural material defects. The NPS of isotropic noise at its source is rotationally symmetric, as shown in Fig. 13.3, but subsequent system components having MTFs that are unequal in the x and y directions may introduce some asymmetry. For example, if a camera film is scanned with a linear sensor array, the original noise source (film granularity) is isotropic, but after scanning, the noise in the scan direction will generally be lower than that in the direction parallel to the linear array because of blur introduced by film translation during the scanning process.

Streaking is most commonly produced by digital writing devices that have heads with linear arrays of marking elements, such as inkjet devices, which contain linear arrays of nozzles that deliver droplets of ink, and thermal dye transfer devices, which contain linear arrays of resistors that heat the donor material. Streaking occurs when the individual elements are imperfectly matched in their characteristics and produce different output for the same input code values. As examples, nozzles delivering more ink in an inkjet printer, or resistors producing more heat in a thermal dye transfer printer, will produce higher output densities.

Because there is relative translation between the medium and the linear arrays, usually achieved by moving the medium past the array, if a uniform image (equal code values at all positions) is sent to the device, the element mismatch leads to one-pixel wide lines of different densities being written parallel to the direction of travel. Except in very low-resolution devices, the individual lines are not visually resolved. Instead, it is clusters of adjacent lines that, by random chance, have a significant excess of high or low-density elements that produce the appearance of streaks. This situation is analogous to that of film grain, where the individual grains of silver halide are not resolved, but instead, unevenly distributed clumps of grains are seen.

Because the distribution of element deviations from the mean is random, not systematic, the streaks vary in both their width and their integrated densities. Hence, a density trace across the streaks (i.e., parallel to the axis of the linear array) yields a sample that is one member of a statistical ensemble constituting a random process. In contrast, a density trace along the streaks (and perpendicular to the array axis) would yield a constant density value. Therefore, streaking may be characterized as being one-dimensional random noise, and its two-dimensional NPS should be zero except along the axis of the linear array. As shown in Fig. 13.3, the NPS along the array axis is broadband in nature, because the spatial configuration of array element mismatch is random.

Banding is usually produced by digital writing devices that write one line at a time in raster fashion, i.e., as the written English language is read, from left to right and then returning to the beginning of the next line. For example, some digital printers direct a beam of light onto a photosensitive or donor medium using a rapidly rotating polygonal mirror. The beam power is modulated in accordance with input code values before striking the polygon. As the polygon rotates, the facet upon which the beam is incident changes angle, sweeping the beam spot across the medium in a straight line. When the polygon has rotated far enough that the beam spot crosses a polygon angle and falls on the next facet, the beam is directed back to the beginning of the line. If the medium is continuously translated at an appropriate rate, when the beam returns to the beginning of the line, the medium has been translated by approximately one line width, and is in the proper position for the next row of code values to be written. If some property of the writer that is correlated with the raster output pattern varies in a cyclical manner, the lines that are written out may vary in density or position in a periodic fashion that leads to the perception of darker and lighter bands across an image, parallel to the direction of travel of the beam spot. In the case of a polygonal mirror, such an effect can be produced if the reflectances of the different facets are not perfectly matched.

If a density trace were taken along the bands, a constant density value would result. In contrast, if a density trace were taken across the bands, the density would vary in a periodic fashion. Frequently, such oscillations are nearly sinusoidal in shape, indicating that a single frequency is dominant. For example, if the facets of a polygonal mirror alternated in having higher and lower reflectances, the banding pattern produced would have a period of two raster lines. In other cases more complex sources of periodic behavior and/or subsequent distortion of the signal lead to more complicated line spectra being produced. The two-dimensional NPS of a banding pattern has power along the axis of periodic variation at a fundamental frequency and its harmonic overtones (2× the fundamental frequency, 3×, etc.), as shown in Fig. 13.3. In some cases, multiple fundamental frequencies of banding may be present in a single device, each deriving from a different source of cyclical variation.

13.3 Establishing a Primary Dimension and Reference Regression

When studying a perceptual attribute, it is often useful to identify a primary dimension of variation that corresponds to a magnitude, and which is expected to influence quality in a monotonic fashion. For example, in the case of banding, two important dimensions of variation are its amplitude and frequency. The amplitude is a measure of the difference between the minima and maxima of a periodic pattern and it may be expected that, at constant banding frequency, greater amplitudes will lead to greater quality loss if the banding is above threshold. Thus, the amplitude and the quality loss should be monotonically related. The frequency variation may be regarded as a secondary dimension that affects the "nature" of the attribute rather than its magnitude. At equal amplitude, the frequency of banding will certainly affect quality loss, but the dependence is not expected to be monotonic because the human visual system is less sensitive to high or very low frequencies than to intermediate ones.

The experimental levels in a psychometric study should include a series varying, to the extent possible, only along the primary dimension. As used here, the term "level" refers to an experimental treatment, e.g., the addition of a particular banding pattern to one or more scenes; it does not imply that the variation must be one of attribute magnitude. The series should contain enough levels that the shape of the quality loss function can be well characterized. To accurately characterize its behavior in the critical threshold region, there should be one level that is very close to threshold and one level not too far above it (at most producing a few JNDs of quality loss). Pilot experiments involving small numbers of images and observers may be used to establish the approximate threshold position. There should also be levels having sufficient degradation that extrapolation will be unnecessary in most cases of practical interest. Although the three fit parameters of the IHIF could be mathematically determined from as few as three levels, additional levels are needed for two reasons: (1) the threshold may vary for subsets of observers and scenes, so extra levels may be needed to determine it accurately for all subsets; and (2) more levels than fit parameters are needed to estimate the confidence interval of the regression. In practice, a minimum of five levels is desirable in the primary dimension series.

An example of a reference regression for the attribute of streaking is shown in Fig. 13.4. The primary dimension corresponds to scaling the streaking NPS without changing its shape, so the streaking pattern is fixed, but its strength (i.e., the magnitude of its deviations from the mean) is varied. The sample levels meet

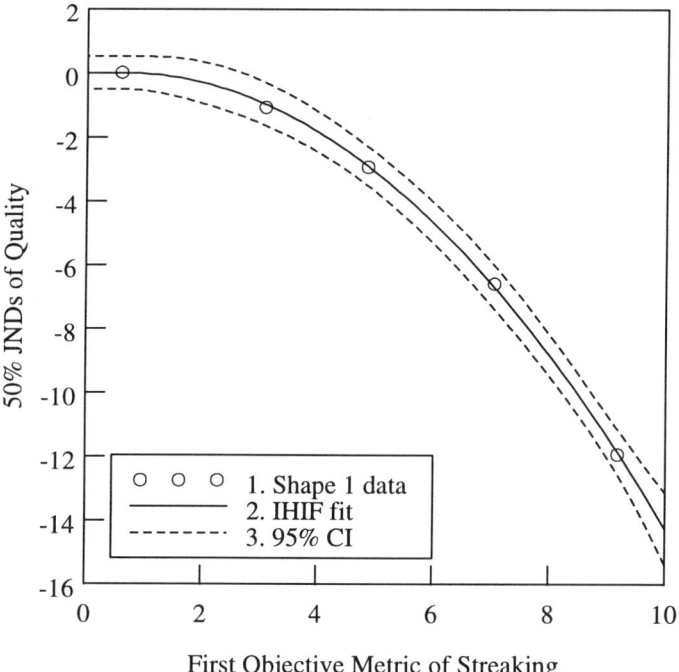

Fig. 13.4 IHIF fit to a quality loss function of streaking for the mean observer and scene. The five levels had the same pattern of streaking and streaking NPS shape, but the magnitude of the pattern and the NPS was varied as the primary dimension of the experimental design.

the criteria stated earlier, with a slightly subthreshold level (leftmost point), a level just above threshold (at ca. −1 JND), and the remaining three levels spaced fairly equally to the maximum level of quality degradation of interest. The objective metric is based upon the NPS of the streaking pattern and an assumed frequency response of the human visual system. The regression fit is excellent, providing a well-characterized reference relationship between the objective metric and quality loss. A fit this good is expected, because the experimental levels were chosen so their quality would be likely to vary monotonically with even a simple measure of magnitude. As long as the quality loss function is monotonic and its curvature is negative (i.e., it is concave downward), it is likely to be fit well by the IHIF. If the curvature condition is not met, it is usually possible to solve this problem by applying a simple transform to the objective metric, such as taking its logarithm.

In some cases, the choice of a primary dimension of variation is somewhat arbitrary, because several candidates may meet the criterion of anticipated monotonic relationship to quality loss. In such cases, if one dimension is of greater interest to product designers because of technological factors, it should be chosen as the primary dimension. The terms primary and secondary are not intended to imply greater and lesser importance, but rather to distinguish the dimension most easily explained by an objective metric and most suitable for establishing a reference regression curve. This curve may be used as a basis against which predictions regarding the secondary dimensions may be quantitatively tested, as explained in the next section.

13.4 Investigating Variations in Secondary Attribute Dimensions

In addition to the primary dimension series, a number of additional experimental levels, varying in secondary attribute dimensions, should be included in the psychometric experiment. Once a reference quality loss function is established for the primary dimension, the data from these secondary levels may be analyzed and used to quantitatively test the predictions of the objective metric for more general variations in the attribute being studied. A prediction might be considered successful if the datum fell sufficiently close to the reference regression curve that it was not, in some sense, significantly different from it. For simplicity, we neglect uncertainty in the individual datum and accept the null hypothesis that the datum is predicted accurately if it falls within the 95% confidence interval of the regression. Because the assessment uncertainty of the individual datum is usually comparable to that of the regression (see Sect. 10.2), this is a somewhat conservative approach.

Figure 13.5 shows an example where the predictions of a streaking metric are clearly unsatisfactory. The objective metric in this figure differs from that of the preceding figure in assuming a relatively wider frequency response for the human visual system. Shape #1 data are from the primary dimension series shown in the previous figure, whereas Shape #2 data are from three levels having much narrower streaking NPS shapes (i.e., lesser streaking bandwidths). The primary dimension regression is acceptable, although the fit is not as good as in the previous figure, as evidenced by the wider confidence interval. The secondary dimension data, plotted as squares, systematically fall outside the reference regression confidence interval, with the magnitude of their quality loss consistently overestimated by the objective metric.

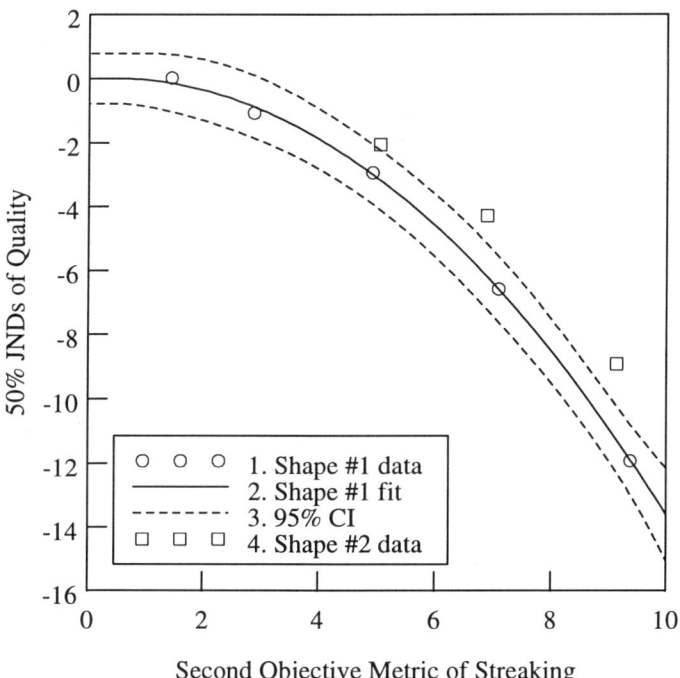

Fig. 13.5 Quality loss functions for primary and secondary series of levels with different streaking NPS shapes. This objective metric does not properly predict the relative impact of the two streaking NPS shapes on quality, as evidenced by the secondary series of data plotting systematically outside the reference regression confidence interval.

Figure 13.6 shows the same y-axis quality loss data as in Fig. 13.5, but plotted against the first objective metric of streaking, from Fig. 13.4, with the narrower assumed frequency response of the human visual system. The secondary dimension data, from the levels with lesser streaking bandwidth, now fall within the very tight primary dimension regression confidence interval. The narrower frequency response of the present metric is clearly superior to that assumed in the previous figure. Still, all three secondary data plot on one side of the regression curve, because the objective metric has very slightly underestimated the magnitude of quality loss associated with the second streaking NPS shape. Given that the broader assumed frequency response led to distinct overestimates of quality loss, and the narrower one to minor underestimates, the latter

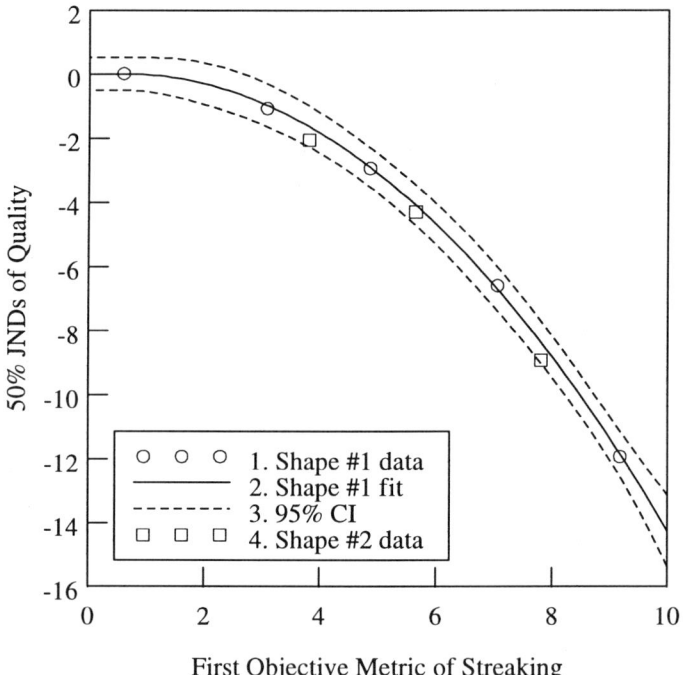

Fig. 13.6 Quality loss data from previous figure plotted against an objective metric based on a different assumed frequency response of the human visual system. The data from the secondary levels, with their different streaking NPS shape, now fall within the reference regression confidence interval of the primary series.

evidently could be broadened slightly, if desired, to remove this small remaining discrepancy empirically.

If an objective metric fails to properly predict quality loss arising from variations along a secondary dimension, it is often possible to modify its design empirically to yield predictions that are accurate. An example of how this might be done is shown in Fig. 13.7, using data from a study of banding. The reference regression is based on a primary dimension of banding amplitude. The primary levels have a constant banding frequency, which is close to the frequency to which the human visual system is most sensitive. The secondary series consists of levels with banding frequencies spanning a factor of sixty, roughly centered

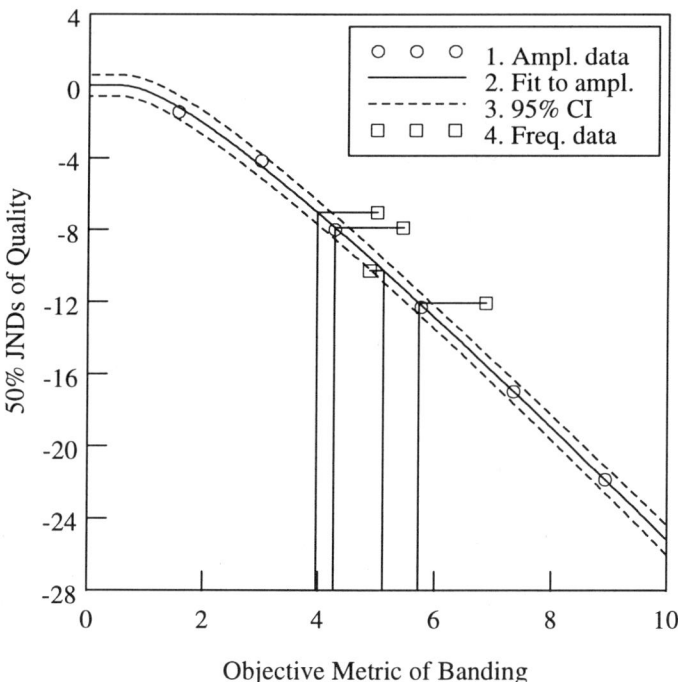

Fig. 13.7 Quality loss function for banding, with the IHIF regression based on a primary dimension of amplitude variations at constant frequency. The data from secondary levels, having different banding frequencies, are not predicted well, but the metric values they should have had may be inferred and used to empirically improve the metric.

on that used in the reference series. As in the case of the streaking metric, the objective metric of banding is based on its NPS and incorporates an assumed frequency response of the human visual system. Because streaking has a broad NPS shape, the streaking metric varied rather slowly as the assumed frequency response was changed. In contrast, because the banding NPS is dominated by a single fundamental frequency, its metric is strongly affected by changes in the assumed visual frequency response, because there is no "averaging" of the metric over a broad frequency range.

Three of the four secondary data in Fig. 13.7 lie outside the reference regression confidence interval, so the assumed visual frequency weighting does not accurately predict the quality loss associated with banding. By mapping the assessed quality loss through the reference regression, as shown by the lines in the figure, the objective metric values required for an exact prediction may be inferred. These values may be converted to aim responses at each banding frequency, allowing empirical modification of the assumed visual frequency response function. For the inferred objective metric values to be precise, the quality loss of the associated level must be large enough that the data do not map onto the weakly sloped portion of the reference regression curve near threshold.

Most objective metric design involves both considerations based on first principles and adjustments based on psychometric data. Typically, the metric is initially designed using what is known from the literature and intuitively plausible assumptions. Then, based on experimental data, empirical modifications are made to improve the predictions of the metric.

13.5 Testing the Limitations of Objective Metrics

It is often impractical to run large enough experiments to ensure that objective metrics are predictive over the full range of all known attribute dimensions. Typically, the range of variation along each dimension is limited to the region of greatest current engineering interest, and some dimensions may not be explored at all because of a lack of immediate practical ramifications. Because market evolution, changes in business conditions, and technological advances can transform cases formerly only of scientific interest into cases of great practical relevance, it is valuable to have some idea of objective metric limitations.

Testing Objective Metrics Using Psychometric Data

One reasonable approach to establishing metric limitations is to check a few extreme cases to determine where an objective metric might fail, without collecting the greater amount of data that would be required to modify the objective metric to make it more predictive. Such an approach allows identification of areas where further future research might be of particular value. An example of this approach is presented in Fig. 13.8 using banding data. It is uncommon for an output device to have banding at multiple frequencies that are each of sufficient amplitude to be of perceptual significance, but it can happen. To thoroughly investigate this phenomenon would require many experimental levels having different combinations of banding frequencies. Instead, just two extreme cases were checked to estimate the potential predictiveness of an

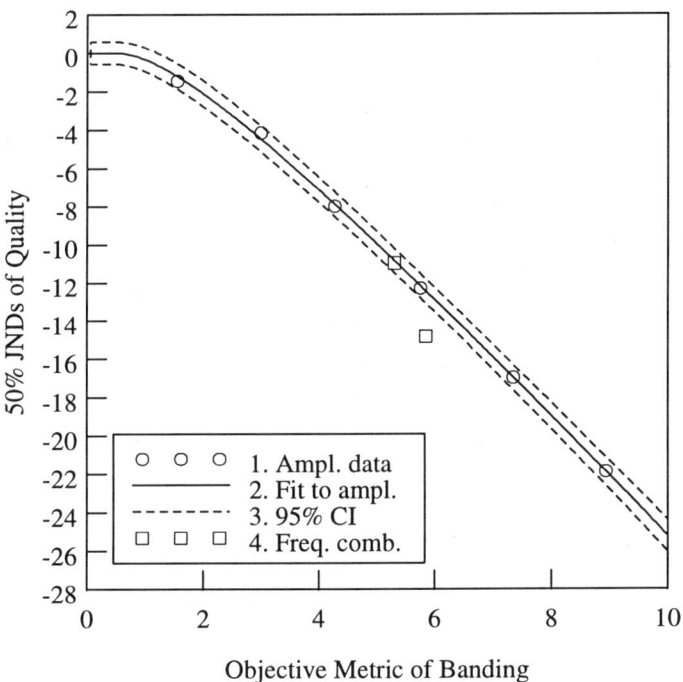

Fig. 13.8 Primary series data and reference regression from the previous figure, with secondary levels involving combinations of banding patterns having different frequencies. The datum from one combination pattern is predicted well, but the other falls well outside the IHIF confidence interval, revealing a potential metric limitation.

existing objective metric of banding. The reference regression in Fig. 13.8 is the same as in Fig. 13.7. The two secondary data plotted as squares are from levels containing combinations of two different frequencies of banding (from the set of frequencies tested in isolation), both at perceptually significant amplitudes. In one case, the two banding frequencies are far apart, and in the other case, they are close together, leading to beat patterns in the spatial domain. The prediction in the latter case is excellent, but in the former case, a modest discrepancy (≈ 2.5 JNDs or 20% error) is evident. These results indicate that at least some predictions in cases involving multiple frequencies may be unreliable. Although there is not enough data to empirically modify the objective metric to improve its predictiveness, at least this potential failure mode is identified and a rough estimate of the direction and magnitude of possible errors has been obtained. This is a good return on the investment of just a few experimental levels.

13.6 Summary

In developing objective metrics, one approach involves initially identifying a primary dimension of attribute variation, usually corresponding to a simple magnitude change in some sense, and determining a reference IHIF regression of the quality loss function in this dimension. The quality range spanned by the primary levels should cover the full quality range of interest, and a minimum of five levels should be included to robustly define the regression shape and allow estimation of regression confidence intervals. Ideally, at least one level should be very near threshold, and one slightly above it, to properly characterize behavior in the critical threshold region.

Subsequently, to test and further refine the objective metric, variations along other secondary dimensions may be evaluated by comparing the calculated or measured objective metric values with those implied by the assessed quality loss and the reference regression. The assessed quality should be far enough from the threshold region that the quality loss function has significant negative slope, so that the implied objective metric value is determined to tight tolerances.

It is difficult to develop an objective metric that is predictive over the entirety of all known attribute dimensions, and the range of properties of practical relevance is subject to change as technology and the marketplace evolve. A reasonable compromise is to carefully test an objective metric over the ranges of greatest current practical interest, and sparingly test cases involving more extreme values or new dimensions of variation to identify potential failure modes and anticipated direction and degree of failure of the metric.

14

A Detailed Example of Objective Metric Design

14.1 Introduction

This chapter presents a detailed account of the entire process of the design and verification of an objective metric of misregistration. The intent of this chapter is to provide a concrete example of the application of various methods and approaches that were described in Chs. 10, 12, and 13. Although a number of other objective metrics will be considered later in Part II, this is the only metric for which all pertinent details are provided. In subsequent discussions, the emphasis will be on selected aspects of objective metric design, or particular challenges faced in experimentation or data analysis.

This chapter is organized as follows. Section 14.2 discusses the experimental design for misregistration, including simulation verification. A potential objective metric is developed in Sect. 14.3 based on a simple model of the perception of misregistration. Verification of the objective metric against psychometric data and empirical adjustment of the metric to increase predictiveness are described in Sect. 14.4. Finally, variability in scene susceptibility and observer sensitivity is quantified in Sect. 14.5.

14.2 Experimental Considerations

Before beginning a psychometric experiment from which an objective metric is to be derived, it is desirable to have one or more candidates for the metric already hypothesized. Consideration of the metric properties helps in the identification of those quantities that will need to be accurately measured or calculated for the test samples, and suggests the types of targets that should be processed with the test samples. Misregistration is a particularly simple artifact to characterize. It can be fully described by determination of the relative positions of a single point in an original scene as mapped to the color records of the final image. For example, it might be measured using a target consisting of a fine cross-hair against a uniform background.

Misregistration is also easily digitally simulated. Color records can simply be translated by integral numbers of pixels or, if finer control is desired, the color records can be resampled for sub-pixel displacements, although in this case care must be taken to avoid aliasing and reconstruction error artifacts (see Ch. 18). Given a knowledge of the digital writer pixel pitch, the induced pixel shifts can be directly converted to distances on the final image, so verification of the amount of misregistration via analysis of a target may seem superfluous; however, double-checking that simulations produce exactly what was intended, is prudent scientific practice.

Selection of test sample properties is a critical aspect of experimental design. Most studies involve several rounds of small pilot experiments prior to the final experiment. In pilot tests, usually only a few scenes at each level are assessed against a quality ruler by just a few observers to obtain rough estimates of their quality. Typically, a primary dimension magnitude series is first analyzed to obtain the approximate dependence of quality on a proposed objective metric. This permits selection of well-placed levels for the final experiment. A few levels with secondary dimension variations are also usually tested to determine whether their effects on quality are significant. If so, the final secondary levels should be chosen so that they fall well above threshold, where the slope of the quality loss function is sufficient to precisely infer the aim objective metric values required for accurate prediction of quality.

In the case of misregistration, the primary dimension was defined to be the shift of the green record relative to stationary red and blue records. Secondary dimensions were shifts in other color records and the geometry of simultaneous shifts in two records.

14.3 Design of an Objective Metric

Misregistration involves relative displacements of different color records, which leads, in small amounts, to apparent unsharpness, and, in larger amounts, to

A Detailed Example of Objective Metric Design

evident colored ghost images, as seen in Sunday comics. Consider the case in which an additive display with red, green, and blue primaries (e.g., a video projector) has no two channels properly aligned. If a white point is imaged against a black field, three closely spaced red, green, and blue points result, as shown in Fig. 14.1. The red, green, and blue points are assigned indices 1, 2, and 3, respectively, and x-y coordinates on an arbitrary grid are assigned to the positions of each of the points. If these three points are sufficiently close together, they will be perceived as forming a single white dot that is slightly blurred. The apparent center of that perceived white dot will be called the visual center. The distance from the visual center to the red, green, and blue points are denoted by d_1, d_2, and d_3, respectively. This coordinate system and these definitions will form the basis for the proposed objective metric of misregistration.

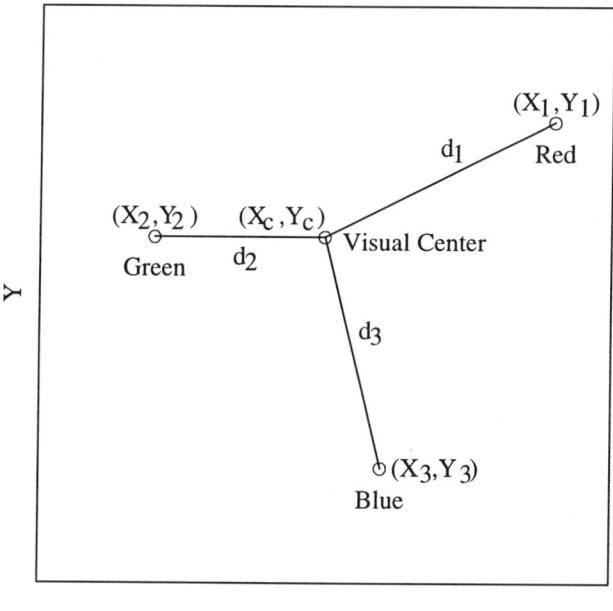

Fig. 14.1 Coordinate system used in developing the objective metric of misregistration. The visual center represents the perceived center of a neutral point formed from three misregistered color records.

It is well known that the different color records have different visibility, with the blue record having particularly little visual impact in high-frequency spatial effects because of the relative scarcity of "blue"-sensitive cones in the human eye. In most cases, the green record has somewhat greater effect than the red, so we may expect that the position of the visual center will be most strongly influenced by the position of the green point and least so by the position of the blue point. If the three points were arranged in an equilateral triangle, then the visual center would be closest to the green point and farthest from the blue. The relative influences of the three color records might be described by a set of three visual weights, with the green weight being the highest and the blue the lowest.

This situation bears a certain resemblance to the problem of rotating bodies in classical Newtonian mechanics. A set of rigidly coupled masses will rotate about what is called the center of mass, which is a point lying within the region bounded by the individual masses and tending to be closer to the heavier masses. If the visual weights were regarded as being analogous to the masses, then one might assume that the coordinates of the visual center could be estimated by an equation like that used to compute the center of mass, as shown in Eq. 14.1:

$$x_c = \frac{\sum_i w_i \cdot x_i}{\sum_i w_i} \qquad y_c = \frac{\sum_i w_i \cdot y_i}{\sum_i w_i} \qquad (14.1)$$

where x_c and y_c are the coordinates of the center of mass, x_i and y_i are the coordinates of the misregistered spots in Fig. 14.1, and w_i is the visual weight of the ith color record. The distance of each point from the visual center R_i is given by Eq. 14.2.

$$R_i^2 = (x_i - x_c)^2 + (y_i - y_c)^2 \qquad (14.2)$$

One might now assume that the severity of perceived misregistration would depend upon the degree of blur induced, which would correlate with the visually weighted distances of the red, green, and blue points from the visual center. To continue our physical analogy, the rate of rotation of a rigid body depends upon its moment of inertia, which is the sum of the products of the component masses times the squares of their distances from the center of mass. By analogy, an objective metric based on a "visual moment of inertia" might be defined by:

A Detailed Example of Objective Metric Design

$$\Omega_d^2 = \frac{\sum_i w_i \cdot R_i^2}{\sum_i w_i} \quad (14.3)$$

where Ω_d is the objective metric in distance units, which might be regarded as a weighted displacement from the visual center. Inclusion of the term in the denominator, which does not appear in a classical moment of inertia, is equivalent to requiring that the visual weights are normalized, i.e., they sum to one. The objective metric, as defined, is essentially a square root of the visual moment of inertia, so that it conveniently has units of distance. Another way of interpreting this metric is as a root-mean-square "visual error" term.

It is expected that a given amount of misregistration, expressed as an equivalent visual displacement in distance units, will have greater impact on quality at closer image viewing distances, where the effect will be more discernible. This suggests that to generalize the objective metric for use at viewing distances other than that experimentally tested (at least to first order), the metric should be converted to units of angular subtense at the eye. Conveniently sized units for such purposes are arc-seconds. Because the subtended angle is small, the angle in radians may be calculated by dividing the visual displacement by the viewing distance, and then simple units conversion yields in turn degrees, arc-minutes, and arc-seconds, as shown in Eq. 14.4:

$$\Omega_a = \left(\frac{\Omega_d}{d_v}\right) \cdot \left(\frac{180 \text{ degrees}}{\pi \text{ radians}}\right) \cdot \left(\frac{60 \text{ minutes}}{\text{degree}}\right) \cdot \left(\frac{60 \text{ seconds}}{\text{minute}}\right) \quad (14.4)$$

where Ω_a is the objective metric in angular subtense units, and the units of viewing distance d_v must match those of the visual displacement Ω_d.

The objective metric of Eq. 14.4 is a reasonable candidate to test against psychometric data, although the visual weights remain to be specified. These may be calculated from first principles for sharpness and noisiness (Sect. 22.5) based on the spectral characteristics of the viewing illuminant and image material colorants or, in the case of self-luminous displays, the emitters (e.g., phosphors). Visual weights for misregistration might be somewhat different from those for sharpness and noisiness because of the distinct nature of the ghost images at greater levels of misregistration, so empirical verification of the visual weights against psychometric data is appropriate.

14.4 Verification of an Objective Metric

Figure 14.2 summarizes the experimental results. The x-axis is the objective metric of misregistration (arc-seconds subtended by the equivalent visual displacement) that is given by Eqs. 14.1–14.4; the y-axis is 50% JNDs of quality change arising from misregistration. The primary dimension involves green record shifts of different magnitudes, which form the basis for the IHIF regression shown. Four secondary levels fall within the regression confidence interval; these correspond to a red shift, a blue shift, orthogonal red and blue shifts (at 90° to one another), and opposite red and blue shifts (at 180° to each other). The positions of these data depend strongly on the assumed visual

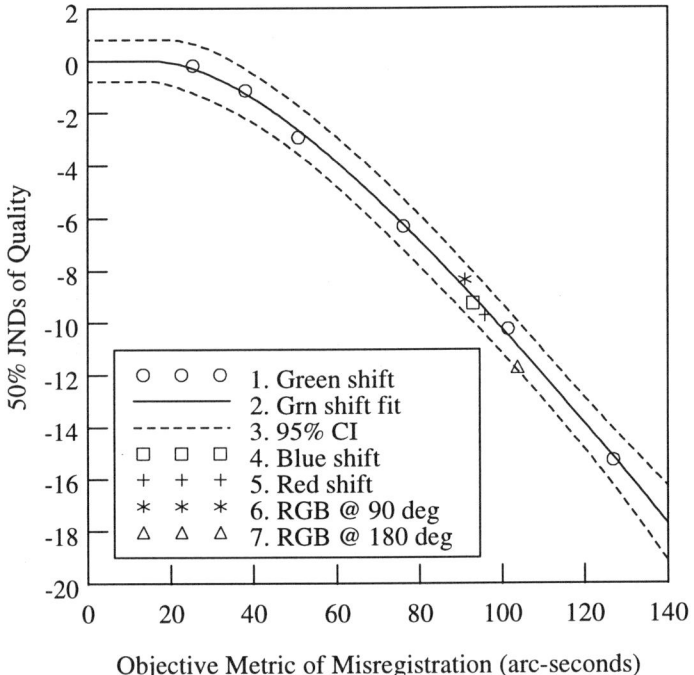

Fig. 14.2 Quality loss function for misregistration, with a primary dimension of green record shift. Secondary levels involve red shifts, blue shifts, and combinations thereof in different relative directions; all variations are predicted accurately by a visual subtense metric.

A Detailed Example of Objective Metric Design

weights, which were optimized through nonlinear regression to minimize discrepancies between predicted and measured values. The red:green:blue proportions obtained were 34:50:16, in fair agreement with the 37:55:08 expected from first principles, if misregistration behaved like sharpness and noisiness (Sect. 22.5). The optimized weights may be rounded off to the mnemonic values shown in Eq. 14.5.

$$w_r = \frac{1}{3} \quad w_g = \frac{1}{2} \quad w_b = \frac{1}{6} \quad (14.5)$$

In this case, the proposed objective metric successfully fit the psychometric data with the adjustment of two parameters, the ratios of the red and blue visual weights to the green visual weight (because the visual weights are effectively normalized in Eq. 14.3, specification of two weights or weight ratios determines the third). Had this not been the case, additional parameters of optimization could have been introduced. For example, the power to which the distances from the red, green, and blue points to the visual center are raised in Eq. 14.3 might be parameterized as in a Minkowski metric, instead of being fixed at two. The value of two is pleasing because the objective metric is essentially an RMS error and the objective metric squared is analogous to a moment of inertia. Nonetheless, there is no obvious a priori reason why the perceived quality loss should depend on displacements in exactly this fashion. If the power parameterization had failed, a substantially different form of objective metric might have been sought. Misregistration is a relatively simple artifact and is successfully described by an intuitively plausible metric, but some other attributes are less amenable in this regard, and may require extensive trial and error and empirical optimization.

14.5 Fitting Scene and Observer Variability

Scene susceptibility and observer sensitivity variability were characterized based on just the primary dimension data to increase robustness of the fits. Based on their average response to the primary dimension levels, observers were categorized into three groups: (1) average (containing all observers); (2) more sensitive 50%; and (3) less sensitive 50%. No correlation was found between expertise or photographic activity of the observer and their sensitivity. The scenes were also ranked based on their mean quality loss across the primary dimension series and then collected into four groups: (1) average (containing all scenes); (2) most susceptible 25%; (3) least susceptible 25%; and (4) middle 50%. IHIF regressions with 3 free parameters were run for all 12 combinations of these groups and the fits studied.

Some minor regression curve crossings were noted as discussed in Sect. 10.4. There were evident differences in asymptotic slopes and thresholds of the subsets of data, so the regressions were repeated with the least variable parameter, threshold radius of curvature, fixed at its value in the IHIF regression of the fully pooled data set. Curve crossings were eliminated by this process with minimal compromise of fit error, so this regression set was accepted. The fit parameters obtained are summarized in Table 14.1, which also includes the numerically determined value of the objective metric for which one JND of quality loss occurs.

As explained in Ch. 10, determination of quality loss functions for different subsets of observers and scenes is useful in application-specific analyses where the observers and/or scenes are expected to differ from the average. The IHIF regressions are also useful in Monte Carlo predictions of the distribution of quality that would be produced by an imaging system, as described in Ch. 28. In such an analysis, many iterations would be executed, and each subset regression would be used with a frequency proportional to its expected occurrence; e.g., the probability of an observer being in the 50% more sensitive group and a scene falling in the 25% most susceptible group would be 50% × 25% = 12.5%. In this way, a representative sample with an appropriate level of variability would be obtained.

Observer Sensitivity	Scene Susceptibility	$\Delta\Omega_\infty$	Ω_r	Ω_{-1}
less sensitive 50%	least susceptible 25%	5.500	20.76	39.9
less sensitive 50%	mean	4.364	16.43	36.7
less sensitive 50%	middle 50%	4.078	17.11	37.2
less sensitive 50%	most susceptible 25%	3.980	11.94	34.4
mean	least susceptible 25%	5.298	18.72	36.3
mean	mean	3.902	15.07	35.2
mean	middle 50%	3.708	15.81	35.5
mean	most susceptible 25%	3.256	10.95	34.3
more sensitive 50%	least susceptible 25%	5.088	16.62	33.7
more sensitive 50%	mean	3.470	13.70	33.7
more sensitive 50%	middle 50%	3.351	14.50	33.8
more sensitive 50%	most susceptible 25%	2.654	9.81	35.9

Table 14.1 IHIF fit parameters for subset regressions; in all cases $R_r =$ 151.6. The quantity Ω_{-1} is the objective metric value at which one JND of quality loss occurs.

A Detailed Example of Objective Metric Design

Figure 14.3 depicts the regression fits for the mean observer and scene and the two extreme cases of more sensitive observers assessing most susceptible scenes, and less sensitive observers assessing least susceptible scenes. This range of variation is smaller than that typically observed with artifactual attributes because of an unusually small degree of variation in observer sensitivity, as mentioned in connection with Fig. 10.7.

The remainder of Part II is devoted to more advanced aspects of objective metric design, but in almost all cases discussed, an approach like that described in this chapter has been followed, at least in the initial phases of the investigation.

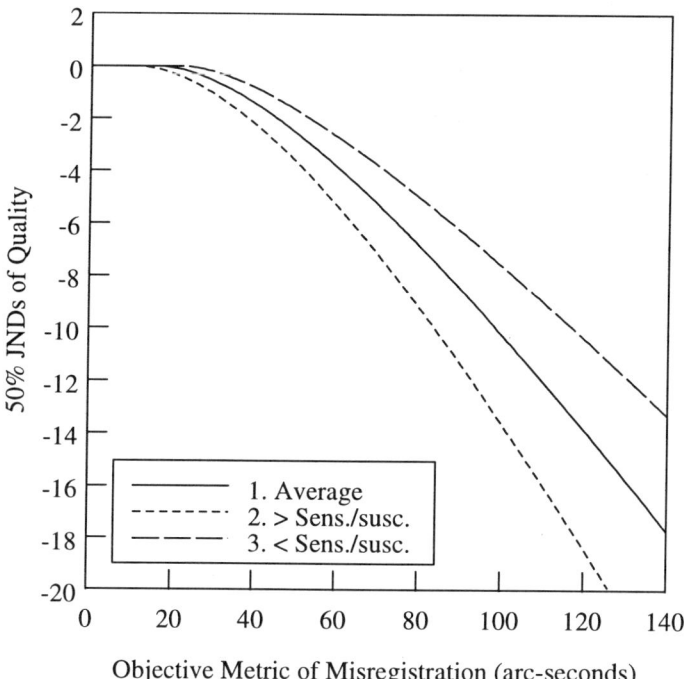

Fig. 14.3 IHIF subset regressions for the average observer and scene and for the more (less) sensitive 50% of observers assessing the more (less) susceptible 25% of scenes. This attribute exhibits somewhat less variability than is typical of artifacts.

14.6 Summary

In this chapter, the artifact of misregistration was used to provide a detailed example of objective metric design. Misregistration is easily measured with simple targets and is readily simulated using digital image processing. A primary dimension of green record shift was selected and pilot experiments were performed with a few observers and scenes to ensure that appropriate amounts of misregistration were added to the test images. Secondary dimensions were shifts in other color records and relative geometry of shifts in two color records. The pilot experiments were helpful in confirming that the secondary levels caused sufficient quality loss to fall in the more strongly sloped portions of the quality loss function, so that the predictiveness of the objective metric could be reliably tested using the approach described in the previous chapter. The tentative form of the objective metric was generated based on a simple model of the perception of misregistration and by analogy with physical phenomena. Visual weights quantifying the impact of misregistration of the different color records were estimated from first principles but then empirically optimized against the psychometric data. Variability in scene susceptibility and observer sensitivity was characterized using IHIF regression fit parameters as described in Sect. 10.4.

The process described in this chapter exemplifies that applied in many other studies of image quality attributes. Misregistration is slightly atypical in being a relatively simple artifact that is well quantified by an intuitively plausible objective metric. Some artifacts are more complex and require considerable trial and error in the objective metric design, and more extensive empirical optimization, to obtain a satisfactorily predictive metric.

15

Weighting Attributes that Vary Across an Image

with Karin Töpfer
Eastman Kodak Company
Rochester, New York

15.1 Introduction

Some artifacts, such as misregistration, are usually global defects, being of similar magnitude and character throughout an image. In this chapter, the discussion will focus on treatment of attributes that vary within an image. Such attributes usually depend on one or more of the following three factors:

1. signal level;

2. location within the image; and

3. orientation (direction).

Most attributes are signal-dependent and, therefore, vary as a function of the visual image density (or, equivalently, lightness) in the final image. For example, film granularity varies with exposure, partly because different exposures lead to the development of different size populations of silver halide grains. Sensors in digital cameras also produce noise that varies with exposure because of the fundamental nature of some of the contributing sources of noise; e.g., shot noise varies with the square root of signal, and fixed pattern noise is proportional to signal (see Sect. 23.3). In comparison, the amplitude of banding produced by a digital printer writing onto silver halide paper is usually signal-independent, but when this constant exposure modulation is rendered by the nonlinear tone scale of the photographic paper, it is scaled by the sensitometric gradient (objective contrast) of the paper, which is highest in the mid-tones but tends to zero as the minimum or maximum densities of the material are approached (analogously, noise power is scaled by the square of the gradient; see Sect. 22.7). Thus, even an artifact that is independent of signal at its source may vary with final image density.

A dependency on location within the imaging field is often caused by variation of lens MTFs with radial position. Often, the best MTF occurs along or near the optical axis of the lens, which usually coincides with the center of the image. Farther out in the field, towards the corners of the image, the MTF usually falls off, unless specific design choices are made to sacrifice performance on-axis for that off-axis. If the MTF of a component does not depend at all upon location within the imaging field, it is said to be shift-invariant, which is one condition for the rigorous application of linear systems theory (Ch. 22). As another example, illumination through a lens often falls off as the fourth power of the cosine of the field angle, causing correlation between location and exposure. Consequently, if a subsequent system component influences an attribute in a signal-dependent fashion, that attribute will also vary with location.

Finally, directional dependencies are particularly notable in digital scanning or writing devices that process one line of the image at a time (i.e., in raster fashion). For example, a document scanner having a linear array sensor (or three sensors for parallel full-color capture) typically relies on page motion to scan across the full field. The page motion imparts an additional degree of blur in the scan direction that is not present in the direction parallel with the linear array(s), leading to a lower MTF in the scan direction. Similarly, when a video display is written one line at a time, there is an extra component of blur in the line direction because the writing spot is on continuously, "blending" the signal values in the line direction.

In this chapter, the treatment of attributes that vary across an image will be described in general terms and the example of signal dependence will be considered in some detail. The chapter is organized as follows. Section 15.2 contrasts the merits of weighting objective quantities compared to quality losses. Section 15.3 briefly considers the treatment of attributes that vary with location and orientation within an image. The tonal distribution, tonal importance, and detail visibility functions needed to characterize the impact of attributes on quality as a function of signal dependence are defined in Sects. 15.4 and 15.5. The method by which the tonal importance function has been measured is explained in Sect. 15.6. Finally, two applications of tonal weighting are described in Sect. 15.7.

15.2 Weighting in Objective versus Perceptual Space

Prediction of image quality change arising from an attribute that varies across an image may be regarded as a weighting problem. If the quality dependence of a particular attribute were known in the case where the attribute were invariant within the image, it seems likely that the quality of an image with variable levels of the same attribute could be estimated by taking some sort of a (not necessarily linear) weighted combination of quality values over the image area. If so, the problem reduces to determining an appropriate weighting scheme.

The scenario described above involves weighting in perceptual space, because it is quality loss values that are being "averaged" in some fashion. As described later in this chapter, we have had excellent success with such weighting schemes carried out in a perceptual space. Often it would be more convenient, however, to perform the weighting in an objective space to simplify computations. For example, suppose that the MTF of one or more imaging system components varied with field position. In the general case, prediction of the sharpness of images produced by the system would require the following steps.

1. The MTFs of each component, at a number of field positions, would be cascaded to yield the overall system MTF at each field position.

2. The value of the objective metric of unsharpness would be computed at each field position from the system MTF.

3. The quality change arising from unsharpness at each field position would be calculated from the objective metric values, using IHIF regressions.

4. The quality changes at each field position would be weighted to obtain a single value.

A great simplification would occur if the component MTFs could be directly weighted to produce a single MTF (per color) for each component, because the calculations in the second and third steps would then only have to be done once (per color), instead of iterating over many field positions. In our experience, for weighting in objective space to yield accurate predictions of quality, there must be an approximately linear relationship between the quality change arising from an attribute and the correlated objective quantity over which a weighted average is taken.

To understand why such a linear relationship might be required, let us assume that overall quality is a weighted arithmetic average impression of the quality of different parts of an image, so that linear weighting in a perceptual space properly emulates the subjective assessment process made by an observer. Suppose that in one half of an image an artifact has an objective metric value of $\Omega_r + \Delta\Omega$ (Ω_r being the threshold value), and so degrades quality by some constant amount $\Delta Q(\Omega_r + \Delta\Omega)$. In the other half of the image the metric value is $\Omega_r - \Delta\Omega$, and the artifact is subthreshold, so that $\Delta Q = 0$. If the two halves of the image contributed equally to the impression of overall quality, the weighted quality loss in perceptual space would be $\Delta Q(\Omega_r + \Delta\Omega)/2$, which is half as large as that in the degraded half of the image. If the objective metric values were averaged instead, the predicted quality loss would be $\Delta Q(\Omega_r) = 0$. This amounts to one part of the image being sufficiently far below threshold to cancel the impact on quality of the artifact elsewhere in the image where it is visible. This result is absurd, because the appearance of the image would not have been significantly different had the subthreshold half of the image been exactly at threshold instead. Having portions of an image free of an artifact can certainly reduce the quality loss associated with that artifact, essentially by dilution, but it cannot render the artifact undetectable elsewhere in the image.

Averaging quality losses and objective metrics yields the same result when these two quantities are linearly related, which occurs when all values either: (1) are below threshold; (2) fall in the asymptotic region of the IHIF; or (3) are so close together that the IHIF can be approximated by a straight line in the region spanned. If an objective metric can be designed in a fashion that the threshold radius of curvature is very small, the shape of the IHIF approaches that of two line segments joined at an angle. If the objective metric is further constrained so that its minimum allowed value is equal to its threshold value, it becomes linearly related to quality loss everywhere. Such an objective metric is defined to be perceptually uniform, and has the advantage that it can be directly weighted. If quality loss functions for different subsets of observer sensitivity and scene susceptibility had different thresholds, slightly different objective metrics, having different minimum values, would have to be defined for each subset, which would be quite inconvenient. Consequently, the opportunities for defining perceptually uniform objective metrics are somewhat limited. Objective metrics of preferential attributes are usually expressed as distances from preferred reproduction positions, and so are necessarily constrained to be positive (see Sect. 20.4); because they additionally often exhibit little curvature, they often are good candidates for perceptually uniform metrics. Examples of two perceptually uniform metrics will be presented later in this chapter.

Even if an objective metric is perceptually uniform, it may be nonlinearly related to changes in the objective measure from which it is computed (e.g., MTF or NPS). In such a case, a weighted average objective metric may be predictive, but a weighted average objective measure may yield a different and misleading objective metric value. Nonetheless, over sufficiently small ranges of objective measure variation, the relationship to the corresponding objective metric may be adequately linear to permit weighting of the objective measures. In many cases, the variation of attributes with orientation and location in the image is small enough that the objective measures may be weighted directly. Signal dependence of attributes is often stronger and probably can only rarely be treated in this fashion.

15.3 Location and Orientation Weighting

Although the bulk of this chapter addresses signal-dependent weighting, a brief description of certain aspects of location and orientation weighting is in order. Some of the sources of location and orientation dependencies were reviewed in the introduction to this chapter and will not be repeated here; instead the focus will be on how these dependencies are modeled.

In orientation weighting, the properties of images and the characteristics of the human visual system are both important. The fidelity of reproduction of edges strongly influences the perception of sharpness. Because images often include man-made objects such as buildings, we may expect that roughly vertical or horizontal edges will be somewhat more common than oblique edges. It is also well known that the contrast sensitivity function of the human visual system is greater in bandwidth along the horizontal and vertical directions than along the diagonals, which is why halftone screens are usually angled, thereby reducing the visibility of their regular patterns. Therefore, it seems likely that greater weighting should be given to orientations closer to horizontal or vertical. In our experience, weighting just these two orientations is sufficient to produce predictive results in nearly all instances. In a few extreme cases, we have obtained slightly better results when a diagonal orientation was included also, but in these cases, a similar improvement was obtained with only two orientations by weighting the orientation having lower quality slightly more.

The simplest approach to field position weighting is to divide the field up into various regions and weight each according to its area. This is probably an appropriate approach for more critical photographic applications; however, in consumer photography, much greater use is made of the center of the image area than its periphery. This might be expected based on compositional considerations and on the properties of point-and-shoot camera viewfinders, which usually show less than the full field of view, and often have explicitly marked central areas for metering and autofocus operations that encourage the photographer to center the main subject.

The spatial distribution of important subject matter in image populations might be measured in several ways. Observers could be asked to view images and their eye movements tracked to determine how much time is spent looking at each area within each image. Alternatively, a representative collection of images could be divided into segments of varying importance by having observers draw outlines of regions of interest and then measuring the spatial coordinates of the region boundaries. Although images having different aspect ratios might be expected to display somewhat different spatial distributions of key subject matter, we have found similar results over a wide range of consumer pictorial images of varying aspect ratio. There is nearly always important subject matter in the center of the image, and rarely any in the corners. To a fair approximation, there is a linear fall-off in probability of occurrence of key subject matter with increasing distance from the center of the image, if an average is taken over all angles. This result may be used to modify simple area-based weights for consumer applications.

15.4 Tonal Distribution and Importance Functions (TDF and TIF)

Three functions might be expected to influence the perceptual weighting of different image tones in observers' assessments of an attribute. These functions, which form the basis for signal-dependent (tonal) weighting, are:

1. the tonal distribution function (TDF), which quantifies the relative frequency of occurrence of different tones in original scenes;

2. the tonal importance function (TIF), which quantifies the relative importance of scene tones at equal occupied image areas; and

3. the detail visibility function (DVF), which quantifies the relative visibility of spatial artifacts at different image densities (with less spatial detail generally being discernable in darker areas).

The first two functions are discussed in this section, and the third is described in the following section.

The TDF is a probability density function (PDF) that describes the distribution of original scene tones associated with a set of photographed images. The combination of the original scene tone distribution and imaging system tone reproduction determines the distribution of visual densities in the final images produced by a system. Scene tones are conveniently quantified in terms of scene density, which is the density of a tone relative to a reference 100% reflecting, diffuse (Lambertian) white. For example, a 20% gray, diffuse reflector in the primary scene illuminant would have a scene density of $-\log_{10}(20/100) \approx 0.7$. The same reflector in a shadowed area would have a higher scene density because of its lower luminance. Specular highlights, such as mirror-like reflections off metal surfaces, are brighter than a diffuse white, and, consequently, have negative scene densities. Use of scene density is convenient because it is easily compared to the resulting visual density in the final image. Scene and image densities are monotonically related and are similar in magnitude, although they do differ systematically in a preferred reproduction (see Giorgianni and Madden, 1998). The TDF may be determined by directly measuring scene tones with a photometer or by photographic photometry, in which images scanned from film or captured by an electronic sensor are analyzed using calibration targets to infer the original scene tone associated with each pixel. To obtain a representative TDF, the measurements from a large number of scenes or images must be analyzed.

Whereas the TDF describes the relative abundance of different scene tones, the TIF characterizes the relative impact of different scene tones on perceived image quality, at equal abundance. It is not obvious a priori what form the TIF might take, but some guesses may be made. Colors frequently associated with primary subjects of photographs or with important compositional elements include skin-tones, foliage, sky, etc. These colors vary considerably in visual density but are mostly mid-tones rather than highlights or shadows, suggesting that the TIF probably peaks in the mid-tones. Furthermore, because main subjects are often uniformly lit for pleasing rendition, extreme luminance values are avoided, again suggesting that mid-tones will have a greater importance than highlights and shadows. The relative importance of highlights compared to shadows may be guessed, also. Because the human visual system adapts to the brightest areas in its field of view, we almost invariably see a high degree of detail in scene highlights, whereas sufficiently dark shadows, especially those of narrow angular subtense, may appear devoid of detail. Therefore, it may appear especially unnatural when highlight detail is compromised in an image, whereas loss of shadow detail at least produces an appearance that is familiar. This suggests that the TIF might be higher in the highlights than in shadows. A discussion of the manner of deduction of the TIF is deferred to Sect. 15.6.

Figure 15.1 shows diagrammatic representations of unnormalized tonal distribution and importance functions. The TDF peaks in the mid-tones, indicating that they occupy a greater fraction of image area than do highlights or shadows. The TIF peaks at lighter tones than does the TDF, indicating somewhat greater importance of tones lighter than the most prevalent mid-tones. These features are consistent with those we guessed earlier.

To review, the TDF describes the abundance of different scene tones, and the TIF characterizes their relative importance at equal occupied area in an image. In the case of tonal clipping, in which shadow and/or highlight detail are lost through poor exposure or dynamic range, these two functions are adequate to

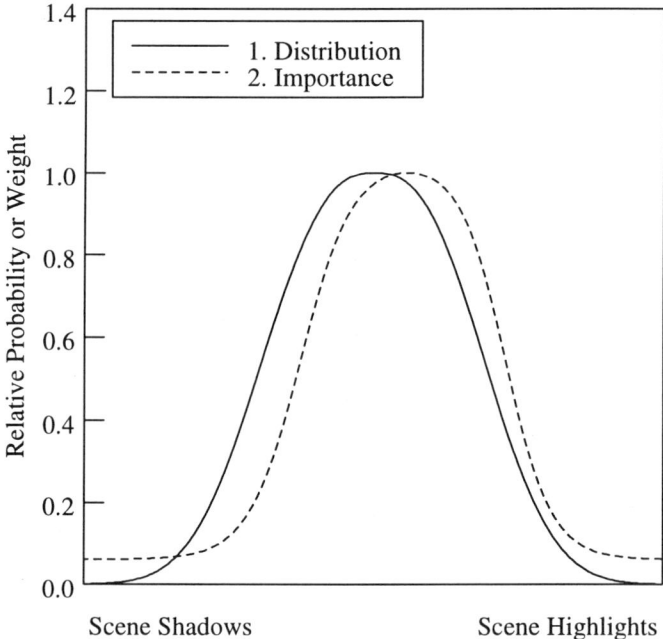

Fig. 15.1 Comparison of relative abundance of scene tones in consumer photography with their importance at equal occupied image area. The tonal importance function (TIF) peaks closer to the scene highlights than does the tonal distribution function (TDF), so tones lighter than the most common ones have the greatest perceptual impact.

predict associated quality loss. In the case of spatial artifact such as noise, a third factor is also of importance, namely, the relative visibility of the artifact as a function of the rendered image density. This dependency is quantified by the detail visibility function (DVF) described in the next section.

15.5 The Detail Visibility Function (DVF)

It is well known that image structure artifacts such as noise, streaking, and banding, at equal objective amounts measured in density space, are less visible (detectable) in darker parts of an image. The detail visibility function (DVF) characterizes this dependence. It is fundamentally a function of the visual density in the rendered image, rather than the scene density, because it is related to visual adaptation.

The DVF is most easily determined by assessing the perceived level of a spatial artifact such as isotropic noise, banding, or streaking in uniform fields. By comparing the ratings of samples varying in both density and objective metric value, an equivalence between objective metric changes and visual density changes may be established. For example, in a study of isotopic noise, Bartleson (1985) found a unit visual density (D_v) change to have the same effect on perceived noisiness as a 0.37 units change in the common logarithm of RMS granularity measured in visual density space (σ_D).

$$\frac{\partial \log_{10} \sigma_D}{\partial D_v} = -0.37 \qquad (15.1)$$

This relationship was found to be closely followed from threshold to suprathreshold levels of graininess. The negative sign indicates that an increase in visual density has the same directional effect as a decrease in granularity. A new objective quantity, the effective granularity (σ_e), may be defined so that equal effective granularities at different visual densities are perceived to have equal graininess. Integrating and exponentiating Eq. 15.1 yields Eq. 15.2.

$$\sigma_e \propto \sigma_D \cdot 10^{-0.37 \cdot D_v} \qquad (15.2)$$

The form of this equation suggests that the expression $10^{-0.37 \cdot D_v}$ be identified as the DVF. Unlike the TDF and TIF, the DVF is an objective transform, not a perceptual weighting. It relates to the detectability of a spatial attribute as a function of visual density, not to the way in which the variation of an attribute across an image is integrated to yield an overall impression of that attribute.

It is sometimes assumed that calculation of RMS granularity measured in lightness space (CIE L*) should yield a quantity that is predictive of perceived graininess at different visual densities. Thus, it is of interest to compare the properties of σ_{L^*} to σ_e of Eq. 15.2. Except at very high densities, L* is calculated from the equation :

$$L^* = 116 \cdot \left(\frac{Y}{Y_0}\right)^{1/3} - 16 \qquad (15.3)$$

where Y is luminance and Y_0 is the reference white luminance (CIE Publication 15.2, 1986). If we choose zero visual density to correspond to the reference white position, then visual density is given by Eq. 15.4.

$$D_v = -\log_{10}\left(\frac{Y}{Y_0}\right) \qquad (15.4)$$

Substituting Eq. 15.4 into Eq. 15.3, and taking the derivative of lightness with respect to visual density yields Eq. 15.5.

$$\frac{\partial L^*}{\partial D_v} = -\left(\frac{116 \cdot \ln(10)}{3}\right) \cdot 10^{-D_v/3} \qquad (15.5)$$

For small noise fluctuations the RMS granularity in visual density space can be converted into lightness space via multiplication by the derivative of lightness with respect to visual density, from Eq. 15.5.

$$\sigma_{L^*} \propto \sigma_D \cdot 10^{-D_v/3} \approx \sigma_e \qquad (15.6)$$

Thus, lightness space granularity σ_{L^*} has the same form as the effective granularity defined by Eq. 15.2, and the constant in the power is only ≈10% different, supporting its use for comparison of granularities at different densities. The fact that the very different tasks of assessing lightness (to define CIE L*) and graininess (in Bartleson's study) yield such similar dependencies on visual density suggests that the DVF may be quite general in nature.

15.6 Determination of the Tonal Importance Function (TIF)

We have estimated the shape of the TIF using two very different experiments. The first experiment involved tonal clipping, which is the loss of detail in highlights and/or shadows because of limited dynamic range or improper exposure. Figure 15.2 (next page) depicts an example of highlight clipping, as occurs in overexposed digital still camera images, in which highlight information is truncated abruptly at the maximum code value. When such an image is rendered so that the mid-tones have a pleasing appearance, the truncated white areas of the image lack detail and appear dull, because the full dynamic range of the display material is not used.

Tonal clipping was digitally simulated by mapping all scene densities above some maximum value to that maximum value (shadow clipping), and all scene densities below some minimum value to that minimum value (highlight clipping). A range of both minimum and maximum values was investigated, and a number of combinations of highlight and shadow detail loss were included. The samples were assessed for overall quality using a quality ruler. The object of the analysis was to seek a TIF such that the fraction of the area under the TIF-weighted TDF that was clipped would be proportional to quality loss ΔQ and would constitute a perceptually uniform objective metric of tonal clipping:

$$\Delta Q \propto \frac{\int_{-\infty}^{D_{s,+}} h_t(D_s) \cdot t(D_s) \cdot dD_s + \int_{D_{s,-}}^{+\infty} h_t(D_s) \cdot t(D_s) \cdot dD_s}{\int_{-\infty}^{+\infty} h_t(D_s) \cdot t(D_s) \cdot dD_s} \qquad (15.7)$$

where D_s is scene density; $h_t(D_s)$ is the TDF; $t(D_s)$ is the TIF; and a ± subscript indicates a highlight/shadow clip point. The DVF is not included in this objective metric because the quality loss is caused by the absence of desired signal detail, not the presence of image structure artifacts.

A range of TIF shapes approximately meeting the criterion of Eq. 15.7 was found, so the behavior of the function was further constrained by requiring that it explain the results of another experiment. This second quality ruler experiment involved the assessment of the overall quality of samples having isotropic noise with a wide variety of density profiles. A TIF was sought such that the quality loss could be expressed as a weighted average of the quality losses at each density, as shown in Eq. 15.8.

Fig. 15.2 Tonal clipping of highlights, causing dull, featureless whites, as occur in digital still camera overexposures.

$$\Delta Q = \frac{\int_{-\infty}^{+\infty} h_t(D_s) \cdot t(D_s) \cdot \Delta Q(\sigma_e(D_s)) \cdot dD_s}{\int_{-\infty}^{+\infty} h_t(D_s) \cdot t(D_s) \cdot dD_s} \tag{15.8}$$

But what does the quality loss at a particular density, $\Delta Q(\sigma_e(D_s))$, actually mean? This quantity is defined as the quality loss that would occur if the perceived graininess and effective granularity at all visual image densities were equal to that at the visual image density to which D_s was mapped. The DVF has been implicitly incorporated into Eq. 15.8 by the inclusion of effective RMS granularity σ_e. Requiring the TIF to simultaneously approximately meet the

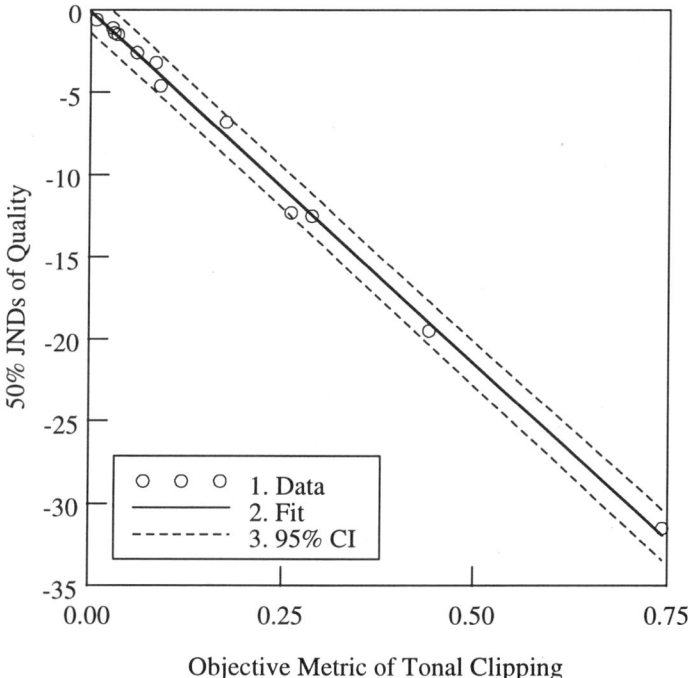

Fig. 15.3 Quality loss function for tonal clipping. The direct proportionality of quality loss to objective metric value indicates that the metric is perceptually uniform.

criteria of Eqs. 15.7 and 15.8 narrows the possible behavior of the TIF to a reasonable degree and leads to a function with the shape shown in Fig. 15.1.

The success of the TIF in explaining the results of the two experiments described is demonstrated in Figs. 15.3 and 15.4. Figure 15.3 (previous page) shows the quality loss function from the tonal clipping experiment with the objective metric defined in accordance with Eq. 15.7. The levels exhibit shadow detail loss, highlight detail loss, and combinations of the two. A proportional relationship between the objective metric and the quality loss arising from tonal clipping is indeed obtained, indicating that the objective metric so defined is perceptually uniform. This quality loss relationship can still be fit with the IHIF; the threshold and radius of curvature at threshold simply assume values of zero.

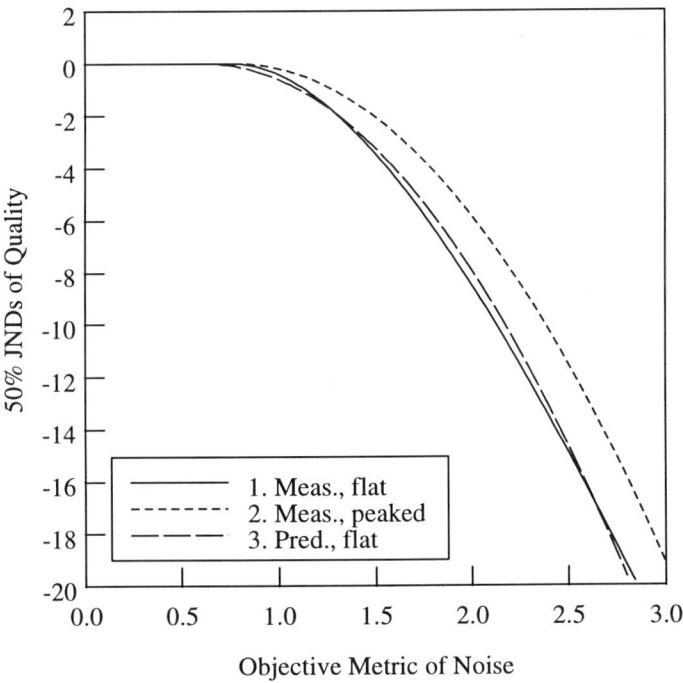

Fig. 15.4 Quality loss functions for noise having two very different dependencies on image density. Curve #3 is a prediction of Curve #1, made from Curve #2, based on tonal weighting, using the detail visibility, tonal distribution, and tonal importance functions.

In Fig. 15.4, results for two different profiles of noise versus image density are compared. The objective metric of noise is linearly related to the logarithm of the RMS print granularity in visual density space and at a mid-tone density. Curve #1 shows the quality loss for a flat noise profile, in which RMS granularity measured in visual density space (σ_D) is independent of visual density. Curve #2 shows the quality loss for a noise density profile that peaks strongly in the mid-tone region, as is characteristic of many systems in which dominant capture stage noise is shaped by output rendering, as discussed in connection with banding earlier in the chapter. The substantial shift between Curves #1 and #2 demonstrates that at matched mid-tone noise (and so matched objective metric), the flat profile leads to noticeably greater quality loss arising from its additional noise in the highlights and shadows. Curve #3 shows the prediction of the quality loss function for a flat noise profile based on the quality loss function for the strongly peaked noise profile and a tonal weighting scheme using the TDF, TIF, and DVF functions. Restated, Curve #3 is a prediction of Curve #1 from Curve #2 based on tonal weighting. The predicted Curve #3 very closely matches the measured Curve #1, demonstrating the success of the tonal weighting scheme.

15.7 General Applicability of the Tonal Weighting Scheme

In this final section, the tonal weighting scheme described above is applied to two new problems to demonstrate its generality. The first example is an extension of the tonal clipping result, but it provides a dramatic confirmation of the applicability of the work in another context. In the tonal clipping experiment described in the previous section, quality ruler assessments were made of digital image simulations of abrupt clipping, which emulates the truncation of highlights that occurs at the maximum code value in overexposures from digital still cameras. Less abrupt clipping of shadows occurs in underexposures of negative film because the response of the film rolls off gradually as its minimum density is approached. By estimating an effective clip point from the sensitometry of the film, a prediction was made of the anticipated loss of quality arising from tonal clipping as a function of ISO exposure to the film. This prediction was tested against the results of an independent experiment involving ratio scaling of analog samples optically printed from differently exposed negatives, including reference samples of known quality. The comparison of the measured data with the predictions from the tonal clipping work are shown in Fig. 15.5 (next page). The x-axis is the logarithm of exposure relative to an ISO normal exposure, so a one-stop underexposure corresponds to −0.3 units. The agreement is exceptional, particularly given the great difference in experimental methodologies.

The second example of an application of tonal weighting is the use of the TDF and TIF to predict quality loss from contouring. The artifact of contouring was discussed in connection with Figs. 10.5 and 11.2, and the objective metric of contouring mentioned below will be considered in detail in Sects. 15.7 and 17.3. A quality ruler experiment was performed using stimuli having different bit depths (numbers of gray levels) in different color records and quantization was performed in both linear (exposure) and logarithmic (density) spaces to change the perceptual uniformity of spacing of the gray levels. In addition, different amounts of noise were introduced prior to the quantization step, to investigate the interaction of noise and contouring. A perceptually uniform objective metric of contouring was sought having the form:

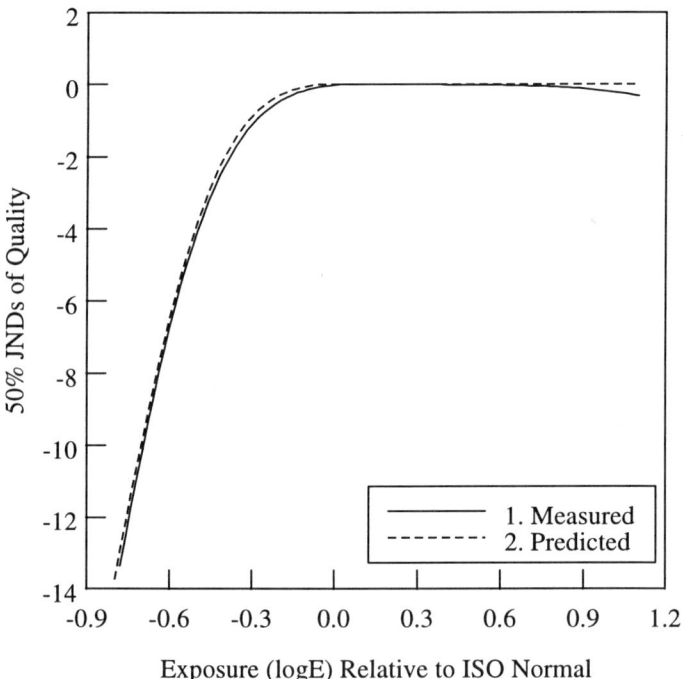

Fig. 15.5 Measured and predicted quality loss as a function of ISO exposure of color negative film (with x = −0.3 corresponding to a one-stop underexposure). The predictions, based on the objective metric of tonal clipping, are in outstanding agreement with the measurements.

$$\Delta Q \propto \sum_{i=2}^{q-1} \chi(\Delta L_i^*, \sigma_{L^*}(D_{s,i})) \cdot \int_{D_{s,i-1}}^{D_{s,i+1}} h_t(D_s) \cdot t(D_s) \cdot dD_s \quad (15.9)$$

where q is the number of quantization levels (e.g., for 8 bits, $q = 2^8 = 256$); $D_{s,i}$ is the scene density at which the ith of $q - 1$ transitions from one quantization level to another occurs; ΔL_i^* is the height of the transition step in the image in lightness units; and χ is a function describing the obviousness of a density step of this height in the presence of lightness space noise σ_{L^*}. Essentially, the quality loss associated with a single quantization transition step has been assumed proportional to the product of the obviousness of the step and the

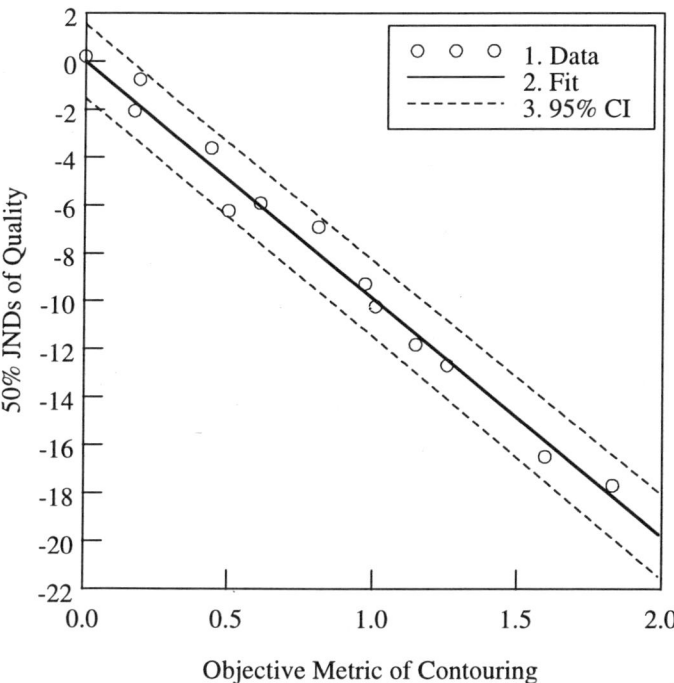

Fig. 15.6 Quality loss function for contouring. The objective metric, which is based on the tonal distribution and importance functions and the height of quantization transitions and pre-quantization noise in ΔL^* units, is perceptually uniform.

relative frequency of occurrence and importance of tones near the step (specifically over the interval from the previous step to the subsequent step). The success of this use of the tonal weighting functions is demonstrated in Fig. 15.6 (previous page). Although there is some scatter in the data, indicating that not all aspects of the impact of contouring on quality have been reflected in the objective metric, a good relationship of proportionality is nonetheless obtained.

Examples such as these, in which relationships derived in one context are predictive in another, are of key importance in developing confidence in the generality and validity of experimental methodology, analysis methods, and theoretical frameworks.

15.8 Summary

Some attributes vary significantly within an image because of dependencies on image density, location within the imaging field, and/or orientation (direction). For example, sensor noise is usually signal-dependent, and so varies with exposure, causing different amounts of noise to be present at different visual densities in the final image. Lens MTFs usually fall off away from the optical axis, so that greater blur occurs farther out in the imaging field, towards the corners of images. Digital scanners and printers that read and write in raster fashion have greater blur in the direction of the line or scan orientations.

The overall quality change arising from an attribute that varies with any of these factors may be calculated by taking a weighted average of quality changes over tones, positions, or orientations. This is referred to as weighting in perceptual space. If objective quantities are approximately linearly related to quality loss, the weighted averaging may be performed in an objective space, i.e., the weights may be applied to the objective quantities rather than to quality loss values, which can simplify modeling calculations significantly. An objective metric that is linearly related to quality loss is described as being perceptually uniform.

In the case of directional variations, an average of horizontal and vertical objective measures (e.g., MTF, NPS) is usually quite predictive of the overall impression of an attribute. Variations over location within the image field may usually be treated similarly, with simple area weighting for critical applications, or with the weights modified to reflect the concentration of key subject matter near the center of consumer images.

Tonal weighting of signal-dependent attributes is somewhat more complex. Larger variations in attribute levels across the image usually require that weighting be done in perceptual space. Tonal weighting is based on three functions. The tonal distribution function (TDF) is a PDF characterizing the relative frequency of occurrence of different scene tones in photographic images. The tonal importance function (TIF) is an inferred function reflecting the relative importance of different scene tones when they occupy equal areas in an image. Both functions peak in the mid-tones, although the TIF peaks closer to the highlights. The detail visibility function (DVF) describes the relative visibility or detectability of spatial artifacts as a function of visual image density. Such artifacts, at equal RMS density deviations, are less visible in darker areas than in lighter areas. The DVF is close to that predicted from the assumption that equal RMS deviations expressed in CIE lightness units should be perceptually equivalent regardless of the visual density. These three functions are useful in predicting the quality changes associated with tonal clipping, contouring, spatial artifacts, and a number of attributes of color and tone reproduction.

16

Analysis of Multi-Attribute Experiments

with Paul J. Kane
Eastman Kodak Company
Rochester, New York

16.1 Introduction

Sometimes the levels of different attributes are correlated because each of the attributes is affected by a common process. For example, blurring an image decreases both sharpness and noisiness, if the noise is suprathreshold. This does not imply that the attributes are perceptually dependent, but rather is an indication that they are physically linked under some circumstances. Covariations of this type can pose difficulties for the researcher who wishes to study the impact of an attribute in isolation, because generation of samples varying only in that single attribute may be difficult or impossible.

This situation may be handled by a multivariate decomposition procedure, in which the multivariate formalism equations are solved for an unknown contributing attribute, given a knowledge of the quality changes of all other attributes, and the overall quality of a sample. In this chapter, the general approach to multivariate decomposition is described in Sect. 16.2, and a practical example involving edge artifacts and the attribute of oversharpening is provided in Sect. 16.3.

16.2 Multivariate Decomposition

The multivariate equations of Ch. 11 provide a mechanism for predicting the overall quality change arising from a set of perceptually independent attributes given the impact on quality of each of the attributes in isolation. If the overall quality of a sample is known, as well as the effect of all but one of the contributing attributes, the equations can be numerically solved for the quality change caused by the unknown attribute.

The overall quality of a sample can be assessed in the usual fashion using a quality ruler. Estimation of the impact of an unavoidable ancillary attribute may be achieved by one of three methods, listed below.

1. A measurement (or calculation) may be performed to determine an objective metric, from which the quality change arising from the attribute may be predicted via a corresponding IHIF regression. This approach will be referred to as mensuration.

2. The degree of the undesired attribute can be assessed by appearance matching to a quality ruler varying in that attribute. Recall from Sect. 7.7 that this process is called projection, by analogy with the projection of an N-dimensional vector onto a particular coordinate axis. Projection may be used to measure the quality loss arising from the attribute directly, or it may be used to infer the associated objective metric, from which the quality change may be predicted with an appropriate IHIF.

3. An experimental level that contains the undesired attribute in isolation may be simulated, and it may be assessed in the usual fashion for overall quality. This approach, called the isolation method, provides a correction for each scene as assessed by each observer, without the need for any objective metric.

In the mensuration and projection methods, the IHIF regression for mean scene and observer is usually used, unless specific information is available regarding the observer sensitivity and scene susceptibility to the attribute. Use of a mean quality change rather than a scene- and observer-specific value may introduce some error into the correction. Compared to the mensuration method, the projection method involves additional assessments to account for the undesired attribute, and so usually involves greater effort, but is preferable in cases where objective characterization of level properties is problematical in some regard. Appearance matching usually has low variability, so only a small number of observers and scenes are typically needed to obtain a reliable estimate of the objective metric value. In the isolation method, the addition of one or more additional experimental levels entails extra simulation and assessment effort, but it directly yields scene- and observer-specific assessments of the impact of the undesired attribute on quality.

All three methods can lead to a decrease in precision of the quality loss values because there is uncertainty in the assessment of the undesired attribute, and when its effect is removed, any error is propagated into the inferred impact of the attribute of interest. By analogy, suppose that a measurement were made with some variance, and an undesired bias in the measurement were estimated independently with the same uncertainty. If the bias estimate were subtracted from the original measurement, the variance of the corrected value (i.e., the difference) would be twice that of the original measurement. The accuracy of the measurement is improved by eliminating systematic error (the bias), but this result is achieved at the expense of precision (the variance).

Eliminating the effect of an undesired attribute involves similar compromises. In the mensuration and projection methods there is usually relatively little imprecision in the estimate of the undesired attribute because those estimates stem from robust objective measurements or calculations, or are based on averages of multiple perceptual assessments. However, the accuracy of these two methods may be compromised by the lack of scene and observer specificity in the corrections. The isolation method is usually less precise because the correction is typically based on a single assessment, but the accuracy is higher because the correction is scene- and observer-specific. The isolation method may be viewed as an "aggressive" correction scheme because it sacrifices precision for accuracy, whereas the other two methods are milder schemes because they largely preserve precision, but do not improve accuracy to as great a degree.

The final choice of method may be dictated by practical considerations. It is not always possible to perform the simulations required by the isolation method because of correlation of ancillary attributes with a required simulation process. For very complex simulation procedures, the objective characterization required by the mensuration method is sometimes difficult to carry out. Finally, use of the projection method may be precluded by the lack of existence of a ruler varying in that attribute. If more than one of these three methods is viable, the best choice may depend on the magnitude of the impact of the undesired attribute. If the impact is small, the needed correction will be minor, and the mensuration or projection methods are likely to provide the best approach. If the impact is large, the isolation method may be required, but decreased precision may be anticipated. Intermediate approaches are also possible; for example, the isolation method can be modified by averaging assessments of the undesired attributes across observers or scenes to improve the precision of the estimates at the expense of their accuracy.

There is no reason why the impacts of the different undesired attributes need to be estimated in the same fashion; any combination of the above three methods may be used when multiple ancillary attributes are present. Once the quality change caused by each of the unwanted attributes has been estimated, and the overall image quality has been evaluated, the impact of the attribute of interest may be computed by numerical solution of the multivariate formalism equations. This is like solving for the leg of a right triangle (attribute of interest) given the other leg (undesired attribute) and the hypotenuse (overall quality). More generally, to continue the vector analogy mentioned in connection with the projection method, it is like finding an unknown component given the length of an N-dimensional vector and the other $N - 1$ orthogonal components.

There will be considerable uncertainty in the deduced magnitude of the attribute of interest if its impact is suppressed by one or more undesired, dominant attributes. Abstracting a small signal in the presence of a large signal is usually problematical even if the signals simply add linearly; the presence of multivariate suppression makes the problem even more difficult. For accurate determination of the quality loss arising from an attribute, the impact of that attribute should be comparable to, if not exceed, that of other attributes influencing quality.

16.3 A Practical Example: Oversharpening

As an example of the application of multivariate decomposition, an investigation of artifacts resulting from excessive boost in sharpening operations will be considered. Digital images are easily sharpened using spatial filtering procedures, and analog images may be sharpened by certain chemical effects in the development process. These digital and chemical processes boost the spatial frequency content of the images in a fashion that makes them appear sharper. However, noise and other artifacts present at the time of the boost operation are usually amplified by it, and a new artifact called oversharpening may be introduced as well.

Oversharpening is an edge artifact that gives the appearance of a halo along sharp edges separating areas of significantly different densities. Figure 16.1 shows a pictorial example of a severe level of oversharpening, which is most obvious along the outlines of the people and props against the light background.

Analysis of Multi-Attribute Experiments

Fig. 16.1 The edge artifact of oversharpening, evident as "halos" along the outlines of the people against the light background.

To elucidate the origin and nature of oversharpening, an example of a spatial filtering procedure called unsharp masking is shown in the spatial domain in Fig. 16.2. The figure depicts an edge from a square wave pattern, one-half period of which is shown. The original edge, Curve #1, is blurred by one or more imaging system components as shown in Curve #2. To compensate for this blur, a three-step digital boost operation is performed, as described below.

1. Each pixel value in the blurred image is replaced with a weighted average of pixels in its neighborhood (including the pixel itself), yielding Curve #3. This operation is known as a convolution, and the weights are arrayed in what is called a convolution kernel.

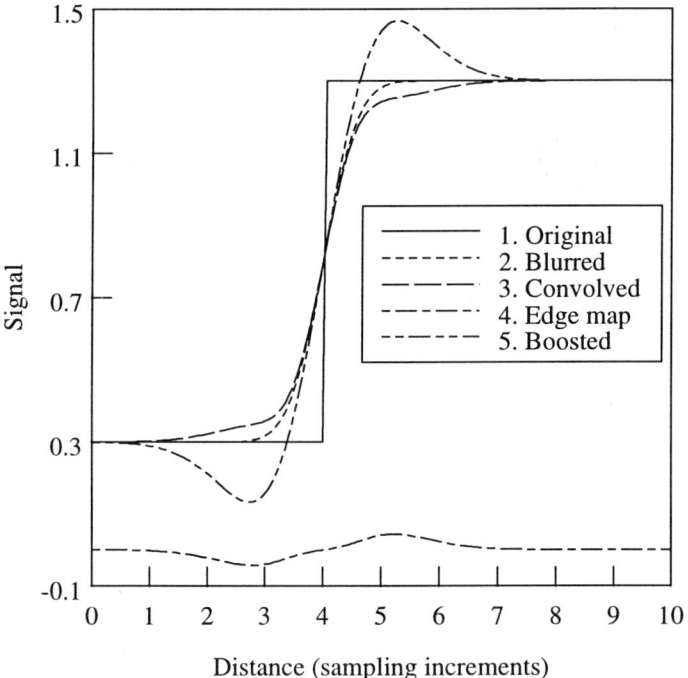

Fig. 16.2 Edge profile followed through an unsharp masking procedure. Although the boost operation increases the signal difference across the edge by undershooting and overshooting the original signal levels (which may lead to a halo-like artifact at the edge), the high slope of the original edge is not recovered.

Analysis of Multi-Attribute Experiments 233

2. The difference between Curves #2 and #3 yields an edge map, Curve #4. The map is close to zero except at edges, which produce characteristic signatures of adjacent positive and negative deviations.

3. The edge map is scaled by a positive gain factor (in this case, having a value of 4.0) and added back to the original image (Curve #2) to produce the sharpened image of the edge (Curve #5).

In the final image (Curve #5), the signal difference across the edge has been amplified, but the high slope of the edge in the original scene (Curve #1) has not been recovered. The boosted profile overshoots the original profile on the high signal side and undershoots it on the low image side. If the unsharp masking gain is excessive, this effect can result in halo-like oversharpening artifacts.

An experiment was performed in which levels exhibiting varying degrees of oversharpening were digitally simulated by blurring original images, which were quite free of noise and other artifacts, and boosting the blurred images via various unsharp masking procedures. The images were visually inspected for presence of artifacts other than oversharpening to ensure that formerly subthreshold artifacts such as image noise were not amplified to suprathreshold levels. The resulting experimental levels varied in both sharpness and oversharpening. It would have been convenient if the undesired variation in sharpness could have been eliminated, leaving only the desired variation in oversharpening. However, to achieve this result would have involved boosting or blurring the final images to a common level of sharpness, which could have changed the degree of oversharpening. The problem is that modification of the spatial frequency content of an image by a sharpening operation influences both sharpness and oversharpening in a correlated manner, so it is not possible to vary the two completely independently. Because of this linkage, the test samples unavoidably varied in sharpness, and multivariate decomposition was required.

In this experiment, two of the three methods for estimating the ancillary attribute (unsharpness) were viable options. The projection method would have required assessment against a quality ruler varying in sharpness, and such a ruler existed. The mensuration method would have involved calculation of an objective metric of unsharpness for which IHIF regressions were known, and such a metric was available. The isolation method would have entailed simulation of samples having the same sharpness as the oversharpened samples, but without any edge artifacts. If an objective metric of oversharpening had already existed (making this experiment unnecessary), such simulations might have been feasible, but without such a metric, the isolation method was not practical. Because the mensuration method was easily applied in this case, it was chosen in preference to the projection method, which would have required additional assessments.

Figure 16.3 shows examples of the types of MTF shapes investigated. The x-axis is cycles/sample, which has a value of one at the sampling frequency and a value of one-half at the Nyquist frequency. Curve #1 shows the MTF associated with the unsharp masking procedure used in Fig. 16.2 if the sample increment were that indicated in the x-axis of that figure. Because spatial frequency content has been amplified, the MTF is greater than unity. By using different convolution kernels and gains, different spatial frequency responses may be obtained. Curve #2 has a higher initial slope, boosting low frequencies more aggressively, but does not produce such high amplifications at higher frequencies. In certain systems under certain viewing conditions, these two responses might yield the same apparent sharpness increase, but the degree of quality loss arising from oversharpening could be quite different.

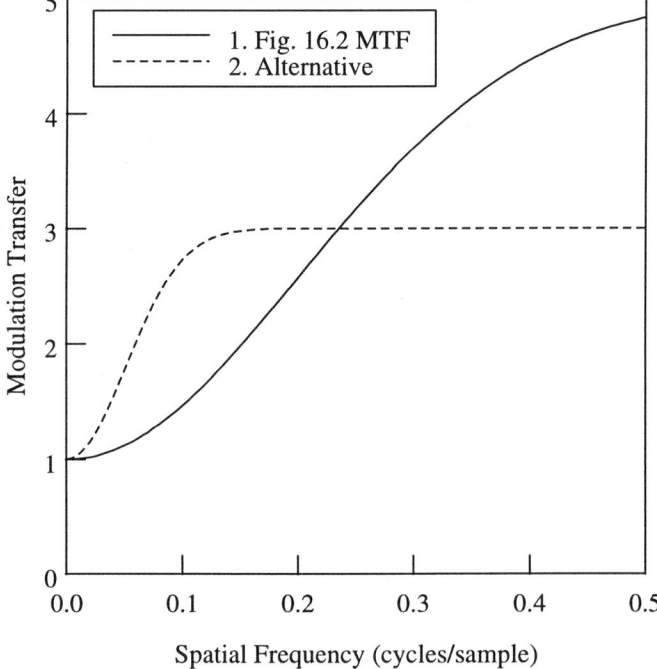

Fig. 16.3 Two possible MTFs associated with digital boosting operations. Curve #1 represents the unsharp masking procedure of Fig. 16.2; Curve #2 has a higher initial slope, indicating that it boosts low frequencies more aggressively.

Analysis of Multi-Attribute Experiments

The primary dimension series involved different gain levels with a single convolution kernel that led to substantial boost at low frequencies, like Curve #2 of Fig. 16.3. A second experimental dimension was the degree of initial blur in the sample, prior to the unsharp masking procedure. The third and final dimension of investigation was the shape of the spatial frequency response of the sharpening procedure. Figure 16.4 shows the primary dimension series regression results. The objective metric of oversharpening was based on a comparison of the final image MTFs that would have resulted with and without boosting operations. The gain series was very well fit by the IHIF. The quality loss values on the y-axis were determined by multivariate decomposition and represent only the effect of oversharpening, with the impact of sharpness removed. It is instructive to consider the leftmost data point, which falls near

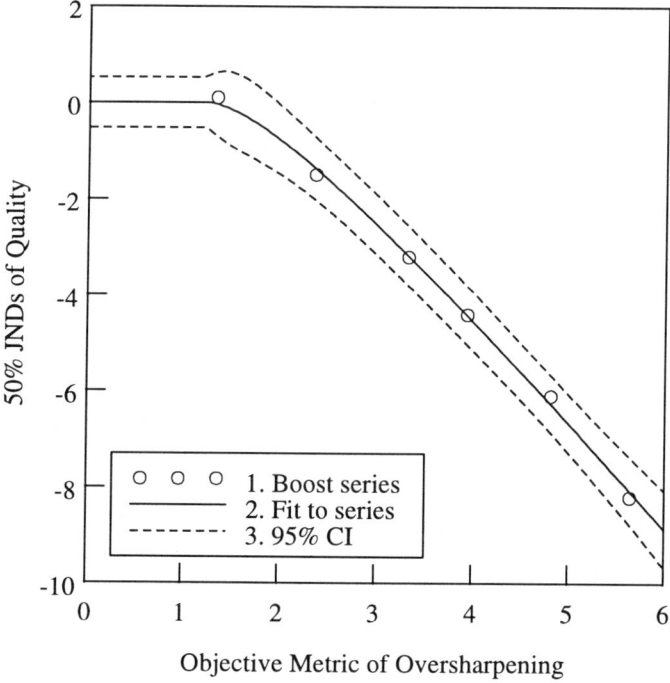

Fig. 16.4 Quality loss function of oversharpening, from a primary series varying in unsharp masking gain. A multivariate decomposition was performed to remove the impact of sharpness changes, which were quantified via the mensuration method.

threshold. Its quality change is very slightly positive (+0.1 JND), although this value is not statistically significantly different from zero. A positive quality change corresponds to an enhancement of quality, and represents a discrepancy in the analysis arising from the multivariate decomposition. If the quality loss caused by unsharpness were slightly underestimated by the mensuration method, and oversharpening were subthreshold, a small quality enhancement arising from oversharpening would be inferred. Samples of each scene at this level were critically examined for edge artifacts by expert observers, who concluded that oversharpening was subthreshold at this level. Consequently, the 0.1 JND discrepancy is indicative of the accuracy of the multivariate decomposition, which would appear to be very good.

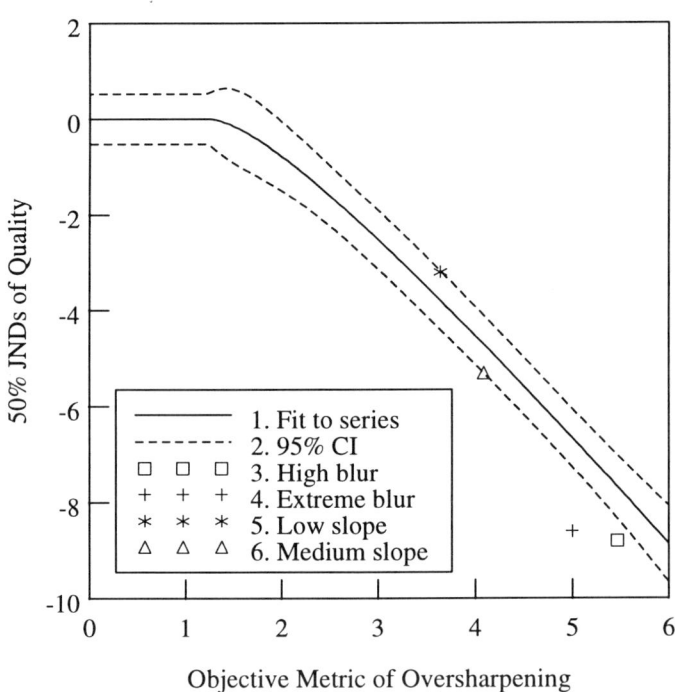

Fig. 16.5 Secondary dimension variations of oversharpening plotted against the primary series IHIF regression. Although the effect of variations in spatial frequency response shape (low and medium slope) are predicted reasonably well, the degradation from oversharpening is underestimated when very blurred samples are sharpened.

Figure 16.5 shows the results from the secondary level series, plotted with the primary dimension IHIF regression. The levels labeled low and moderate slope had spatial frequency response shapes with less aggressive low-frequency boost than in the primary series, like Curve #1 of Fig. 16.3 (lower initial slope) compared to Curve #2 (higher slope). These low and moderate slope data fall just within the IHIF regression confidence interval, so the objective metric of oversharpening successfully predicted the effect of these variations in spatial frequency response. The fact that the low and moderate slope points fall on separate sides of the high slope regression, further suggests that there is not a systematic error in the objective metric, because the data are not monotonically ordered by their slope.

The high and extreme blur data fall outside the confidence interval, and the extreme blur is misestimated to a greater degree, suggesting a modest systematic error as the amount of initial blur increases. Because the amounts of blur applied in these levels were larger than routinely encountered in practice, and because the resulting discrepancies were not too severe, the objective metric was deemed to be satisfactory, but with known limitations.

In this chapter, we have reviewed how to analyze experiments in which undesired but perceptually independent attributes unavoidably vary in the test samples. In the next chapter, treatment of perceptually dependent attributes will be considered.

16.4 Summary

It is not uncommon for an attribute being studied to be accompanied by unavoidable variations in another attribute, perhaps because of the fashion in which the first attribute is generated. Even if two attributes both vary as a function of a common process, they still may be perceptually independent, and so subject to analysis based on the multivariate formalism. In such cases, a multivariate decomposition can usually be used to isolate the effect of the attribute of interest.

To accomplish a multivariate decomposition, the images are assessed for overall quality, and then the quality losses that would be produced by the undesired attribute variations alone are determined by one of three methods:

1. measurement or calculation of the objective metric value, from which the quality change is computed using IHIF regressions (mensuration);

2. assessment of the level of the undesired attribute (as opposed to overall quality) against a quality ruler based on that same attribute (projection); or

3. assessment of overall quality of additional samples containing only the undesired attribute (isolation).

The mensuration method is very efficient, but is not applicable to attributes for which objective metrics are not available. The projection method requires the existence of a calibrated quality ruler varying in the undesired attribute. The isolation method requires that it be possible to simulate the undesired attribute without producing ancillary artifacts. The isolation method is particularly powerful because scene- and observer-specific corrections may be determined, but it involves a considerable number of additional image quality assessments. All three methods reduce systematic error (bias) in the estimates of the impact of the attribute of interest, but at the expense of noise in the inferred values, with this effect being particularly strong in the isolation method. As the impact of the undesired attributes increases relative to that of the attribute of interest, this becomes the method of choice.

Once the quality changes caused by undesired attributes are quantified, the multivariate equation is then solved numerically for the missing quantity, which is the quality loss that would have occurred had the attribute of interest been present alone. This technique works well as long as an undesired attribute, at the levels present, does not dominate the quality loss. If an unwanted attribute is the chief contributor to quality loss, multivariate suppression effects cause the deduced quality loss arising from the attribute of interest to have substantial uncertainties.

17

Attribute Interaction Terms in Objective Metrics

17.1 Introduction

Chapter 11 discussed the challenge of predicting multivariate image quality from a knowledge of the impact of each of the contributing attributes in isolation. Pairs of attributes were classified as perceptually dependent or independent based on whether the level of one attribute affected the perception of the level of another attribute. It was stated that in practice, most image quality attributes, if defined carefully, are approximately perceptually independent at most practical levels. In this chapter, two examples of interactions between perceptually dependent attributes are considered; these are: (1) the impact of isotropic noise on streaking, a weak interaction discussed in Sect. 17.2; and (2) the effect of pre-quantization isotropic noise on contouring, a strong interaction discussed in Sect. 17.3. In both cases, the objective metrics of the affected attributes can be extended so that the perceived level of the affected attribute in the presence of the affecting attribute may be predicted. This observation leads to some interesting deliberations in Sect. 17.4, which suggest that the multivariate formalism may be even more generally applicable than initially proposed.

17.2 The Weak Interaction of Streaking and Noise

As discussed in Sect. 13.2, streaking is a type of noise that is random in one dimension and deterministic in the other dimension, producing the appearance of parallel streaks of varying width and density. These streaks may be much more visible than a two-dimensionally random pattern of isotropic noise at the same RMS noise level because of the more regular and extended pattern. It seems plausible that addition of isotropic noise to an image containing streaking might help to visually break up the extended streaking pattern and render it less noticeable, thereby decreasing its impact on quality. This effect would be of particular interest if the quality loss arising from the additional isotropic noise were more than offset by the reduction of quality loss caused by streaking. It would be very convenient if the quality loss from streaking in the presence of isotropic noise could be predicted from the same IHIF regressions as in the noiseless case, by simply modifying the objective metric of streaking to take

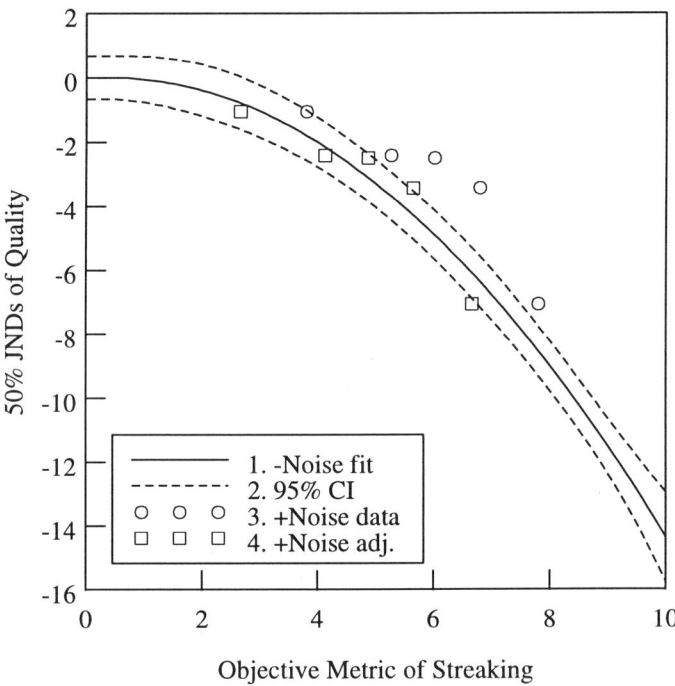

Fig. 17.1 Effect of visually substantial noise on quality loss arising from streaking. Incorporating a weak noise interaction term into the objective metric of streaking shifts the data for levels with noise (circles); the adjusted data (squares) are in good agreement with the IHIF regression based on levels without noise.

Attribute Interaction Terms in Objective Metrics

into account the visual masking of streaking by isotropic noise. In fact, this is possible, as described below.

Figure 17.1 presents the results of a quality ruler experiment on streaking, which was described in Sect. 13.4. The IHIF regression is based on variations in both magnitude and NPS shape of streaking. In addition, new data based on samples with a fixed, perceptually substantial amount of added isotropic noise are shown. An experimental level containing the isotropic noise alone, without streaking, was assessed in accordance with the isolation method, allowing multivariate decomposition to determine the quality loss arising from streaking, in the presence of isotropic noise, on a scene- and observer-specific basis. The five noise points plotted as circles all lie above the regression curve, indicating that the streaking had less effect on quality in the presence of noise, thereby demonstrating the perceptual dependence of streaking on noise. However, so much isotropic noise was required to produce a noticeable interaction effect that the impact of streaking on quality was multivariately suppressed by the quality loss caused by isotropic noise. Evidently, the interaction of these attributes is rather weak, and, at least under these circumstances, adding isotropic noise to reduce the quality loss from streaking is undesirable, because the quality loss caused by the added noise will more than offset its benefits.

The suppression of streaking by noise caused significant uncertainty in the multivariate decomposition, leading to some scatter in the streaking plus noise data. Within the precision of the data, the effect of the addition of a constant amount of noise to varying magnitudes of streaking noise can be explained by a fixed shift along the objective metric axis, as demonstrated by the five noise points plotted as squares. This allows the streaking metric to be extended to cases in which noise is present by including in the objective metric a subtractive term that depends on the RMS granularity. Our goal of maintaining the IHIF regression by modifying the objective metric of streaking has thus been met.

17.3 The Strong Interaction of Contouring and Noise

Contouring is an artifact caused by an insufficient number of gray levels being supported in a quantization operation. Contouring is especially visible in slowly varying areas, such as clear blue sky (which slowly becomes a deeper blue from the horizon to the zenith) or background sweeps in studio photography. In such slowly varying areas, contouring causes a stair-step effect because the image densities change in discrete steps rather than smoothly and continuously.

Fig. 17.2 The artifact of contouring, evident as discrete steps of density change that create a pattern of arcs in the sky (see text).

Attribute Interaction Terms in Objective Metrics 243

Fig. 17.3 Visual masking of the contouring level of Fig. 17.2 by isotropic noise introduced prior to the quantization step.

A pictorial example of contouring is shown in Fig. 17.2. The contouring is most obvious in the clear sky, which should slowly and smoothly increase in density from the horizon to the zenith. Instead, discrete density transitions produce a series of stepped arcs with fixed densities. The curvature of the arcs is probably caused by camera lens illumination ($\cos^4\theta$) fall-off (see Sect. 15.1), which, in a quantized uniform field would produce circular contours. Figure 17.3 was simulated exactly as Fig. 17.2 except for the addition of a visually modest quantity of isotropic noise prior to the quantization stage. Because of visual integration over a number of pixels, the noise breaks up the contouring pattern quite effectively, without the isotropic noise seriously degrading quality itself. The reduction of contouring at the expense of isotropic noise is a good exchange in this case, significantly improving the overall quality of the image.

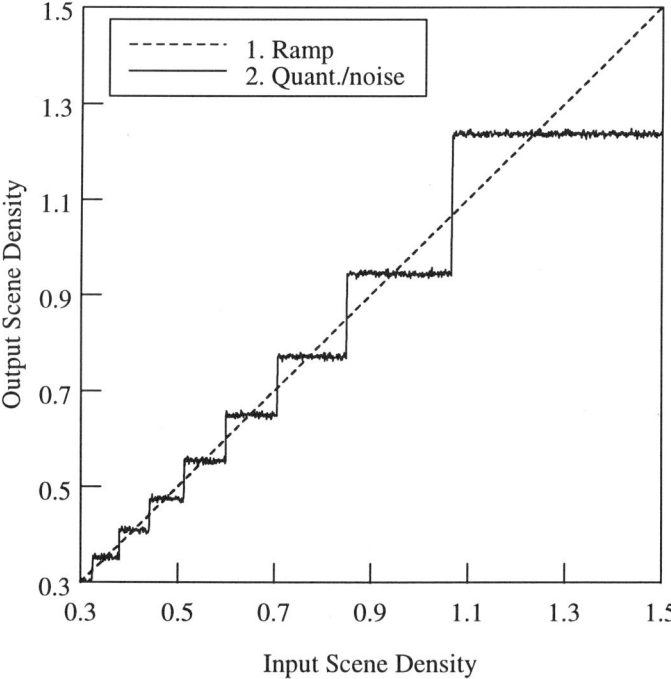

Fig. 17.4 Effect of linear (exposure space) quantization and post-quantization noise on a ramp target. Noise arising after the quantization event does little to obscure the stair-step effect of contouring.

Attribute Interaction Terms in Objective Metrics 245

To understand in detail how noise can mask contouring, consider the reproduction of a "ramp" target that varies from white to black, at a constant rate of density change per unit distance. Suppose that we capture an image of the ramp using a system with high spatial resolution and a variable linear exposure quantization rate, which can be changed from a coarse bit depth (e.g., 5 bits, supporting $2^5 = 32$ gray levels) to an arbitrarily fine bit depth. The system is designed such that the quantization acts like a nearest integer operation rather than a truncation operation (i.e., a signal equivalent to 13.8 will be rounded up to 14 rather than truncated to 13). The system also allows a fixed amount of isotropic noise to be added either prior to, or subsequent to, the quantization operation; these will be referred to as pre- and post-quantization noise.

First, we turn off the noise altogether, set the bit depth to a very high value so that the system is essentially continuous, and capture an image of the ramp. The output values of the image are calibrated against the scene densities of the ramp so that the input densities can be inferred from the output values. Having made this calibration, we obtain Curve #1 of Fig. 17.4, in which the inferred output scene density is equal to the input scene density, and the scene density profile is indeed a ramp from white to black, going from left to right.

Next, we set the bit depth to a low value, recapture the ramp, add post-quantization noise, and infer the scene density profile. Because the output can now only have a small number of discrete values, the ramp becomes a stair step pattern, as shown in Curve #2. To make the amounts of noise shown in the profiles perceptually meaningful, we have averaged the values of pixels over a neighborhood of the size that could just be resolved by the human visual system. The steps are smaller in the highlights than in the shadows because the quantization has occurred in linear exposure space, as is common in digital cameras. Equal increments in linear space become unequal steps in logarithmic (scene density) space, because the step heights are a higher percentage of the signal where the signal is lower (towards the shadows). The added post-quantization noise does not change the overall shape of the stair steps, but simply adds small fluctuations to them.

In our final capture, the post-quantization noise is replaced with an equal amount of pre-quantization noise, leading to the results shown on the next page in Fig. 17.5, Curve #3. Provided for comparison are the ramp (Curve #1) and noise-free stair steps (Curve #2). The latter were not shown in Fig. 17.4 because they would have been completely obscured by the post-quantization noise trace. The profile is now dramatically changed, with the stair steps rounded off or even eliminated, with greater effect occurring where the step heights are smaller compared to the fixed noise (i.e., in the highlights, at low scene density). What has happened is this: near a step edge, the noise caused the output to toggle between two possible levels, creating rapid spatial fluctuations. When multiple pixels are spatially averaged by the eye, the fluctuations are smoothed out, and intermediate densities are perceived, obscuring the stair step pattern. Essentially, excess spatial resolution permitted some visual averaging of the output, increasing the number of effective output levels. This same effect is used to advantage in digital halftone systems.

The quality ruler experiment on contouring, described in Sect. 15.7, included variations with: (1) different numbers of quantization levels in different color records; (2) quantization performed in different spaces (linear and logarithmic); and (3) a fixed level of noise added prior to quantization (post-quantization

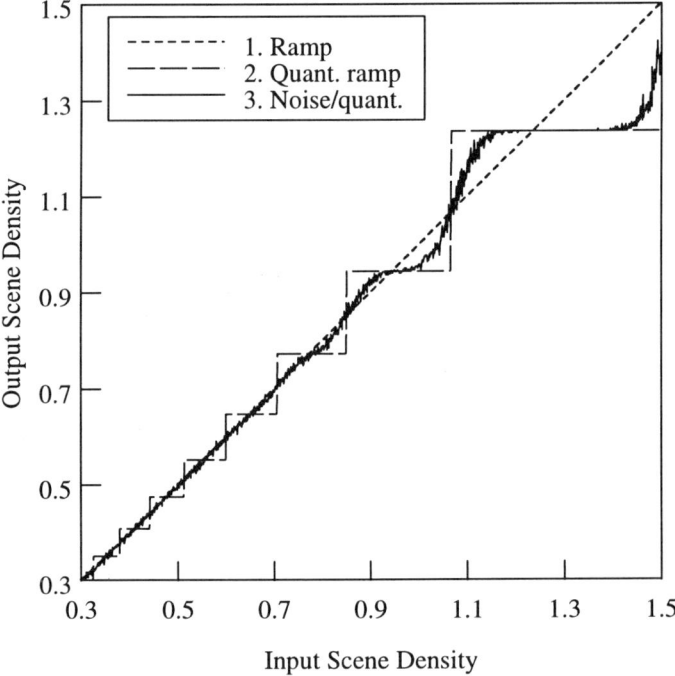

Fig. 17.5 Same as Fig. 17.4 except that the noise has been added prior to the quantization step. The pre-quantization noise smoothes out the stair-step appearance because of visual integration over multiple pixels, reducing the quality loss arising from contouring.

noise was not investigated because the impact on perceived contouring was so small). The impact of the isotropic noise on quality in isolation was estimated by the mensuration method, based on objective measurements and corresponding IHIF regressions. So little pre-quantization noise was required to reduce contouring that the impact on quality of the isotropic noise was minimal, and negligible multivariate suppression of the contouring occurred. The perceptual dependence of contouring on noise is therefore classified as a strong interaction.

Figure 17.6 shows the quality loss regression for contouring, previously depicted in Fig. 15.6, which is simply a straight line because the objective metric is perceptually uniform. Also shown are four data points, plotted as circles, which

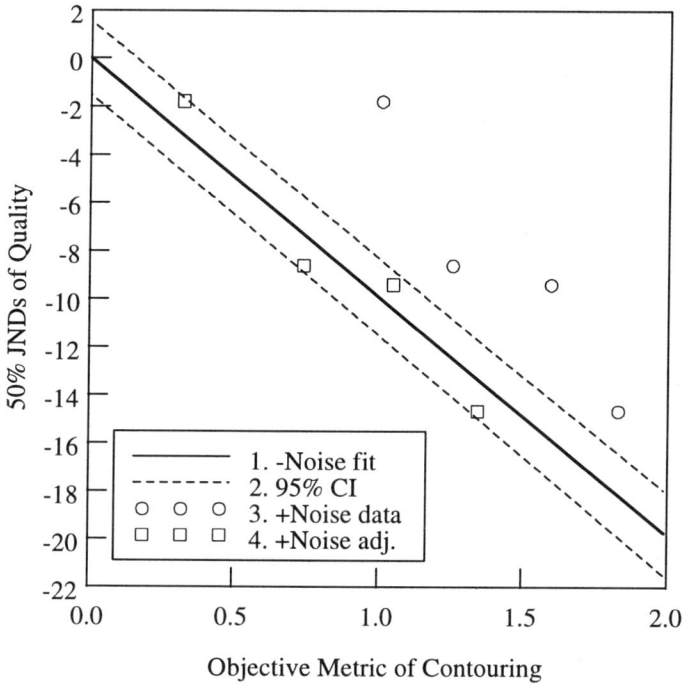

Fig. 17.6 Effect of visually minor pre-quantization noise on quality loss arising from contouring. Incorporating a strong noise interaction term into the objective metric of contouring shifts the data for levels with noise (circles) into agreement (squares) with the IHIF regression based on levels without noise.

are based on the levels with isotropic noise added prior to quantization. The quality change caused by the noise in isolation was estimated by the mensuration method and a multivariate decomposition performed to yield the quality change arising from contouring in the presence of the noise. These data all plot above the regression line, indicating that the quality loss arising from contouring has been mitigated by the addition of the noise. Unlike the case of streaking, in Fig. 17.1, a simple shift of these data along the objective metric axis cannot bring all the points within the regression confidence interval; a more complicated correction for noise within the objective metric of contouring is required.

The objective metric of contouring was given previously in Eq. 15.9, which is repeated here.

$$\Delta Q \propto \sum_{i=2}^{q-1} \chi(\Delta L_i^*, \sigma_{L^*}(D_{s,i})) \cdot \int_{D_{s,i-1}}^{D_{s,i+1}} h_t(D_s) \cdot t(D_s) \cdot dD_s \qquad (15.9)$$

In this equation, q is the number of quantization levels (e.g., for 8 bits, $q = 2^8 = 256$); $D_{s,i}$ is the scene density at which the ith of $q - 1$ transitions from one quantization level to another occurs; ΔL_i^* is the height of the transition step in the image in lightness units; and χ is a function describing the obviousness of a density step of this height in the presence of lightness space noise σ_{L^*}. The term being summed over the step transitions is the fractional area under the tonal distribution function (TDF) times the tonal importance function (TIF) that lies between the preceding and following step transitions (a measure of the number and importance of tones near the transition), multiplied by a contouring function χ that quantifies the obviousness of the transition.

In the absence of pre-quantization noise, χ is just a thresholded function of the transition height in CIE L* lightness units. As shown in Fig. 17.5, the pre-quantization noise rounds off or even smoothes out a step transition according to the RMS variation of the noise compared to the height of the step. Accordingly, the contouring function χ is extended to account for pre-quantization noise by, prior to thresholding, subtracting a term from the step height that depends on the ratio of the RMS noise variation to the step height. So, instead of subtracting a noise interaction term from the objective metric as was done for streaking, a noise correction depending on step height was subtracted from each term in the objective metric sum. This approach accounts for the fact that the degree of masking provided by a given amount of noise depends on the height of a step compared to the noise level. The revised objective metric of contouring is satisfactorily predictive, yielding the adjusted data plotted as squares in Fig. 17.6, which lie just within the regression confidence interval.

The masking of contouring by pre-quantization noise is the strongest perceptual interaction we have studied. In our experience, even weak interactions are rare among artifactual attributes, although they are slightly more frequent between preferential attributes of color and tone reproduction. These observations imply that the multivariate formalism is widely applicable and that studies of interactions will primarily serve to characterize second-order effects.

17.4 A Reconsideration of Perceptual Independence

The success of the approach adopted in this chapter, namely, the modification of objective metrics to account for attribute interactions, brings up an interesting philosophical point. The reduction in the appearance of, and quality loss arising from, streaking and contouring that is caused by the addition of isotropic noise has been considered to be an attribute interaction, because the presence of an affecting attribute (noise) has changed the perception of an affected attribute (streaking or contouring). Yet, the objective metrics of these affected attributes have been modified so that they are predictive of the affected attributes even in the presence of noise. Furthermore, the quality changes arising from contouring or streaking in the presence of noise that are predicted by the extended objective metrics may be combined according to the multivariate formalism to predict overall quality (this result is guaranteed because the objective metrics are adjusted to predict the outcome of a multivariate decomposition). Thus, the quality changes deduced by multivariate decomposition and the modified objective metrics behave just as if they were associated with perceptually independent attributes. These attributes might be called effective streaking and contouring, indicating that they include the effects of visual masking and other perceptual phenomena. Effective streaking and contouring defined in this way could be regarded as perceptually independent of noisiness, although their objective metrics would have to contain terms depending on isotropic noise to be predictive.

Let us review our reasoning to determine if this alternative viewpoint offers new insights. In Sect. 11.2, a simple test was given for determining if two attributes were perceptually independent. Samples of three series were prepared, one series having different levels of the first attribute, a second series varying in the second attribute, and a third series containing combinations of these levels of the two attributes. The samples of the third series were then rated against the first two series for appearance of each attribute, i.e., a projection assessment was performed. If the samples of the third series were matched to the samples having the same constituent levels of the attributes, the attributes were deemed independent. This procedure seems straightforward, but implicit in this definition of perceptual independence is the assumption that combining the two attributes in the samples of the third series somehow did not alter the degree of either attribute, so they should match up with samples of the first two series having received the same objective treatments (e.g., having patterns with the same streaking NPS added to them, or being quantized to the same number of levels in the same space). This might be called the metrocentric viewpoint, because some objective measure (e.g., streaking NPS, or profile of a ramp exposure), upon which an objective metric could be based, is viewed as determining the image attribute.

Suppose instead we adopt what shall be called the psychocentric viewpoint. The results of the three-series experiment are now interpreted differently. The perceptual attributes are assumed to be defined in a fashion that they are perceptually independent, implying that their effects on quality combine in accordance with the multivariate formalism. Consequently, the apparent interaction revealed by the projection assessments on the third series is ascribed to a failure of an objective metric (or other objective quantity) to accurately predict the appearance of one attribute in the presence of the other. For example, an objective metric of contouring might reflect the size of quantization steps in CIE L^* units and their locations in scene density units. The metrocentric viewpoint implicitly assumes that the size and scene tones of the quantization steps are sufficient to specify the degree of contouring, and if these quantities are equal, the contouring is identical. In the psychocentric viewpoint, this specification is deemed incomplete, because pre-quantization noise affects the appearance of contouring, but has been omitted from the objective metric. Essentially, the three series test has become a screening method for identifying objective metric inadequacies rather than perceptual interactions. If the objective metric of contouring were extended to include pre-quantization isotropic noise, and the third series were simulated to produce contouring metric values equal to those of the first and second series (rather than using the same simulation processes as in the first two series), the apparent interaction would disappear.

At this point, a loose analogy with the mathematical constructs of matrices and eigenvectors may be helpful. Suppose that pertinent objective metrics without interaction terms are collected into a vector. It is desired to derive from this a new vector containing the quality changes caused by each associated attribute. From this new vector, a multivariate sum may be computed and the overall image quality predicted. We imagine the vector conversion to be accomplished by multiplication by a transform matrix, the elements of which are allowed to be generalized operators. If the attributes are perceptually independent in the metrocentric sense, the matrix will be diagonal, because each objective metric by itself is sufficient to predict quality change arising from that attribute. However, for each attribute that affects another, one off-diagonal matrix element will be non-zero, so that the prediction of quality change caused by the affected attribute will also depend on the objective metric of the affecting attribute. In this metrocentric arrangement, the starting vector contains metrics without interaction terms, but the transform matrix will not be diagonal if interactions exist.

In the psychocentric viewpoint, the transform matrix is diagonalized by replacing the original objective metric vector with a new vector, the elements of which are unchanged in the absence of interactions, but are combinations of the old elements where an interaction does exist. As a result of this process, the matrix has been simplified by being diagonalized, but the objective metric vector elements have been made more complex in the process of accounting for interactions. This process is analogous to finding the eigenvectors of a matrix.

The psychocentric viewpoint may be combined with the multivariate formalism to produce a unified paradigm of image quality, the tenets of which might be stated as follows.

1. A set of N or more arbitrary attributes that fully describe an N-dimensional subset of overall image quality can be combined to form a set of exactly N orthogonal attributes that span the image quality space.

2. These N orthogonal attributes, expressed in terms of their impact on quality in isolation, combine to yield N-dimensional overall image quality according to a universal relationship, the multivariate formalism.

Although the differences between the metrocentric and psychocentric viewpoints are interesting, and despite the elegance of the paradigm described above, which viewpoint is adopted does not really influence the nature of the experiments and analyses that must be carried out. Regardless of viewpoint, stronger interdependencies must be identified and quantified, and the objective metrics associated with affected attributes must be extended through incorporation of interaction terms. What is most important is that stronger interactions are uncommon, but when they do occur, such extensions of objective metrics are feasible.

17.5 Summary

The multivariate formalism is intended to predict the overall quality of an image based on the contributions of multiple attributes, assuming that they are perceptually independent. In some cases, attributes may interact perceptually, with the level of one attribute affecting the perception of another. In such cases, it is useful to extend the predictiveness of the objective metric of the affected attribute by incorporating a correction based on the affecting attribute. The quality change predicted from such extended metrics via IHIF regressions will lead to accurate multivariate formalism calculations despite the attribute interactions.

This approach is exemplified in this chapter by the extension of objective metrics of streaking and contouring to account for visual masking by noise. The dependence of streaking on noise represents a weak interaction; addition of perceptually large amounts of noise results in rather modest reductions in perception of streaking, so that the effect of the interaction may be difficult to measure because of multivariate suppression. In contrast, the appearance of contouring is so effectively reduced by pre-quantization noise that the addition of visually modest amounts of isotropic noise can actually lead to an improvement in overall image quality. In our experience, strong interactions are rare, so that relatively few calibrated experimental investigations of interactions are required in practice.

It is argued that apparent perceptual dependence can equally well be regarded as a failure to identify the fundamental nature of the attribute and design an accordingly complete objective metric. This psychocentric viewpoint assumes that it is always possible to define perceptually independent attributes that directly combine according to the multivariate formalism, although their corresponding objective metrics may have to contain interaction terms to be adequately predictive.

18

Attributes Having Multiple Perceptual Facets

with Paul J. Kane
Eastman Kodak Company
Rochester, New York

18.1 Introduction

In the preceding chapter, a psychocentric viewpoint was proposed, in which the existence of orthogonal perceptual attributes combining according to the multivariate formalism is assumed. Even if this viewpoint is valid, however, there is no guarantee that these orthogonal attributes can easily be identified, nor that correlating objective metrics can be readily designed. In nearly all our studies, identification of orthogonal image quality attributes has been reasonably straightforward and design of predictive objective metrics, while often challenging, has ultimately proven feasible. In a few instances, however, a fully satisfactory treatment was not developed during the initial investigation. In this chapter, the most notable example of this type is described so that the reader can appreciate some of the difficulties that may be encountered in practice.

The artifact complex described in this chapter occurs during the reconstruction of an analog signal from a digital signal, as is required to display a digital image. Section 18.2 provides a very brief background discussion of sampled systems and discusses aliasing, primarily to distinguish it from the artifacts related to reconstruction, which are the subject of Sect. 18.3. Finally, Sect. 18.4 addresses the issue of how many independent perceptual attributes are represented by the reconstruction artifacts complex.

18.2 Sampling, Aliasing and Reconstruction

A detailed explanation of sampling theory is far beyond the scope of this work; the interested reader is referred to Wolberg (1990) and the works cited therein for a discussion of this topic from the viewpoint of digital imaging applications. An extremely simplified discussion of the basic aspects of sampling and aliasing follows.

When an analog image, such as that produced by a lens at the capture plane in a camera, is sampled at discrete, regularly arrayed points, as by an electronic sensor in a digital camera, the resulting digital signal frequency spectrum

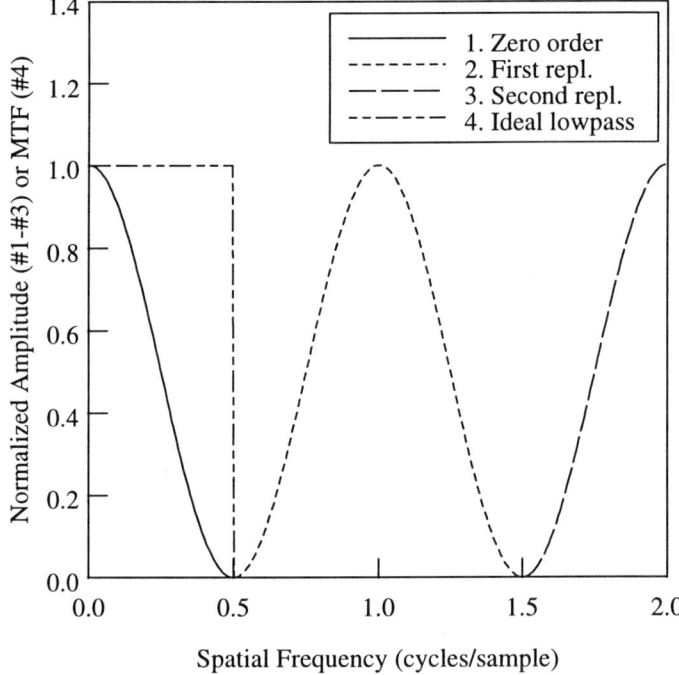

Fig. 18.1 Spectrum of a sampled signal having no aliasing. Because there is no overlap between the zero-order spectrum and higher order replicates, the original analog signal can be recovered by applying the ideal low-pass filter of Curve #4.

consists of the analog (zero-order) spectrum replicated at multiples of the sampling frequency, as shown in Fig. 18.1. The x-axis of this figure is spatial frequency in cycles per sample, with a value of one corresponding to the reciprocal of the spatial separation between sampling points (i.e., the pixel pitch). Although only the range 0–2 cycles/sample is shown, the pattern shown extends symmetrically from minus infinity to plus infinity. The y-axis shows MTF or frequency spectrum amplitude, the latter normalized to unity, for convenient comparison with MTFs. In this example, the original analog signal had no power above one-half cycle per sample, which is referred to as the Nyquist frequency. Consequently, the replicate spectra centered on multiples of the sampling frequency (... −2, −1, 0, 1, 2 ... cycles/sample) do not overlap and only the zero-order and first and second replicate spectra have power in the range of 0–2 cycles/sample shown in the figure.

Because there is no spectral overlap between the zero-order spectrum and the replicates, an ideal low-pass filter having unit response below the Nyquist frequency and zero response at higher frequencies (Curve #4) could be applied to the sampled signal to isolate the zero-order spectrum from the replicate spectra and thereby recreate the original analog signal. The process of converting a digital signal to an analog signal is called reconstruction. The process of reconstruction must occur before an image can be viewed, because human vision is an analog phenomenon. Writing a digital image to a hardcopy medium or displaying it on a video monitor are both forms of reconstruction.

Image resampling may also be regarded as incorporating a reconstruction step. For example, to quadruple the number of pixels in a digital image requires a 2× interpolation (up-sampling) in each direction. One way of doing this would be to form new columns of pixels alternating with the original columns and having values equal to the average of the two adjacent original pixels. The same procedure would be repeated in the row direction to complete the operation. Similarly, to reduce the number of pixels in an image, for decreased storage space or faster transmission of the image, alternating rows and columns could be eliminated; this procedure is referred to as a 2× down-sampling. Although both operations just described can be carried out in purely digital form, it is often more convenient to regard the operation as consisting of a reconstruction step, yielding an analog signal, followed by a resampling at twice (interpolation) or one-half (down-sampling) the original sampling frequency to yield a new digital image. This paradigm is particularly helpful when trying to understand non-integer interpolation or down-sampling.

Figure 18.2 shows a second example of a sampled signal, but in this case, the original signal has a relatively greater bandwidth (0.75 cycles/sample instead of

0.5 cycles/sample), so that the spectral replicates overlap. This condition is known as aliasing and is common in digital photography because the analog image at the camera capture plane usually contains frequencies higher than the Nyquist frequency of the electronic sensor unless special precautions are taken. In essence, power at frequencies above the Nyquist limit in the original analog signal has been remapped to lower frequencies, below the Nyquist limit, because of replicate overlap. When the signal is reconstructed, even the use of an ideal low-pass filter (Curve #5) cannot isolate the zero-order spectrum from its replicates.

Figure 18.3 shows a simple example of aliasing in the spatial domain. The x-axis is the index of the sampled point, i.e., the samples are taken at integer

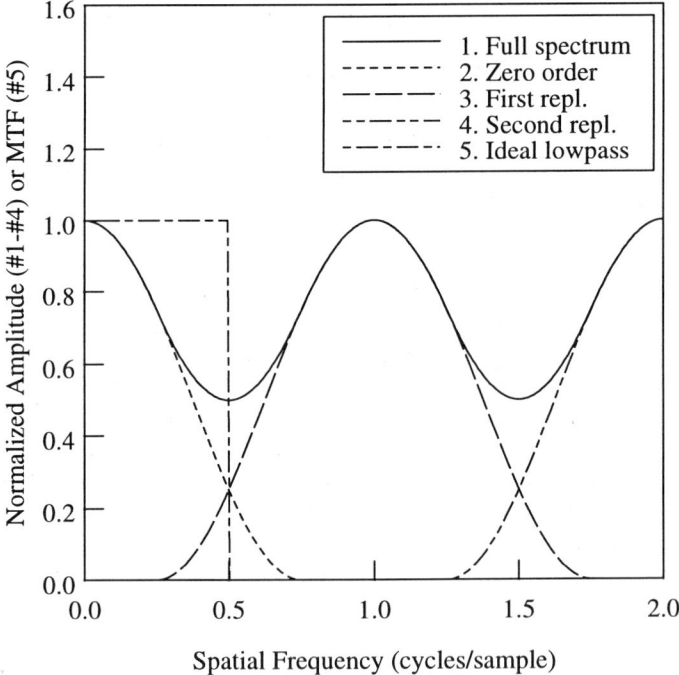

Fig. 18.2 Spectrum of a sampled signal that is aliased. Because the zero-order spectrum overlaps the first replicate spectrum, the original analog signal cannot be recovered by applying the ideal low-pass filter of Curve #5.

Attributes Having Multiple Perceptual Facets 257

values of x; the y-axis is signal amplitude. A pure sine wave with frequency 0.75 cycles/sample is sampled and reconstructed with an ideal low-pass filter. The resulting signal consists solely of an aliased sine wave from the first spectral replicate, which occurs at 1 − 0.75 = 0.25 cycles/sample. Consequently, the reconstructed sine wave has 3× the period of the original sine wave.

Aliasing can be prevented by limiting the bandwidth of the signal prior to sampling. In some digital cameras, this is done with an optical anti-aliasing filter, which is usually constructed from a birefringent material that produces multiple images shifted from one another by fractions of a pixel. This introduces a controlled degree of blur, reducing high-frequency content of the analog image. Because the birefringent filter does not have ideal characteristics, some lower frequency attenuation occurs as well, leading to a loss of sharpness. The

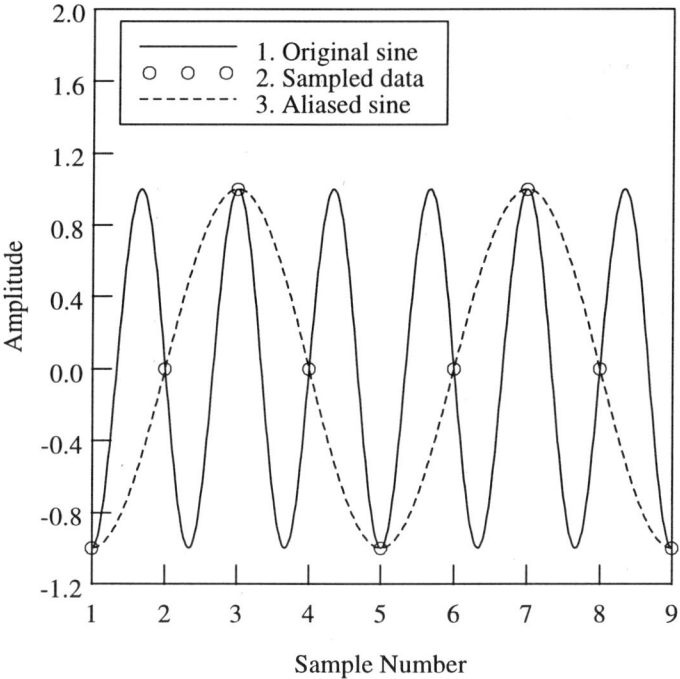

Fig. 18.3 Aliasing of an insufficiently sampled sine wave. The original frequency of 0.75 cycles per sample is aliased to 1 − 0.75 = 0.25 cycles/sample, tripling the period in the reconstructed sine wave.

best compromise between these conflicting requirements, namely, retention of sharpness and avoidance of aliasing, is dependent upon the characteristics of the full imaging system. Similar considerations apply in selecting a digital spatial pre-filter to limit bandwidth prior to down-sampling to avoid aliasing.

18.3 Reconstruction Artifacts

If the analog signal being sampled is sufficiently narrow in bandwidth (i.e., has no power above the Nyquist frequency, which is one-half the sampling frequency), aliasing will not occur and perfect reconstruction is possible by using an ideal low-pass filter having a block frequency response. When aliasing

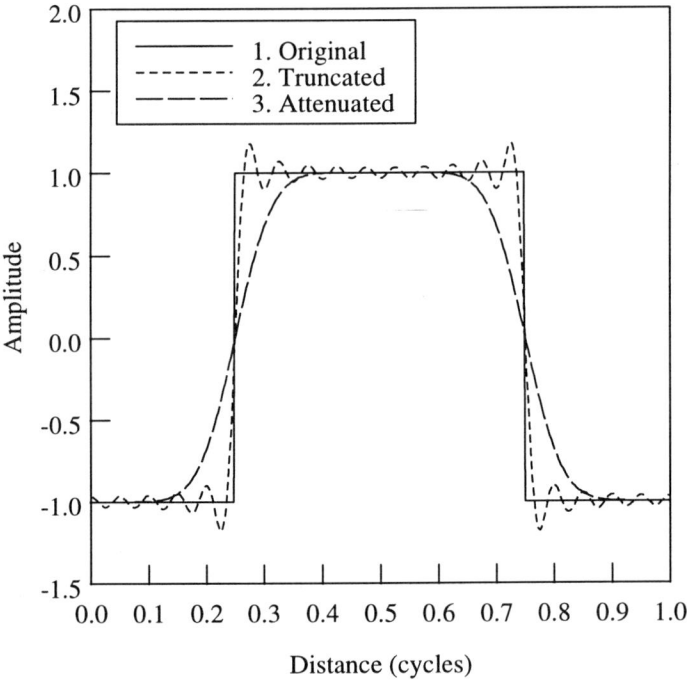

Fig. 18.4 One period of an original square wave (Curve #1) and two reconstructions from its Fourier series. Abruptly truncating the Fourier series produces ringing (Curve #2), whereas attenuating the series more slowly eliminates the ringing but blurs the edges (Curve #3).

has occurred, application of even an ideal low-pass filter will fail to perfectly reconstruct the original signal because the zero-order spectrum cannot be isolated. It is often assumed that a spectral response similar to that of the ideal bandpass is still desirable in cases where aliasing has occurred, but this is not necessarily the case. The ideal low-pass filter allows some aliased spectral content to persist, and truncates the spectral content of the original signal above the Nyquist frequency, but there is no guarantee that it achieves the optimal balance of these two opposing and undesirable effects.

Figure 18.4 shows the implications of spectral truncation in the spatial domain. An original square wave, one period of which is shown in Curve #1, may be expressed as an infinite Fourier series. If the Fourier series is truncated by omitting all terms above a particular harmonic, and the truncated series transformed back to the spatial domain, the modified square wave shown in Curve #2 results. The ripple artifact is referred to as ringing, and is evident as density oscillations near the edges. This ringing differs from the halo effect discussed in connection with oversharpening (see Fig. 16.2), in that the latter effect involves only a single excursion (overshoot or undershoot) of density on each side of an edge. Ringing in the spatial domain is caused by abrupt spectral truncation in the frequency domain, and so may be ameliorated by use of a more gradual attenuation. Curve #3 shows the result of multiplying the terms of the Fourier series of the square wave by a weighting function that diminishes slowly with increasing frequency, and transforming the results back to the spatial domain. The ringing has been eliminated, but at the loss of high-frequency detail, leading to rounded edges and a potential loss of perceived sharpness. The analogue of the weighting function in the spatial domain is called a windowing function, properties of which have been extensively studied (Harris, 1978).

The Fourier transform of the ideal low-pass filter frequency response yields the spatial form of the ideal reconstruction (or interpolating) function, which is referred to as a sinc function. The sinc function dies out slowly (like reciprocal distance); therefore, many original pixel values are required in the computation of each reconstructed value, leading to a computationally inefficient process. In practice, less spatially extensive reconstruction functions with efficient computational implementations are sought, which provide some level of approximation to the ideal low-pass frequency response. The windowing functions mentioned above provide one means of limiting the spatial extent of the reconstruction function without excessive ringing. Windowing causes the MTF of a sinc reconstruction to change from a block response, with an infinitely sharp transition at the Nyquist frequency, to a smoother response that falls off more gradually. Other reconstruction methods such as linear interpolation and cubic convolution have roughly similar responses. To the extent that such responses are less than one in the filter pass-band (below the Nyquist frequency), the reconstructed image will be blurred. To the extent that their response is greater than zero in the filter stop-band (above the Nyquist frequency), spurious high-frequency power may be passed, even in the absence of spectral overlap.

The passage of first and higher order replicate power because of imperfect stop-band behavior is referred to as reconstruction error and is demonstrated in Fig. 18.5. The spectrum of a sampled image with no aliasing is shown in Curve #1. An ideal low-pass filter cutting at the Nyquist frequency, 0.5 cycles/sample, could isolate the zero-order spectrum and permit perfect reconstruction of the signal. Instead, this figure shows the effect of the simplest of all reconstruction

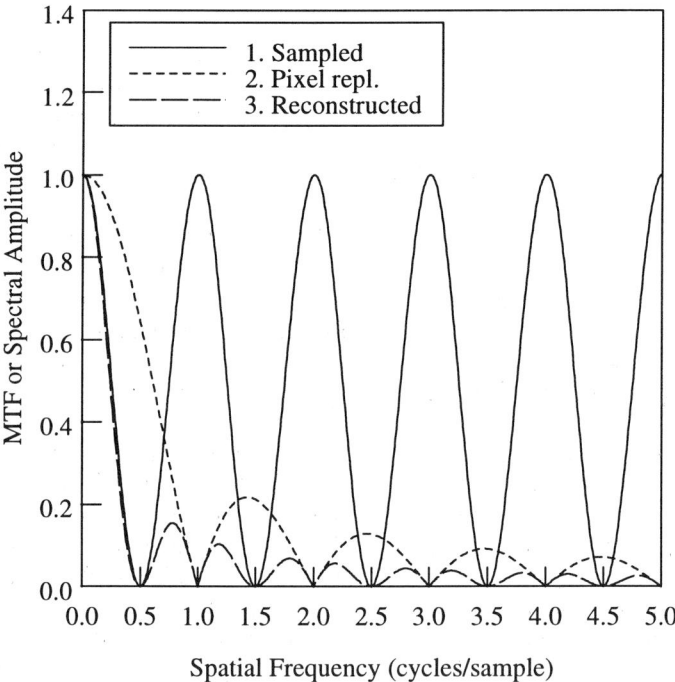

Fig. 18.5 Reconstruction error from pixel replication, viewed in the frequency domain. An original sampled signal (Curve #1) filtered by the pixel replication MTF, Curve #2, yields the spectrum of Curve #3, which contains extensive spurious high-frequency content, despite the fact that the original signal was not aliased.

methods, referred to variously as sample and hold, boxcar integration, or pixel replication. The first two names usually are applied when a signal is encoded as a temporal sequence of values, called a bit stream, which, for example, might drive an output device that writes in a raster fashion. In the temporal domain, the method consists of applying the most recent value in the bit stream until a new value becomes available.

If the signal is processed as a two-dimensional array, as in a digital image interpolation, the term pixel replication is applicable. For example, to effect a 2× interpolation, each pixel in the original image is replaced by a 2 × 2 array of identical pixels, quadrupling the number of pixels in the image. Although

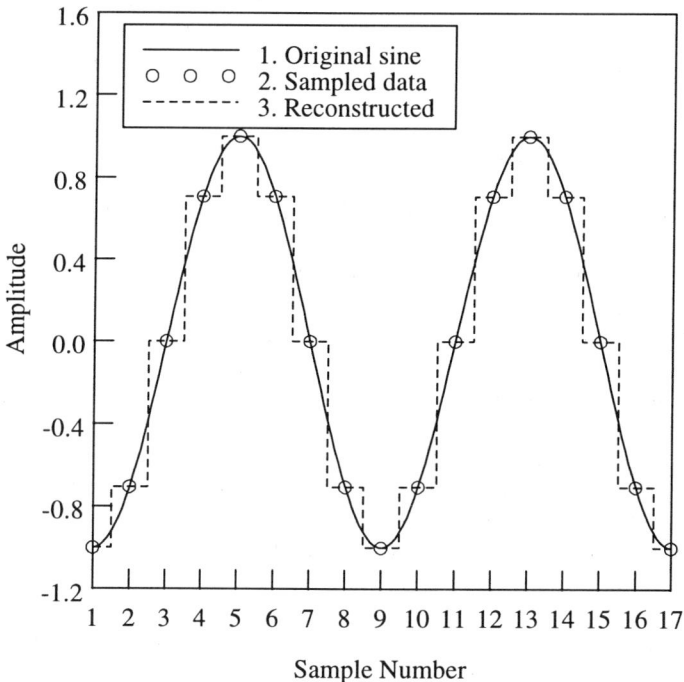

Fig. 18.6 Reconstruction error from pixel replication, viewed in the spatial domain. The reconstructed signal contains a number of sharp edges that were absent in the original unaliased sine wave signal; these edges arise from the spurious high-frequency power produced by reconstruction error.

computationally very efficient because no multiplications or additions are required, pixel replication has quite a bit of loss in the pass-band and very poor rejection in the stop-band, as shown in the MTF of Curve #2. In fact, even though the original analog signal was not aliased, quite a bit of the power from not only the first, but even the higher order replicates is passed, as shown in Curve #3, which is the cascade (frequency by frequency product) of Curves #1 and #2. In some regards, this effect is the opposite of aliasing, in that content at low frequencies in the original analog signal has been reproduced at higher frequency in the reconstructed signal.

Figure 18.6 (previous page) shows the spatial manifestations of spurious high-frequency content in one dimension. The x-axis is the index of the spatial sample, i.e., samples are taken at integer values on the x-axis. The y-axis is signal amplitude. A pure sine wave (Curve #1) has been sampled without aliasing (circles) and reconstructed using replication of nearest neighbor pixels (Curve #3). Although the original sine wave signal varied smoothly, the reconstructed image is like a series of ascending and descending stair steps, and has many sharp edges not present in the original signal, which correspond to the spurious high frequency content associated with the reconstruction error.

The two-dimensional analogue of this effect is shown in Fig. 18.7 (see following pages), where a slanted edge is reconstructed using pixel replication, resulting in a jagged edge. This artifact, called pixelization, is a manifestation of reconstruction error (just as ringing is a manifestation of spectral truncation). For comparison, the results from two other reconstruction processes are also shown in Fig. 18.7. In one case, a 15-element Hamming windowed sinc function has good pass-band and stop-band characteristics but cuts so sharply at the Nyquist frequency that the zero-order spectrum is truncated severely, leading to ringing. In the other case, reconstruction is by cubic convolution, a popular method in image processing, which has an MTF intermediate between that of the other two methods. It reduces both the pixelization and ringing substantially, but at the expense of poorer pass-band characteristics leading to lower sharpness.

In summary, selection of an optimal reconstruction function involves balancing the quality losses arising from at least three artifacts:

1. pixelization (from reconstruction error);

2. ringing (from truncation of the signal spectrum); and

3. blur (from spatial averaging of the signal).

Attributes Having Multiple Perceptual Facets

Pixelization depends mostly on reconstruction stop-band characteristics, blur on pass-band properties, and ringing on the abruptness of the transition between the pass-band and stop-band. The balance between performance in the stop-band and pass-band may be modified by changing bandwidth. More abrupt transitions simultaneously allow better stop-band and pass-band behavior, but may lead to ringing in the presence of aliasing and often involve computations that are more extensive. The best compromise is a function of the amount of aliasing present and so is system-dependent.

18.4 Perceptual Attributes Associated with Sampling Artifacts

Except in the case of some synthetic imagery, digital imaging systems generally involve an equal number of sampling and reconstruction operations, and a minimum of one of each type of operation, because the original scene and final displayed image are both analog in nature. The preceding discussion of sampling and reconstruction, while cursory, has shown that a variety of artifacts may be associated with analog-to-digital and digital-to-analog transformations. These include aliasing, pixelization from reconstruction error, ringing from spectral truncation, and blur. In this section, we will focus on the perceptual aspects of the first three artifacts, some of which pose a particular challenge.

Aliasing is usually most evident in pictorial images as low-frequency patterns, often wavy or (in motion imaging) scintillating in appearance, in areas that should have high-frequency detail. A classic example is the crawling pattern sometimes seen in finely patterned suits worn by television news announcers. Aliasing is usually a distinctive artifact that can be successfully treated as an independent attribute and is readily correlated with an objective metric. Ringing, which appears as density oscillations parallel with edges, and pixelization, which transforms diagonal edges into stair step patterns, would seem at first to be equally distinct attributes. However, inspection of pictorial images produced by a wide variety of reconstruction methods reveals that a continuum of artifacts from ringing to pixelization may be produced by MTFs ranging from nearly rectangular (like an ideal low-pass filter) to slowly diminishing (like pixel replication), respectively. In all cases, edges are distorted in some fashion, but only in the extreme response shapes are the terms ringing or pixelization fully descriptive of the appearance. Pictorial examples of aliasing, pixelization, ringing, and a condition intermediate between pixelization and ringing are shown in Figs. 18.8–18.11, respectively.

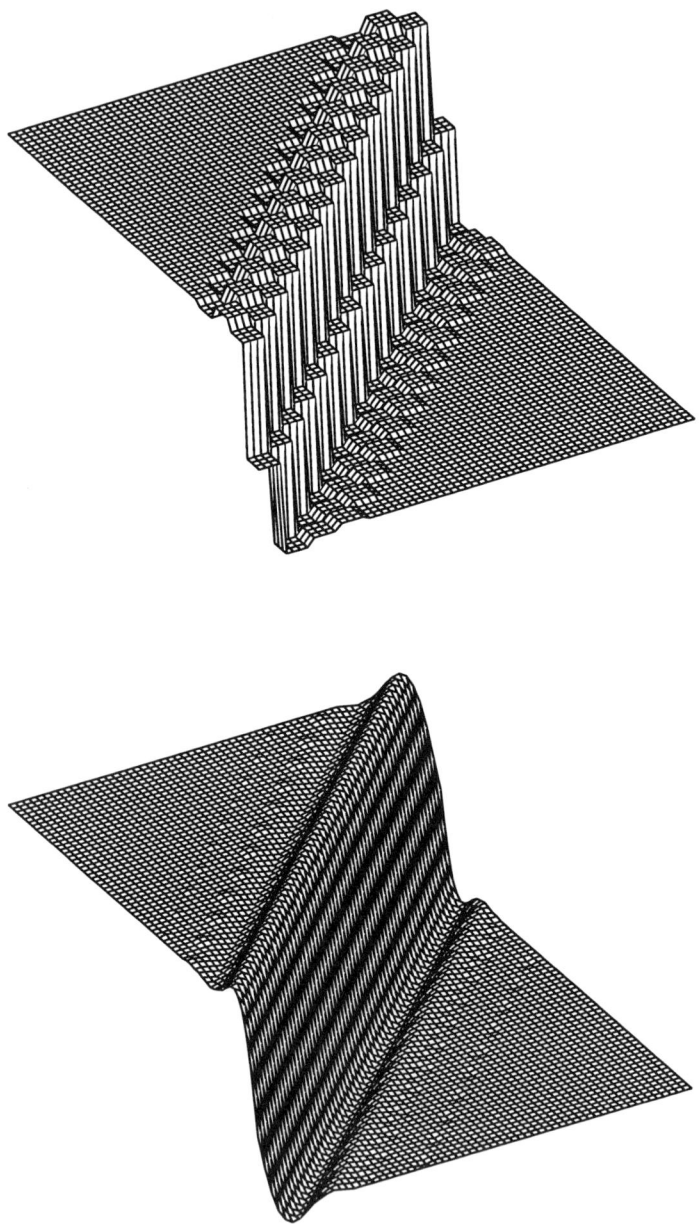

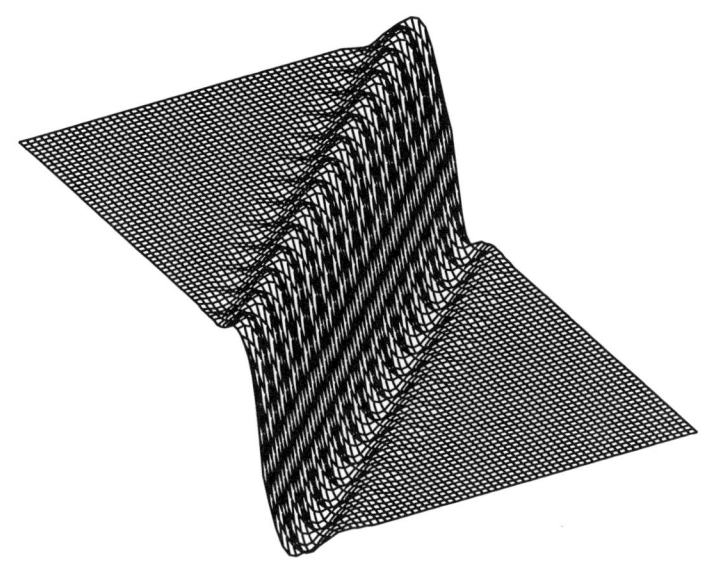

Fig. 18.7 Reconstruction infidelity in two-dimensional images of a diagonal edge. In these images, a slanted edge has been down-sampled without pre-filtering, and reconstructed using three different methods. Pixel replication (opposite page, top) is the computationally fastest method, but it leads to the most severe pixelization from reconstruction error, which is evident in the jaggedness of the edge. Windowed sinc reconstruction (opposite page, bottom) is a slower technique that attempts to approach the ideal low-pass response, but suffers from edge ringing caused by spectral truncation of the signal, especially when aliasing is present. The ringing is characterized by the oscillations on either side of the edge. Cubic convolution (above) is a relatively fast technique that produces edge artifacts that are intermediate in appearance between the extremes of pixelization and ringing, and lead to less quality loss on average.

Fig. 18.8 The artifact of aliasing, most obvious as the wavy low-frequency pattern in the black and white vertically striped building.

Attributes Having Multiple Perceptual Facets 267

Fig. 18.9 Pixelization from reconstruction error, evident as stair step diagonal edges, notably in the shadow cast in the opening.

Fig. 18.10 Ringing from spectral truncation, evident as oscillations along edges, notably around the light at the end of the lamppost.

Attributes Having Multiple Perceptual Facets 269

Fig. 18.11 Reconstruction infidelity of intermediate appearance, exhibiting edge artifacts but neither distinctive ringing nor pixelization.

Essentially, the apparent intergradation of ringing and pixelization is a problem in classification. How many independent attributes are represented by this complex of artifacts? In principle, this question can be answered by determining whether the effects of proposed attributes combine in accordance with the multivariate formalism. In the present case, ringing and pixelization could be hypothesized to be two distinct attributes, and the assessed overall quality of combinations of the two could be compared with multivariate predictions. This approach requires application of one of the three standard methods for estimation of the impact of the individual attributes (mensuration, projection, or isolation). Isolation is impractical because the two attributes cannot be independently simulated over the entire continuum, but only at the extreme positions. Performing two sequential reconstructions separated by a resampling might appear to offer the potential for simulating arbitrary combinations of the two artifacts, but the resampling and second reconstruction are likely to substantially modify the contribution of the first reconstruction, compromising the approach. Projection might be possible, if extreme series could be made with differing amounts of quality loss, e.g., by varying resolution at the time of reconstruction. However, projection involves appearance matching, and it has already been stated that a continuum of appearances intermediate between pixelization and ringing are produced, so the projection assessment might be difficult or impossible. This leaves mensuration, which requires the development of objective metrics for each of the two artifacts. We adopted this approach in our work on this artifact complex.

Experimental levels, ranging widely in quality, including some tending strongly towards either ringing or pixelization, were simulated from high-resolution images by: (1) variable pre-filtering; (2) reducing the number of pixels by different down-sampling factors; and (3) increasing the number of pixels back to the original number using a selection of interpolation methods. Pre-filtering controlled the zero-order spectrum shape and affected both pixelization and ringing. The down-sampling factor affected the resolution at which the reconstruction operation was performed and determined the frequencies to which the artifacts were mapped in the final image, with higher down-sampling factors producing lower-frequency, and thus more visible, artifacts. The interpolation method affected the pass-band, stop-band, and transition properties of the reconstruction MTF. Multivariate decomposition was used to remove the impact of unavoidable variations in other image quality attributes, which included aliasing, blur, and noise. Aliasing varied with pre-filter MTF and down-sampling factor. Blur varied with down-sampling factor, pre-filter MTF, and interpolation MTF. Noise varied primarily because of aliasing, which maps high-frequency noise to low frequencies where it is more visible, in the same manner that high-frequency signal is corrupted. The impact of aliasing and blur were estimated by mensuration, but noise was quantified by projection, because of the complexity of aliased noise calculations.

Next, objective metrics of pixelization and ringing were developed based on the down-sampling factor series at the extreme reconstruction frequency responses. These permitted estimation of the impact of ringing and pixelization in samples made with reconstruction methods having intermediate frequency responses, if these descriptors could be meaningfully applied to such samples. Attempts to predict the quality change associated with intermediate reconstruction MTFs, using the multivariate formalism, were not very successful. At a given down-sampling factor, larger variations in quality were predicted with changes in reconstruction MTF than were actually observed. In fact, the range of quality change arising from reconstruction MTF was smaller than that caused by rather modest changes in down-sampling factor. At the extremes, where the distortions closely approximated either ringing or pixelization, the quality loss was greater; evidently intermediate positions, with a balance between ringing and pixelization tendencies, were preferred, but not dramatically so. Recognizing that pixelization and ringing might not be separate attributes, as evidenced by their continuous intergradation, we also attempted to combine the two trial objective metrics into a more global metric, but without compelling success.

It has already been mentioned that when series of samples at the same down-sampling factor were laid out in the order of their spectral properties, a continuously changing appearance from pixelization to ringing was observed. Nonetheless, there was an impression of overall infidelity that was much the same across all the samples. In aggregate, we took the modeling results and our anecdotal observations to suggest that a third and pervasive facet of reconstruction quality related to resolution was still not properly represented in our objective metrics, even though both the ringing and pixelization metrics were defined in a fashion that was strongly resolution-dependent. We assumed that the missing facet did not manifest itself as a distinctive artifact and so had not been recognized earlier.

Therefore, we defined a third trial objective metric called the limiting pixel subtense to attempt to capture this facet of reconstruction quality. The limiting pixel subtense is computed as follows. The stage of the imaging system having the fewest independent pixels is identified. Often this is the capture stage, but if the image is down-sampled after capture to reduce storage space or transmission time, or for display at lower resolution, the limiting stage may occur later in the imaging chain. One of these limiting pixels is projected to the final image plane to determine its size there, which is equivalent to dividing the final image into the limiting number of pixels and calculating the resulting pixel size. This size is converted to angular subtense at the eye. The limiting pixel subtense was intended to be a measure of the potential visibility of the fundamental "building block" available to the reconstruction process or processes.

Because we did not know how to generate samples having quality affected only by limiting pixel subtense, and not by ringing or pixelization, no method of estimation of the impact of limiting pixel subtense in isolation was available. By this time, however, it seemed likely that we were dealing with an artifact complex that represented a single perceptual attribute with a multifaceted appearance, so that the multivariate formalism was not required for predictions of quality loss arising from the global attribute, which will be called reconstruction infidelity. Instead, we looked for a combination of the three trial

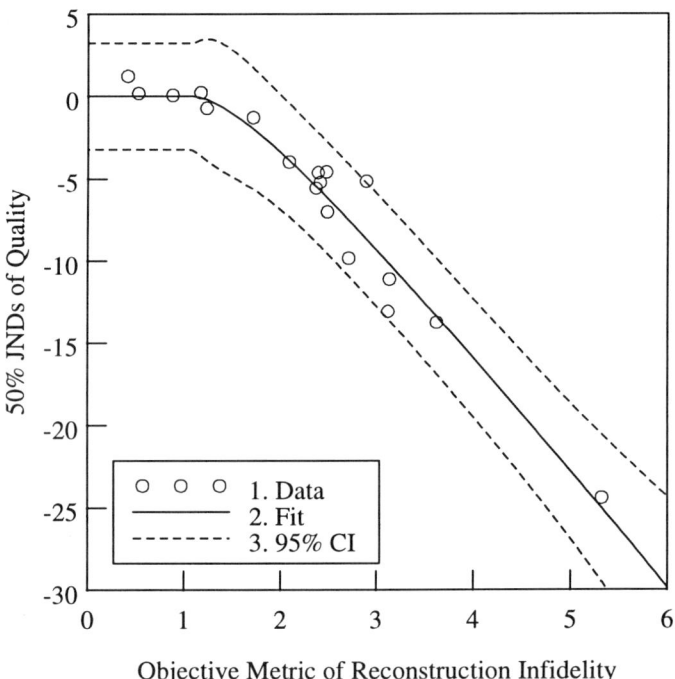

Fig. 18.12 Quality loss function of reconstruction infidelity. The objective metric has terms reflecting pixelization, ringing, and limiting pixel subtense, but still does not accurately predict the effects of all the variations in pre-filtering, down-sampling factor, and interpolation that were tested, as evidenced by the poor fit.

objective metrics that was predictive of quality loss arising from reconstruction infidelity over the wide range of experimental levels tested. In this endeavor we were moderately successful using a weighted sum of the ringing, pixelization, and limiting pixel subtense component metrics, as shown in Fig. 18.12. The greater dispersion of data about the IHIF regression line and the correspondingly larger regression confidence interval than observed in other studies presumably reflect the incompleteness of the final objective metric, which evidently still does not fully explain all the effects of variations in pre-filtering, limiting resolution, and interpolation method. Given the complexity of the phenomena involved in the reconstruction process, this degree of predictiveness may not be unreasonable.

We also built combination objective metrics based on all possible subsets of the three component objective metrics, and found that omission of any of the three component metrics substantially reduced the predictiveness of the combination objective metric, indicating that all three captured important and distinct aspects of reconstruction infidelity. Limiting pixel subtense was the best single predictor, however, further supporting the view that ringing and pixelization are merely extremes of a continuum of appearances corresponding to the single attribute of reconstruction infidelity.

Experimentation, analysis, and modeling of the sort described in this section should permit the development of satisfactory first-order treatments of most attribute complexes.

18.5 Summary

Aliasing results from inadequately sampling an analog signal relative to its bandwidth and leads to replacement of high-frequency patterns in the original signal with low-frequency artifacts that are often wavy or, in motion imaging, scintillating. Reconstruction error is caused by inadequate suppression of spurious high-frequency content in the digital signal when it is converted to an analog signal (reconstructed), e.g., so it can be displayed. It is often especially noticeable as a jagged structure along a diagonal edge (pixelization). Spectral truncation is another reconstruction defect caused by abruptly truncating the signal frequency spectrum, and is manifested as ringing, i.e., density oscillations parallel with strong edges.

Aliasing may be reduced, or eliminated, by band-limiting the analog signal prior to sampling, through use of an optical anti-aliasing filter. Reduction of aliasing usually degrades sharpness because suppression of high-frequency content in the signal is invariably accompanied by some attenuation of lower frequencies. Reconstruction error is ameliorated by reducing passage of high-frequency content in the reconstruction stop-band, but this usually is associated with attenuation of low-frequency content in the pass-band, leading to blur, as in the analogous case of anti-aliasing filtration. Simultaneously minimizing pass-band attenuation and maximizing stop-band rejection leads to an abrupt transition in the frequency response near the Nyquist frequency. This abrupt cut-off leads to spectral truncation of the signal if the analog signal has sufficient bandwidth for significant spectral overlap to occur. Because correlations between aliasing, reconstruction error, spectral truncation, and degree of blur are quite complex and system-dependent, full characterization of the attributes involved and development of associated objective metrics is very valuable for optimization of sampling and reconstruction processes in digital imaging system design.

It is not always straightforward to identify perceptually independent attributes corresponding to a set of interrelated artifacts. In the case of reconstruction, what might have been interpreted as two distinct attributes, ringing from spectral truncation, and pixelization from reconstruction error, proved to be better described as a single attribute with multiple perceptual facets. This conclusion was reached based on a variety of evidence:

1. qualitative observation of the continuous nature of the artifact complex in images;

2. failure of the multivariate formalism to quantitatively predict the quality of combinations of the artifact extremes; and

3. predictiveness of an objective metric dominated by a term that is not specific to either of the extreme artifact manifestations.

Intergrading artifact complexes and associated attributes having multiple perceptual aspects pose a particular challenge in image quality research, but using a combination of qualitative observation of images, multivariate analysis of psychometric data, and design of multicomponent objective metrics, it should be possible to develop reasonably predictive treatments.

19

Image-Specific Factors in Objective Metrics

19.1 Introduction

Section 10.4 described our usual method for quantifying variability in observer sensitivity and scene susceptibility, which involves specification of different IHIF fit parameters for different subgroups of observers and scenes. This permits selection of quality loss functions that are appropriate for particular applications. For example, in setting specifications for a critical professional application, the IHIF fit parameters for a subgroup of more sensitive observers and more susceptible scenes might be employed. It is also possible to statistically sample from the sets of fit parameters in Monte Carlo analyses so that scene and observer variability are reflected in performance modeling (which will be discussed in Ch. 28).

An alternative approach to describing susceptibility or sensitivity variability is the incorporation of scene- or observer-dependent factors in the objective metric itself. If factors of both types were included in an objective metric, then a single set of IHIF fit parameters could suffice to predict the quality change arising from an attribute, for all subgroups of observers and scenes. In such an approach, if an objective metric value is quoted, its associated subgroup must be explicitly identified.

Incorporating scene or observer variability into an objective metric usually entails considerable extra effort. The nonlinear relationship between response functions for different subgroups are readily handled by the IHIF, but building such nonlinearity into an objective metric can be quite complex. The effort generally is not justified, except in cases where two criteria are met:

1. the quality change arising from an attribute is strongly affected by scene and/or observer characteristics; and

2. these characteristics can be conveniently quantified and their effect modeled in the objective metric.

The artifact of redeye, which is the subject of this chapter, meets both of these criteria. As described in Sect. 19.2, the quality loss caused by redeye depends strongly on scene-specific characteristics as well as on imaging system properties, thus meeting the first criterion. Furthermore, predictions of scene susceptibility can be made based solely on the size of the pupil in the final image, so the second criterion is met as well, as demonstrated in Sect. 19.3.

19.2 Origin of Redeye and Factors Affecting Its Severity

When light impinges on the human eye at an angle not too far off the optical axis, it may propagate through the pupil, be reflected back from the fundus, and exit the pupil at approximately the same angle it entered. The fine blood vessels in the fundus color the reflected light red. In most ambient lighting, only a very small fraction of the illumination is sufficiently close to the axial angle to produce the red reflection, and so it is not evident. If most or all of the illumination is nearly axial, however (as may be the case in flash photographs), the red reflection can become visible in an image, leading to an artifact called redeye. In more extreme cases, the pupils of the subject appear to glow, and a substantial loss of quality results.

A similar effect can be observed when certain animals are viewed at night in the headlights of a car or by flashlight, although the reflected color varies by species. If an observer moves the flashlight farther away from his or her eyes, or steps out of the car, so that the angle between the headlights, the animal, and his or her head increases, the intensity of the reflection will diminish dramatically. Similarly, in a flash photograph, the angle between the flash tube, the subject, and the camera lens strongly affects the severity of the redeye, with detectable redeye rarely occurring if this angle exceeds three degrees. For this reason, some compact cameras are designed with a flash that extends away from the lens, either by popping up out of the camera body or, more effectively, by being mounted on the lens cover assembly so that when the cover is flipped open, the flash is deployed approximately one camera height above the lens. This approach is especially effective at shorter camera-subject distances, when the extra separation can increase the flash-subject-lens angle above three degrees.

The human pupil contains the pigment melanin, which gives it its dark color. The amount of melanin in the pupil correlates with skin and hair pigmentation, so that darker-skinned and darker-haired individuals typically have higher melanin concentrations in their pupils. Melanin attenuates the light propagating through the pupil on both the entry and exit passes, and so higher melanin content leads to reduced redeye severity. Although there is some variation of pigmentation within races, it is still the case that a similar statement can be made for races, namely, that darker complexioned races on average exhibit less redeye. For example, a photographic test involving more than one hundred subjects and conducted under controlled conditions yielded the following percentages of detectable redeye under one particular set of conditions.

	African-American	Asian	Latin American	Caucasian Adult	Caucasian Youth
Detectable redeye (%)	9	15	41	70	82

Table 19.1 Percentage of occurrence of visually detectable redeye for different demographic groups of varying race and age, at one particular combination of ambient light level, flash-lens separation, camera-subject distance, pre-flash energy, and angular magnification.

These data also show the effect of age on propensity towards redeye. This effect is caused by a greater responsiveness of pupil size to changes in light level at younger ages. At high light levels, the human pupil approaches a diameter of about 2 mm, regardless of the age of the subject. However, the degree of dilation of the pupil at lower light levels decreases with age, with maximum size varying from about 9–10 mm in youth to perhaps 7 mm in old age. Allowing a person to fully adapt in a very dark room, and then taking a flash photograph might allow a crude estimate of the individual's age. The iris diameter is about 12 mm throughout life and could provide a convenient internal reference in the measurement. Given the variability of aging phenomena and the time required for adaptation, however, this test would be unlikely to replace the checking of identification cards at establishments serving liquor!

All else being equal, images with redeye are observed to suffer greater quality loss as the size of the pupil in the final image increases. This effect might be caused by either or both of the two factors listed on the next page.

1. As the pupil size in the image increases, the artifact of redeye becomes more noticeable to the observer.

2. As head size (and, on average, pupil size) increases in an image, it becomes more likely that a person is the principal subject of the photograph, so artifacts in his or her face might more strongly influence image quality.

One mechanism for reducing the severity of redeye at lower light levels is to "pre-flash" the subject just prior to the photographic exposure. A pulse of light will cause the pupil to contract briefly, with the maximum contraction occurring around 0.7 seconds after the pulse. A drawback of pre-flash is that facial expressions can change between the time the shutter button is pressed and the time the photograph is actually captured, because of the delay required for the pupil to react.

The pupil size in an image, or, more precisely, the angular subtense of the pupil in the viewed image, may be predicted from the absolute pupil size, the camera-subject distance, and what is called the angular magnification of the system. Angular magnification is proportional to camera lens focal length and printing magnification, and inversely proportional to viewing distance, as discussed in connection with Eq. A5.13 in Appendix 5.

In summary, the intensity or saturation of the red reflection varies strongly as a function of subject pigmentation and flash-subject-lens angle (which depends on camera-subject distance and flash-lens separation). The absolute pupil size (measured in life, not in an image) is a function of subject age, ambient light level (which controls adaptation), and pre-flash energy delivered to the subject (which varies with the flash power and the camera-subject distance). Finally, the pupil size in the final viewed image, expressed as a subtended angle, depends on the absolute pupil size, camera-subject distance, and system angular magnification (which is a function of camera lens focal length, printing magnification, and viewing distance). The relationships among these factors are complex, involving correlations, which must be accounted for in the prediction of redeye severity.

19.3 Design of a Scene-Specific Objective Metric of Redeye

Most of the artifacts discussed in this book are produced by the imaging system, relatively independently of the scene characteristics. Consequently, these artifacts may be characterized by propagating test targets (such as slanted edges or uniform fields) through an imaging system; such targets serve as proxies for real scenes, but are much simpler to analyze. Scene content usually has a second-order effect, producing moderate levels of susceptibility variation, as we have seen in previous chapters. In contrast, of the factors that influence redeye, only three (flash-lens separation, pre-flash power, and angular magnification) are properties of the imaging system per se, whereas four other factors (subject distance, light level, race, and age) are properties of the scene, and each of the latter properties significantly affects the quality loss caused by redeye. Even if a redeye test target or fixture could be designed, it would probe only the influence of system properties, whereas most of the variation in redeye severity results from scene properties. Consequently, inclusion of scene-dependent factors in an objective metric of redeye is highly desirable.

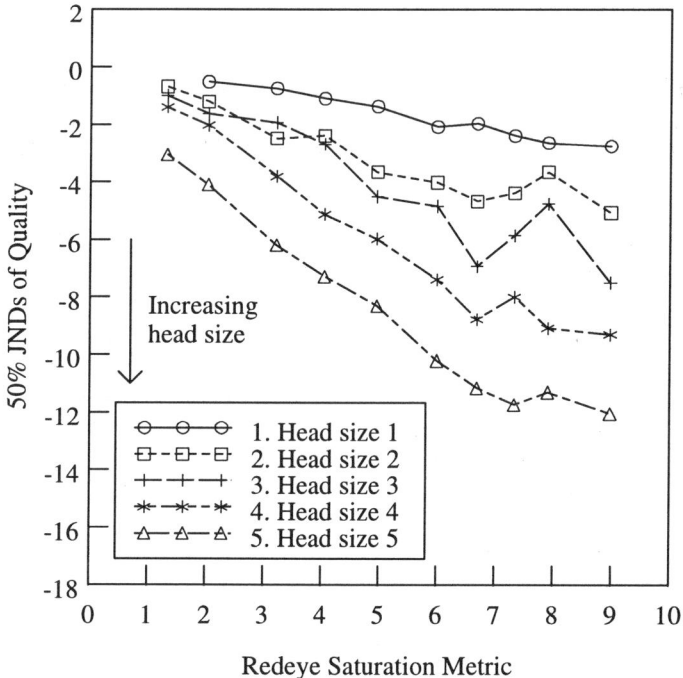

Fig. 19.1 Quality loss arising from redeye at five image head sizes, at constant pupil to iris ratio. The objective metric reflects the impact of the red saturation of the pupil, but the subject head size, a scene factor, has a fully comparable effect on quality loss.

Figure 19.1 (previous page) demonstrates how one scene-dependent factor, camera-subject distance, strongly influences the quality loss arising from redeye. The y-axis is mean quality loss from redeye in JNDs, pooled over observer but not scene. The x-axis shows an objective metric that satisfactorily explains redeye quality loss caused by variations in the red saturation or intensity (see below), thereby accounting for variation arising from pupil pigmentation and flash-lens separation. Only data for a fixed pupil to iris diameter ratio (i.e., a constant absolute pupil size in life) is shown in this figure, eliminating variation from subject age, ambient light level, and pre-flash power. Consequently, the only factors not accounted for are angular magnification and camera-subject distance. Under the restrictions noted above, the head size in the image depends only on the ratio of these quantities. This ratio spans a range that produces head sizes from that of a full-length group shot (Curve #1) to that of a close-up portrait (Curve #5). The figure, therefore, shows the variation in quality loss that would occur if a fixed angular magnification system (fixed focal length camera lens and single print size) were used at five camera-subject distances approximately spanning the range encountered in consumer flash photography. The resulting variation in quality loss at a given red saturation position is roughly a factor of five, with greater quality loss being associated with larger head sizes. This level of scene variation exceeds that of nearly all other artifacts studied, demonstrating the value of incorporating scene-dependent factors into an objective metric of redeye.

One advantage of a scene-dependent objective metric is that predictions can be made for individual images (for some specified observer sensitivity). To fully realize this advantage, the objective metric should be calculable solely from measurements made on a given pictorial image, without requiring knowledge of the conditions that led to the creation of the image. For utility in modeling, however, it must also be possible to predict the outcome of these measurements based on scene and system properties. Therefore, two separate experiments were performed to fully characterize the phenomenon of redeye.

1. A psychometric study was carried out to develop an objective metric of redeye that was based on quantities that could be conveniently measured in pictorial images.

2. A photographic experiment was conducted to empirically correlate these image measurements to scene and system variables, enabling prediction of quality loss arising from redeye based on these fundamental factors.

Measurements considered in the design of the objective measurement included: (1) pupil size, which is expected to affect the visibility of redeye; (2) head size,

Image-Specific Factors in Objective Metrics

which is expected to affect the importance of subject in the image and the quality loss at a given level of visibility; and (3) red saturation or intensity of the pupil, which is expected to affect visibility. All three of these can be readily measured from digital image data, e.g., from a scan of a hardcopy image.

The samples used in the psychometric test were produced by digital image simulation, which conveniently allowed a range of pupil sizes and saturations to be investigated in a well-controlled fashion. The results of this experiment, pooled over observer but not scene, are shown in Fig. 19.2. In addition to containing all the data from Fig. 19.1, this plot includes data from a range of pupil to iris ratios. The objective metric of redeye on the x-axis is an extension

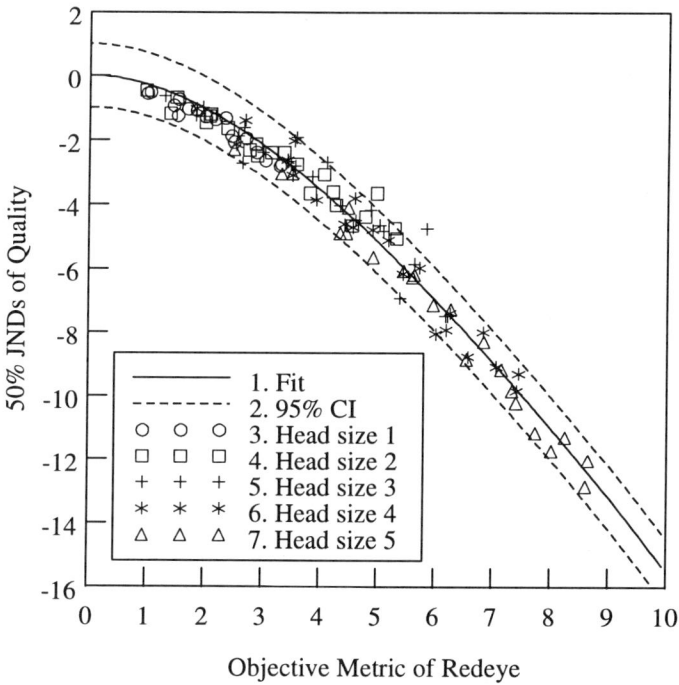

Fig. 19.2 Quality loss function of redeye against an objective metric that accounts for both pupil size and saturation in the final image. The dramatic scene susceptibility variability of Fig. 19.1 has been fully explained by this objective metric, so that the five head size traces of Fig. 19.1 have collapsed into a single well-defined relationship.

of the red saturation metric of Fig. 19.1 that includes the effect of pupil size. Further inclusion of the correlated effect of head size failed to increase predictiveness of the metric and so was omitted for simplicity. This suggests that redeye is perceived as an artifact almost independently of the importance of the human subject in the image. The scene susceptibility variability of Fig. 19.1 has been successfully incorporated into the objective metric of redeye, a remarkable result. This permits quantitative prediction of the mean quality loss caused by redeye in a particular image, based solely on two simple measurements from the image itself.

In the photographic experiment, flash pictures were taken of over one hundred subjects under controlled conditions and the pupil size, head size, and red saturation of the pupil were measured for all images having visually detectable redeye. The factors that varied in the experiment were subject age, subject race, ambient light level, flash-lens separation, and pre-flash energy. Variations in camera-subject distance and system angular magnification were not included because their effects are optical in nature and can be rigorously modeled from first principles. The results of the photographic experiment were linked with those of the psychometric experiment to permit full Monte Carlo modeling of image quality loss from redeye as a function of scene- and system-dependent factors, specifically age, race, light level, flash-lens separation, pre-flash energy, camera-subject distance, and angular magnification.

Although incorporation of scene- or observer-dependent factors in objective metrics involves extra research effort, the additional resource expenditure may be worthwhile in cases where variability from such factors is large. If an objective metric includes scene-specific factors that may be calculated from available data, then predictions of quality can be made for individual images (assuming some particular observer sensitivity). This is convenient when generating demonstrations of image quality effects because the desired quality levels may be directly simulated, without repeated assessment and modification cycles. Furthermore, the ability to calculate image-specific objective metrics can be of considerable utility in deciding what type of digital processing would optimize the quality of a particular image, e.g., in a digital photofinishing operation or on a home computer. If information is available about the observer preferences, via some type of customer or user profile, then objective metrics with observer-specific factors are analogously useful in customizing digital image processing. In the next chapter, scene and observer preference-dependent objective metrics of color and tone reproduction are described.

19.4 Summary

Variations in scene susceptibility and observer sensitivity to an attribute are usually characterized by different IHIF fit parameters, which may be statistically sampled in Monte Carlo calculations, or selected based upon the application. In some cases, though, it is feasible and desirable to incorporate scene- or observer-dependent factors directly into the objective metric. Despite the extra research effort involved, this approach may be advantageous when the degree of scene or observer variability is substantial compared to the level of variation from system-related characteristics. Particular value is derived from such objective metrics if their scene- or observer-dependent factors can be predicted from easily measured or readily available data.

The development of a scene-dependent objective metric is exemplified by studies of the redeye artifact. The severity of this artifact is influenced at least as much by scene-dependent parameters (e.g., subject age and race, camera-subject distance, and ambient light level) as it is by system parameters (e.g., flash-lens separation, pre-flash energy, and angular magnification). An objective metric of redeye that did not include scene-dependent factors would not be very predictive because the variation arising from scene effects is so large. Development of a scene-dependent metric, based on simple measurements made from specific images, led to very successful predictions of quality loss from redeye for a wide range of scene and system characteristics. Further work relating the outcomes of these measurements to system and scene parameters permitted construction of full Monte Carlo models for predicting redeye quality loss distributions based on all known pertinent parameters, a powerful capability.

20

Preference in Color and Tone Reproduction

with Karin Töpfer
Eastman Kodak Company
Rochester, New York

20.1 Introduction

Color reproduction refers to the relationship between colors in an original scene and the corresponding colors depicted in an image of that scene. Tone reproduction is the relationship between the lightness (or visual density) of neutral tones in scenes and images thereof; it does not refer to the chromatic character of black and white images, as in the term "sepia toning". Although tone reproduction is a subset of color reproduction, it is so important perceptually that it is convenient to treat it as a distinct entity. Although the impact of color and tone reproduction on overall image quality is undisputed, some readers may doubt whether its perceptual attributes, which are preferential in nature, can be quantified in the same rigorous fashion as artifactual attributes. Several aspects of this issue have already been addressed in Chs. 1, 4, 9, and 10. For review, the observations made previously are briefly summarized below.

In Sect. 1.2, an artifactual attribute was defined as one that normally degrades quality when it is detectable in an image, whereas a preferential attribute is one that is generally evident, but for which a distribution of optimal positions exist, depending upon the scene content and the observer's personal taste. It was explained in Sect. 4.2 that preference could be quantified in the same units, JNDs, as artifactual attributes, by defining a JND of preference to be a stimulus difference producing the same outcome in a paired comparison experiment as would one JND of an artifactual attribute. For example, stimuli differing by one 50% JND, whether for an artifactual or preferential attribute, will be chosen in the proportions of 75%:25% in paired comparisons. In the case of preference, the term just "noticeable" difference is a bit misleading, because two stimuli that are evidently different in appearance to essentially all observers may still each be preferred by an equal number of these observers, corresponding to equal quality, and a stimulus difference of zero JNDs of quality. Although the JND terminology may be imperfect, expressing degree of preference in terms of JNDs of quality is a very powerful tactic, because it permits the characterization and prediction of overall image quality of samples affected by both artifactual and preferential attributes in a consistent and accurate manner.

In Sect. 4.3, the dependence of quality on a preferential attribute was described in terms of the convolution of two functions:

1. the preference distribution, which describes the relative frequency with which different values of a preferential attribute are deemed to be optimal by a group of observers assessing a selection of scenes; and

2. the quality loss function, which quantifies the dependence of JNDs of degradation on the deviation from the optimal position for a single observer and scene.

As discussed in Sect. 10.3, the selection of observers is particularly critical in experiments addressing preferential attributes, because it is important to obtain a representative preference distribution. In Ch. 9, the IHIF was formulated in such a way that it can equally well describe the quality loss functions of artifacts or preferential attributes, if the objective metric of a preferential attribute is defined so that it reflects some kind of a (necessarily positive) distance from an optimal position. In the case of color and tone reproduction attributes, this distance metric is often three-dimensional, reflecting the nature of the perception of color. The threshold parameter in the IHIF may be interpreted as the minimum distance in color space, from the optimal position, that leads to a measurable loss in quality. Alternatively, an artifactual attribute may be viewed as a special case of a preferential attribute, in which the preference distribution is equivalent to a delta function located at the detection threshold.

This chapter is organized as follows. After defining selected attributes of color and tone reproduction quality (Sect. 20.2), we will describe differences between preferential and artifactual attributes in terms of: experimental design (Sect. 20.3), definition of objective metrics (Sect. 20.4), psychometric data analysis (Sect. 20.5), and visualization of results (Sect. 20.6).

20.2 Definition of Color and Tone Attributes

Definitions and physical correlates of selected attributes of color and tone reproduction quality are summarized in Table 20.1. These definitions employ terminology frequently used in color science, colorimetry, and color appearance modeling. For a better understanding of these subjects, we recommend monographs by Hunt (1976), Hunt (1991), Giorgianni and Madden (1998), Fairchild (1998) and Berns (2000). The terms lightness and chroma refer to the CIE 1976 L* a* b* (CIELAB) metrics L* and C*, respectively (CIE Publication 15.2, 1986). More saturated colors typically have higher C^*/L^* ratios.

Attribute	Definition	Physical Correlate
Color Balance	Overall shift of an image away from a neutral rendition, producing a color bias or cast	CIELAB (a*, b*) coordinates of mid-tone neutral patches
Density Balance	Overall shift of an image towards lighter or darker tones than in the preferred rendition	Lightness values of mid-tone neutral patches
Memory Color Reproduction	Relationship of reproduced versus original colorimetry of familiar colors such as skin, sky, and foliage	CIELAB coordinates of color patches surrounding the prototypical memory color
Contrast	Relationship between original scene lightness perceived by a photographer and final image (reproduced) lightness perceived by an observer	Gradient (derivative) of tone scale (see definition below) and the maximum and minimum lightness values achievable
Relative Colorfulness	Extent to which the hues of non-neutral colors are apparent in a reproduction compared to the original scene	Chroma ratios of reproduced and original colors of non-neutral patches, plus available color gamut volume
Detail in Saturated Colors	Ability to discriminate lightness differences in saturated colors	Lightness of patches at different hue and chroma positions that fall on a reference gamut boundary

Table 20.1 Selected attributes of color and tone reproduction.

The tone scale function, which is listed as the physical correlate of contrast in Table 20.1, is defined as the mapping from original scene lightness, as perceived by the photographer, to reproduced lightness, as perceived by the observer viewing the final image. This tone scale function is often recast in terms of equivalent densities (relative to the reference 100% white) rather than lightnesses, so that it is analogous to the familiar characteristic curve (i.e., the H and D, D log E, or D log H curve), which is used to quantify the mean channel signal (MCS) response of imaging media and systems. The definition of the MCS response of imaging system components and their use in models of signal and noise propagation will be described in detail in Ch. 22.

The attribute of contrast is primarily correlated with the gradient (slope) of the tone scale, which may be evaluated over a range or at an individual point, this latter case yielding an instantaneous gamma (γ) value. The dynamic range of the reproduction, i.e., the ratio between the maximum and minimum achievable image luminances, also affects perceived contrast.

Absolute colorfulness refers to the degree to which the hue of a color is evident. Neutral colors possess zero colorfulness, whereas monochromatic light exhibits high colorfulness. The appearance of colorfulness is influenced by illumination level, the same object appearing more colorful in bright light than in dim light. Relative colorfulness, as defined here, is the extent to which the hues of reproduced colors are apparent, compared to those of the original colors viewed under similar circumstances. Relative colorfulness is quantified by the ratio of distances from the neutral axis (i.e., the chroma values) of the reproduced and original colors. It is, therefore, a measure of chromatic amplification and is a color analogue of neutral contrast. Just as higher contrast systems are often subject to greater tonal clipping (Sect. 15.6), systems of higher relative colorfulness may sacrifice detail in saturated colors. For example, at high levels of colorfulness, the texture of a saturated sweater might be lost, or subtle variations in the petals of a vivid flower might not be reproduced.

Alternative definitions and classifications of perceptual aspects into different numbers and types of attributes are certainly possible. For example, what is defined as the single attribute of contrast above might be divided into distinct attributes corresponding to mid-tone contrast and dynamic range. As another example, the defined attributes of colorfulness and detail in saturated colors might be regarded as merely being aspects of a single attribute of chromatic rendition. The goal of any classification scheme should be to identify perceptually independent attributes that combine according to the multivariate formalism and together explain most of the variation of overall image quality arising from different practical color and tone reproduction positions. Because some of the elements of color and tone reproduction are linked with one another, it is not always easy to generate samples that allow definitive diagnoses of the perceptual behavior of the individual elements. In some cases, it may be possible to explain psychometric data by more than one classification scheme, especially if the range of stimuli tested is limited.

20.3 Experimental Design Considerations

Studies of color and tone attributes often involve variations that are aligned with one or more of three perceptual dimensions of color, namely, lightness, hue, and chroma. Absolute hue position and hue differences may be quantified using the CIELAB metrics of hue angle, denoted h^*, and hue distance, denoted ΔH^*, respectively. Whereas the chroma difference ΔC^* is a radial component in the CIELAB coordinate system, ΔH^* is a tangential component, such that the vector sum of ΔC^*, ΔH^*, and ΔL^* is equal to ΔE^*, the overall color difference (Berns, 2000; CIE Publication 15.2, 1986). Some studies require modifications in all three dimensions, whereas others involve variations primarily in a single factor (except for levels that test for interactions). For example, a tonal reproduction study might just involve variations in lightness rendition, and a study of colorfulness only changes in chroma profile, but an investigation of memory color reproduction would be likely to entail variations in all three dimensions.

To exemplify certain aspects of experimental design, the case of a memory color reproduction study will now be considered. The general experimental approach, discussed point by point subsequently, is as follows.

1. The three-dimensional colorimetric distribution of real world examples of the memory color is characterized and is used to select representative test scenes and diagnostic color patches for objective metric definition.

2. The preferred reproduction of the centroid of the memory color distribution is estimated through preliminary experimentation.

3. The experimental design is specified in terms of variations in hue, chroma, and lightness about the estimated preferred position.

4. The test stimuli are simulated using local color manipulation in a fashion that minimizes the introduction of color artifacts.

First, the colorimetric properties of the memory color are determined to aid in scene and diagnostic patch selection. The scene set should contain objects having original color positions that span the three-dimensional distribution of real world examples of the memory color. It is convenient to specify a single prototypical color that falls near the center of mass (centroid) of this distribution, which forms the basis for the color manipulations that generate the experimental stimuli. The shape of the memory color distribution may vary with geographical location and other demographic factors; e.g., skin-tones vary considerably among races (the term skin-tones will be used to mean the range of colors represented in human skin). In such cases, it is preferable to define multiple prototypical colors and to study them independently.

Second, the preferred reproduction of a prototypical color is estimated through preliminary experiments, which are guided by information available in the literature. This preferred reproduction may vary with observer demographic characteristics such as age, gender, geographical location, and ethnic background. Consequently, it may be important to select observers in a fashion that a representative or inclusive sample is obtained. Correlation of prototypical color and preference should not necessarily be assumed; e.g., races with similar skin-tone colorimetry may still have different preferred reproduction positions.

Third, the experimental design is specified in terms of deviations from a center point that corresponds to the reproduction of the prototypical color at its estimated position of preference. In our experience, fractional factorial or central composite designs (Montgomery, 1984) are suitable for three-dimensional color studies. It is important to have at least five levels in each dimension to quantify the preference distribution shape and the asymptotic behavior of the quality loss function surface at large deviations from the optimum. An example of a modified central composite design in terms of lightness, hue, and chroma deviations is shown in Fig. 20.1. Solid lines connect the six axial points, with their intersection representing the central point of the design. The off-axis points are divided into inner and outer cubes to better define the quality loss function curvature near the origin. The vertical lines attached to the inner design points illustrate the ΔL^* distance from the lightness reference position.

Fourth, stimuli representing the design positions are prepared. All memory color studies, including the classic analog studies by Hunt et al. (1974), are based on local manipulations of color space in the vicinity of the memory color. If the color shifts of the memory color are too large, the image will look inconsistent or unnatural. Psychometric evaluations of such images are expected to be of limited value in practice. Therefore, it is important to select an appropriate range of color variations. It is also critical to perform all image simulations in such a manner that color reproduction artifacts are minimized. This can be achieved by designing a smooth, gradual transition from the unaffected regions to the

Preference in Color and Tone Reproduction 291

centroid position, where the maximum shifts of lightness, hue angle, and chroma occur. We have found it advantageous to introduce shifts in a scene color representation space, so that the subsequent step of rendering for output smoothes transitions without introducing artifacts such as tonal clipping. This is particularly important for local lightness manipulations. If desired, image segmentation can be used to prevent incidental modification of other objects having similar colorimetry; e.g., wood and other natural colors can be isolated from the shifts applied in skin-tone studies.

The extension of the considerations discussed above to experiments other than those involving memory color reproduction is mostly straightforward. For example, in a study of tonal rendition, the scenes chosen should represent the

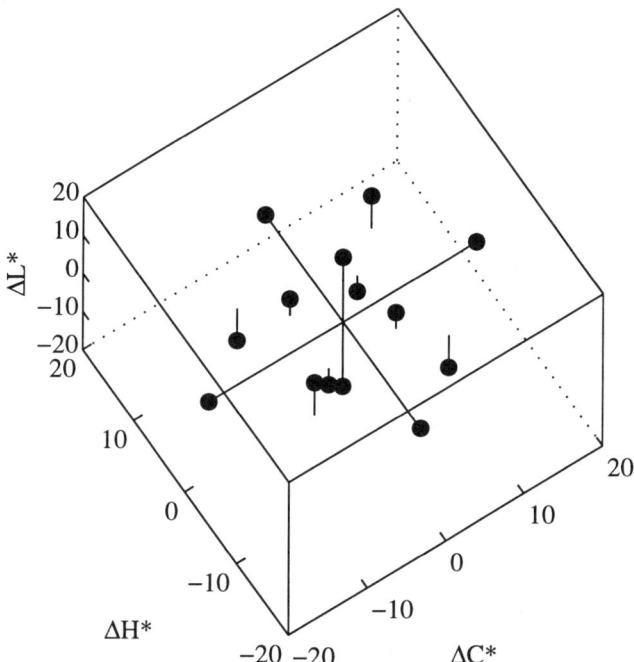

Fig. 20.1 Modified central composite design used in memory color studies. The test levels differ in lightness, chroma, and hue from a central point that approximately coincides to the preferred reproduction of the prototypical memory color.

range of scene lightness distribution shapes encountered in practice; similarly, scenes used in an investigation of colorfulness should have chroma distributions spanning those found in actual imaging applications. As discussed in Sect. 20.5, if dependencies of quality on scene or observer characteristics are found, the differences should be modeled, so that predictions can be made for distributions of the governing properties that differ from those represented in the experiment.

20.4 General Form of Color/Tone Objective Metrics

A number of examples of objective metrics that characterize artifactual attributes have been described in recent chapters. This section will address aspects of the design of objective metrics that pertain to preferential attributes. Specifically, we shall consider objective metrics that correlate with color and tone reproduction quality and are based upon colorimetric measurements of reproductions of targets consisting of various color patches.

Most of our objective metrics, Ω, for color and tone attributes are formulated according to the following general equation:

$$\Omega^{n_c} = \sum_{i=1}^{P} w_{i,p} \cdot \left(\sum_{j=1}^{V} w_{j,v} \cdot \Delta v_{i,j}^{n_c} + \text{additional terms} \right) \quad (20.1)$$

$$\left(\text{where } \sum_{i=1}^{P} w_{i,p} = \sum_{j=1}^{V} w_{j,v} = 1 \right)$$

where n_c is either one or two, depending on whether the metric is linear or a root-mean-square (RMS) quantity; $w_{i,p}$ is the patch weight given to the ith of P patches; $w_{j,v}$ is the variable weight given to the jth of V variables characterizing some aspect of color and tone reproduction; and $\Delta v_{i,j}$ is the difference of the values of the jth such variable between that of the preferred position of the ith patch and that of its actual reproduction. Additional terms, when present, usually reflect interactions between variables or higher order effects; rarely is more than one additional term needed to fit psychometric data well. Some effects of additional terms will be seen in the quality contours of Sect. 20.6.

Examples of variables corresponding to the Δv are listed in the rightmost column of Table 20.1. Dependencies of objective metrics of each attribute in the table are discussed below. A particularly simple case is the RMS color balance metric, which has $\Delta v = \{\Delta a^*, \Delta b^*\}$, and is based on just a few, equally weighted gray patches of moderate reflectance. In this case, Eq. 20.1 reduces to the following form.

$$\Omega^2 = \frac{1}{P}\sum_{i=1}^{P}\left(w_{a*}\cdot(\Delta a*)^2 + (1-w_{a*})\cdot(\Delta b*)^2\right) \quad (20.2)$$

It might seem that only the value of chroma should matter, but the hue of the color shift has a significant effect. Specifically, deviations occurring along the b* axis, which approximately corresponds to natural daylight illuminant variations (from bluer in shade to yellower at sunset), usually are tolerated to a greater degree than those along the a* axis. The case of contrast mismatch, in which the color balance has complementary hue in the shadows and highlights, can be treated via a generalized color balance metric in which neutral patches span the entire output dynamic range (rather than just mid-tones) and the patch weights $w_{i,p}$ are unequal, reflecting tonal weighting (Sect. 15.4).

In the case of density balance, the physical correlate listed in Table 20.1 is the lightness values of mid-tone neutral patches, so there is only one variable in the objective metric ($V = 1$), which is $\Delta v = \Delta L*$, and the P patches are each mid-tone neutrals. In the case of memory color reproduction, all three perceptual dimensions of color are pertinent and so $\Delta v = \{\Delta L*, \Delta C*, \Delta H*\}$. Different sets of patches are used for each memory color, but in all cases, the patches span the three-dimensional distribution of real world examples of the memory color. The weight of each patch may reflect its relative abundance in this distribution. The objective metric of detail in saturated colors has $\Delta v = \Delta L*$; the patches represent various hue and chroma positions that fall on the gamut volume boundary surface of a reference system having excellent color gamut.

The contrast metric has $\Delta v = \{\Delta\gamma, \Delta L^*_{min}, \Delta L^*_{max}\}$, where the latter two terms reflect potential deficiencies in minimum and maximum lightness values achievable based on output media constraints. These deficiencies can lead to hazy blacks and dull whites, respectively. The patch weights $w_{i,p}$ reflect tonal weighting considerations as described in Sect. 15.4. Application of tonal weighting in objective, rather than perceptual, space is reasonable in this instance because the objective metric is nearly perceptually uniform.

Finally, the objective metric for colorfulness is based on two variables, namely, the mean and variance of the distribution of chroma ratios between reproduced and original colors. The former measures the average degree of chromatic amplification and the latter the degree of consistency thereof (with the preferred value being zero). Equation 20.1 can be put in a form encompassing this objective metric if the patch weights are set to $1/I$, additional terms are omitted, and the Δv are defined as follows: (1) the difference between the chroma ratio and its preferred value, to represent the mean; and, (2) the square of the difference between the chroma ratio and the mean value thereof (over the entire patch set), to represent the variance. The set of patches for which the metric is calculated has a distribution of original chroma values similar to that found in a large sample of consumer images, to ensure that the sample is representative.

Given the above considerations, a test target for prediction of color and tone reproduction quality should include at least the following: (1) a series of closely spaced neutral patches spanning the range from minimum to maximum lightness, for characterizing color balance, density balance, and contrast; (2) a similar series of patches (or continuous sweeps) at various hues for quantifying detail in saturated colors; (3) clusters of patches spanning the three-dimensional distribution of real world examples of each memory color; and (4) a set of patches providing a representative distribution of chroma positions for predicting colorfulness. In addition, inclusion of patches of problematical colors, such as infrared reflectors, is desirable. The patches corresponding to particular objects should have spectra (not just tristimulus values) closely matching that of the objects they represent. Finally, for diagnostic purposes, and to support additional metrics, patches regularly arrayed within the volume of real world surface colors should be included to ensure coverage of all regions of color space, with particular emphasis on mid-tones and nearly neutral colors, because of their abundance in photographed scenes.

Before concluding this section, a brief discussion of the implications of inclusion of preferred positions (optima) within objective metrics is warranted. Each term in the generalized objective metric of Eq. 20.1 depends on the deviation of a variable from its preferred value, hence the metric is defined relative to an optimal state. Such an approach is not required for artifactual attributes because their preferred level is always the same, namely, at or below threshold. Optima may be determined by a combination of empiricism and theory, and in some cases, they may vary with observer and scene characteristics (see next section). Application of color appearance models can serve to define positions for which the reproduction appears similar to the original scene, accounting for factors such as adaptation level (perceived luminance contrast is higher at higher light levels), surround conditions (perceived luminance contrast is lower in a dark surround, such as during slide projection), and so on. Empirical adjustments to account for preference are needed, because accurate recreation of the colorimetry of the scene generally does not constitute the preferred reproduction (Buhr and Franchino, 1994, 1995). Once such an adjustment is known, the color appearance models allow estimation of how the optimum will change as a function of viewing conditions, which permits application of an objective metric to viewing conditions differing from those in which it was derived, through translation of the preferred reproduction.

20.5 Fitting Psychometric Data

As described in Sect. 10.4, quantification of quality loss as a function of objective metric value for various subsets of observers and scenes is relatively straightforward in the case of artifacts. Because observers agree the desired state of an artifact is an absence of visibility, the primary differences between observers is simply that of sensitivity to the artifact, and the primary difference between scenes is their susceptibility to the same. Consequently, to characterize the variability of responses (quality changes in JNDs arising from the attribute being studied), it is usually sufficient to simply classify observers into more or less sensitive groups and scenes into more or less susceptible groups, based upon the responses obtained. From all possible combinations of observer and scene groups (e.g., more sensitive observers viewing less susceptible scenes), subsets of data may be defined for separate regressions. The group containing all scenes and the group of all observers are included when making these combinations, so in the situation described above, nine subsets would result.

The combination containing all observers and all scenes corresponds to the fully pooled data set, the mean response of which is attributed to that of the average observer and scene. The fully pooled data is fit using the IHIF (Eq. 9.4), which has three regression parameters (threshold, radius of curvature at threshold, and asymptotic JND increment). Subset data is also fit with the IHIF, but usually, for robustness, one of the three fit parameters is fixed at the value obtained in the regression of the fully pooled data (see Sect. 10.4). Because the objective metric value does not vary between subsets, all variability in the data is modeled through use of different IHIF fit parameters.

Quantification of quality changes arising from preferential attributes is substantially more complex. Because the objective metrics are composed of terms that depend on preference, and because preference may vary with subset, the definition of meaningful subsets may require a more sophisticated approach. Furthermore, the regressions involve more fit parameters and so must be carried out with greater care. We have found the following approach useful in defining subsets of observers and scenes and determining how to fit their data adequately using the fewest fit parameters:

1. identify groups of observers and scenes with similar behavior, using cluster analysis;

2. characterize the identified groups, to the extent possible, by looking for common features, such as similarity of scene content;

3. test whether readily identified factors, such as observer demographics, affect the responses in a significant manner, as determined though analysis of variance (ANOVA);

4. using analysis of variance and cluster analysis results, classify observers and scenes into groups and form subsets for regression; and

5. determine how to model the subset data adequately using the fewest fit parameters possible, based on variation of the fit parameters across subsets and sequential F-tests.

These points will now be considered in detail. Cluster analysis is used to identify groups of observers or scenes having or producing correlated responses across the test stimuli. A number of different methods for cluster analysis are available in standard statistical packages (SAS Institute, 1989). A selection of these methods are applied to the data and those clusters independently identified by several different methods are deemed valid groups. Next, common features within each group are sought, so that the groups may be characterized in terms that facilitate the assessment of their significance in specific imaging applications. For example, in studies of both foliage and blue sky, two clusters of scenes were identified. In one cluster, objects exhibiting the memory color constituted a portion of the main subject of the photograph, as in scenic landscape images. In the second cluster, the main subject (e.g., a group of people) did not contain the memory color, which was evident only in the ancillary content (e.g., the background). In both the foliage and sky cases, more rapid quality loss with deviation from the preferred reproduction was observed when the memory color was a part of the main subject of the image. As another example, in a relative colorfulness study, we detected differences in preference between scenes that contained objects of well-known color, such as fruit, and those that did not. In the latter case, observers, unconstrained by familiarity, preferred more colorful renditions.

An alternative and complementary approach to classifying variability of responses is the analysis of variance. In this technique, potential influencing factors must be identified first, after which their impact on response is tested for statistical significance (whereas in cluster analysis, distinct response groups are first determined, after which the analyst must try to identify the underlying factors causing the response difference). For example, analysis of variance was used to determine whether race, gender, or experience of the observer were significant factors in the evaluation of skin-tone reproduction quality. Race was found to be highly statistically significant, whereas the other two factors were not.

If either scenes or observers have failed to be classified into groups by analysis of variance and cluster analysis, the same groupings as in artifact studies may be adopted, i.e., more and less sensitive observers and more or less susceptible scenes. Having identified the groups of observers and scenes that will be used, subsets of data corresponding to combinations of the scene and observer groups are defined for purposes of regression analysis, just as in the case of artifacts.

In addition to the three fit parameters in the IHIF, the objective metric of a preferential attribute (unlike an artifactual one) may contain several fit parameters related to the optimal position. To improve the robustness of the resulting regressions, it is desirable to use the fewest fit parameters that accurately model the data. It is therefore useful to consider how many fit parameters might be contained in Eq. 20.1. In the absence of interaction terms, the P patch weights $w_{i,p}$ and the V variable weights $w_{j,v}$ could constitute $P-1$ and $V-1$ fit parameters, respectively (the last weight in each case is determined because the weights must sum to one). Furthermore, the definition of the optima used in computing the $\Delta v_{i,j}$ could add as many as $P \cdot V$ additional fit parameters. In practice, a number of constraints are placed on the values that these parameters have relative to one another.

First, because the patch weights $w_{i,p}$ are normally specified based on considerations such as appropriate sampling of color space and tonal weighting, they are generally not used as regression fitting parameters. Second, for a given variable Δv, the optimal values for different patches are assumed to be related to one another in a deterministic fashion, based on the expected global properties of preferred reproductions. Consequently, the optimum for each variable Δv normally depends on only one fit parameter even though the preferred value for each individual patch may be different. Therefore, a total of $2 \cdot V + 2$ fitting parameters are typically available, V from optima, $V-1$ from the variable weights $w_{j,v}$, and 3 from the IHIF. In a memory color reproduction metric, $V=3$, so eight fit parameters are likely to be employed in the regression.

As in the case of artifacts, all fit parameters are used in modeling the fully pooled data; however, to increase robustness and consistency of subset regressions, as many of the parameters are fixed at their fully pooled values as is possible, while still fitting the data satisfactorily. Because of the increase in number of fit parameters, the task of deciding which parameters should be fixed and which should vary is more complicated in the case of preferential attributes than in those of artifacts, where each of the 3 fit parameters could simply be fixed, one at a time, and the regression fits examined visually.

One approach to this problem is to allow all parameters to vary in a trial regression of each subset, and to tabulate the resulting, unconstrained values of each parameter by subset. If a parameter is observed to change little between subsets, it can probably be fixed at the fully pooled value. It is also possible that the parameter varies with scene group but not observer group, or vice versa, in which case an average value by group may be used. In many instances, it is obvious by visual inspection of the resulting fits whether fixing a parameter was appropriate, but in borderline cases, a statistical criterion should be adopted to make a decision. For example, a sequential F-test may be employed to determine if freeing up a previously fixed fit parameter results in a sufficient reduction in the sum of squared errors to be warranted.

In our studies, certain patterns have emerged regarding which parameters may be fixed and which need to vary with subset. In all memory color studies, the variable weights $w_{j,v}$ for lightness, hue, and chroma were constant across subsets, whereas the IHIF regression parameters always changed. In a foliage reproduction study, the optimum lightness, hue, and chroma of the reproduction varied depending on the observer group but not as a function of scene. In a study of blue sky reproduction, observers preferred a darker rendition when the sky was an important element in the composition, as in a scenic photograph. In a skin-tone reproduction study, observers' sensitivity to the deviation of the skin-tone reproduction from the preferred position differed significantly with the observer's ethnic background. Furthermore, each ethnic group had distinct preferences regarding the optimum lightness, hue, and chroma reproduction of their skin-tone, relative to its actual colorimetry.

20.6 Quality Contours of Preference

Visualizing a quality loss function of three color dimensions is difficult. Perhaps the easiest way to do so is though the use of contour plots, which are like topographic maps of quality in two-dimensional slices through color space (Berns, 2000). A fixed value is chosen for one color dimension, thereby defining a color plane, and points having equal quality are joined by lines that show how the quality changes within the plane. Typically, the surface will look like a rounded hill having a peak of zero JNDs at the preferred reproduction position, and dropping off in all directions from there. From Eq. 20.1, for a root-mean-

Preference in Color and Tone Reproduction

square metric with $n_c = 2$, and in the absence of interaction terms, we expect such contours to be ellipses with the major and minor axes parallel to the color dimension axes.

This behavior is indeed observed in Fig. 20.2, which shows the quality loss contours of chroma and hue for foliage reproduction at the preferred lightness position, as assessed by more and less sensitive observers (solid and dashed lines, respectively). As expected, the contour lines of the more sensitive observers are closer together because they require a smaller stimulus difference to produce a given quality change than do less sensitive observers. In addition, the preference of more sensitive observers is shifted to lower chroma, indicating that they prefer less colorful foliage than do less sensitive observers.

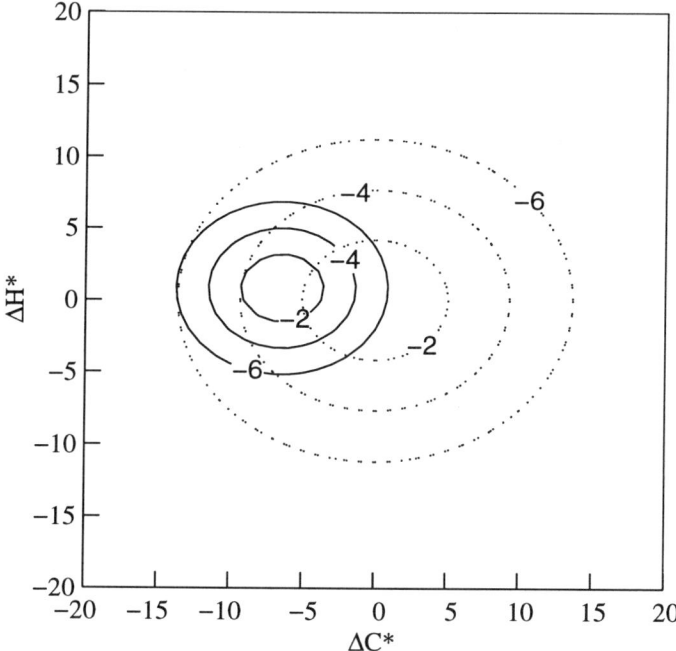

Fig. 20.2 Quality loss contours of chroma and hue differences for foliage reproduction at preferred lightness. More sensitive observers (solid lines) prefer a less colorful reproduction of foliage than do less sensitive observers (dashed lines).

In some cases, rotated or asymmetric contours are encountered, indicating the necessity for additional terms in the objective metric defined by Eq. 20.1. Examples of two such cases are now considered. Figure 20.3 shows quality loss contours for blue sky reproduction as a function of chroma and hue differences from the optimum position. The lightness was held at its preferred position. The major and minor axes of the ellipses are rotated slightly off the hue and chroma difference axes (perhaps because of perceptual non-uniformity of CIELAB space in the blue-purple region), so an interaction term is required in the objective metric. The ratio of hue to chroma difference weighting in the associated objective metric is higher than that for foliage, reflecting the low tolerance for hue deviation from the optimum in blue sky.

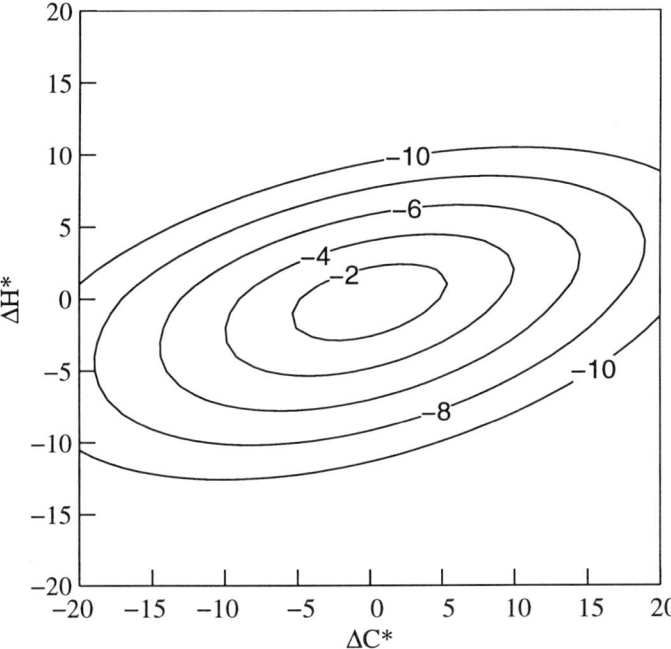

Fig. 20.3 Quality loss contours of hue and chroma differences for blue sky reproduction at optimum lightness. The ellipses are rotated relative to the hue and chroma difference axes, necessitating the inclusion of an interaction term in the objective metric.

Preference in Color and Tone Reproduction

Figure 20.4 shows quality loss contours for Caucasian skin-tone reproduction as a function of lightness and chroma differences from the optimum position, with hue reproduction held fixed at its preferred value. The contours are approximately elliptical, but the major and minor axes are not parallel to the lightness and chroma axes. In Fig. 20.4 the major axes, along which quality falls off most slowly, approximately coincide with a line of constant chroma to lightness ratio. In the 1976 CIE L* u* v* (CIELUV) system, saturation is defined as C*/L*, whereas in the CIELAB system (used in Fig. 20.4) this correspondence is only directionally correct. With this caveat noted, it still appears that changes in saturation are tolerated less than commensurate changes in chroma and lightness having the same colorimetric (ΔE^*) difference.

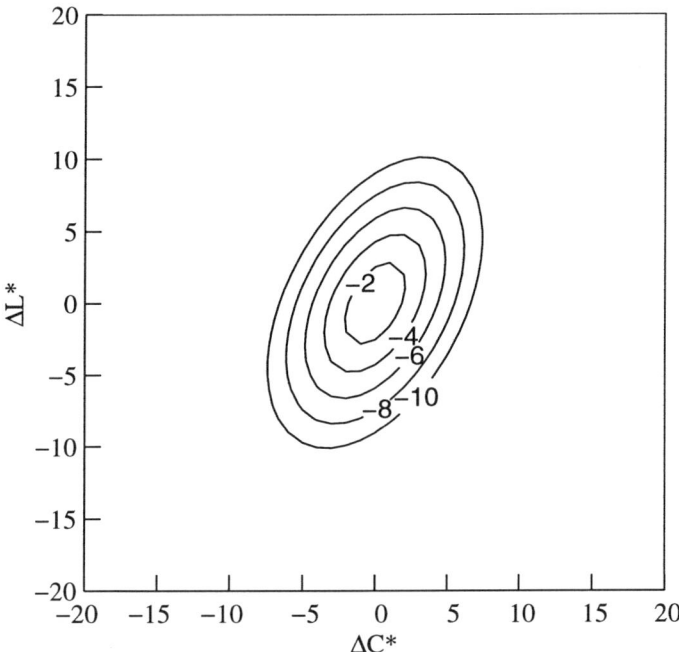

Fig. 20.4 Quality loss contours of lightness and chroma differences for Caucasian skin-tone reproduction at optimal hue. The major axes of the ellipses, which correspond to the minimum quality degradations, roughly coincide with a line of approximately constant saturation.

Our last example, shown in Fig. 20.5, depicts the −5 JND quality loss contours, and the 10% of peak contours of the preference distributions, for Caucasian skin, blue sky, foliage, and overall image color balance in (a*, b*) space at preferred lightness. The results for color balance are shown centered on a neutral position for convenience, even though all colors in an image are shifted. The contours of preference and quality loss for a given color are similar in size and shape. This result implies that positions preferred only 10% as often as the most preferred position (averaged over a set of scenes and observers) would be perceived (by the average observer, viewing the average scene) as being ≈5 JNDs inferior to the peak quality. Even small CIELAB deviations from the preferred position of skin-tone reproduction lead to rapid quality loss, whereas there is considerable

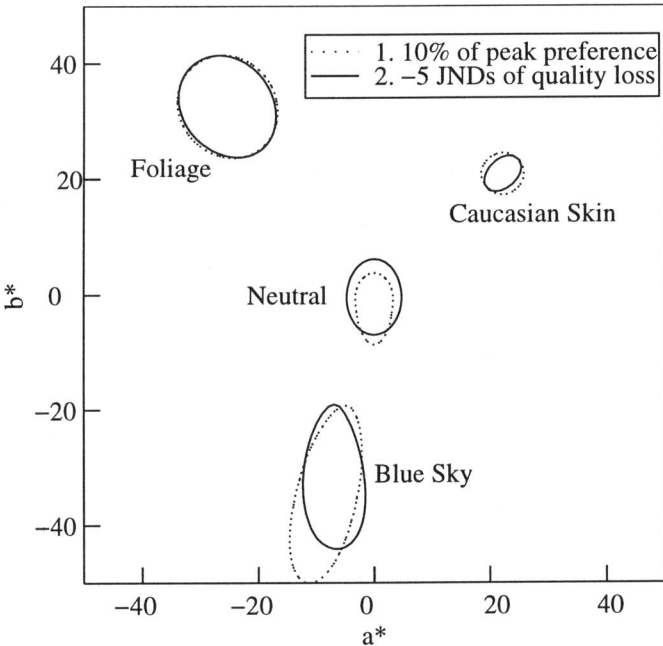

Fig. 20.5 Comparison of −5 JND contours of quality loss, and 10% of peak preference contours, of Caucasian skin, blue sky, foliage, and neutral gray (as a proxy for the color balance attribute), . The contours of quality loss and preference for a given color are similar to one another, but there are large variations between the four colors.

tolerance for larger variations about the optima for blue sky and foliage. The large variation in size and shape of the contours between these four colors demonstrate that cognitive and other high level processes are important in image quality evaluations. Such processes are not reflected in colorimetry, which characterizes visual capture only, nor are they addressed by more advanced color appearance models (Berns, 2000).

20.7 Summary

Preference is an important element in the evaluation of the effects of color and tone reproduction on image quality. Despite the preferential aspect of these attributes, the framework of image quality modeling developed for artifactual attributes is still applicable, with certain extensions. The primary complications arising in modeling of color and tone reproduction quality are threefold.

1. Color perception is three-dimensional, leading to more complex objective metrics, experimental design, data analysis, and visualization of results.

2. Objective metrics quantify the deviations of the actual reproduction from a preferred reproduction, the latter of which is determined at the cost of additional experimentation and increased difficulty of regression analysis.

3. Color and tone attributes are linked by system constraints that make it difficult to simulate them in isolation, so that it may be challenging to perform experiments that definitively identify the best set of perceptually orthogonal attributes for use in the multivariate formalism.

Selected color and tone attributes are defined in Table 20.1. Objective metrics of color and tone reproduction are based on the colorimetry of reproductions of color test targets. The general form of such a metric, as given in Eq. 20.1, is a weighted sum over color patches of deviations from the preferred position of one to a few weighted variables that represent the perceptual dimensions of importance, such as lightness, chroma, and hue. Scene and observer variability are reflected not only in the IHIF fit parameters, but also in additional fit parameters characterizing the preferred reproduction position. Selecting representative sets of scenes and groups of observers is critical in developing predictive color quality metrics because of differences in preference depending upon the type of scene and the background of the observer.

Although simple color difference metrics and color appearance models contribute to the objective metric definitions, they alone are not sufficient for quantifying image quality because of cognitive and other high level processes that are important in image quality evaluations. The wide range of shapes and sizes of contours of quality loss and preference distributions for color balance deviations from the preferred rendition of skin-tones, blue sky, foliage, and overall images substantiate this conclusion.

21

Quantifying Color/Tone Effects in Perceptual Space

with **Karin Töpfer**
Eastman Kodak Company
Rochester, New York

21.1 Introduction

The previous chapter discussed the design of objective metrics correlating with individual color and tone reproduction attributes. This chapter will demonstrate that the ability to predict color and tone reproduction quality can be substantially extended by transformations applied in perceptual space. The reader will recall that an operation is said to be carried out in perceptual space if its inputs and outputs are quality changes expressed in JNDs. Examples of operations carried out in perceptual space include the multivariate formalism (Ch. 11) and the weighting schemes applied to attributes that vary across an image (Ch. 15). These contrast with the extensions of metrics, such as the incorporation of terms reflecting attribute interactions (Ch. 17), which are accomplished in objective space. It is not always obvious whether an observed effect should be modeled in objective space, by generalizing a metric, or in perceptual space, through some quality transformation. A rough guideline might be that lower level visual processes such as masking are often amenable to treatment in objective space, whereas higher level cognitive effects are better emulated in perceptual space.

In this chapter, three examples will show how transformations in perceptual space can be successfully employed to model different aspects of color and tone reproduction quality, as well as to account for varying characteristics of image populations encountered in different imaging applications. These examples are briefly summarized here and are treated in detail in separate sections.

1. The degree to which the reproduction of a particular color affects the quality of a given image depends upon the importance of objects of that color in the composition of the image, which, in turn, is correlated with the extent of occurrence of that color in the image. In Sect. 21.2, this effect is demonstrated by examining the influence of the subject head size on the impact that skin-tone reproduction has on image quality. This information, in combination with expected head size distributions, allows prediction of the overall effect of skin-tone reproduction on quality in different applications (e.g., portrait versus consumer photography).

2. The degree to which certain color and tone defects are evident in an image and the extent to which they influence perceived quality is affected by whether a suitable reference image is available for comparison. In Sect. 21.3, this effect is demonstrated by comparing quality assessments of images having different levels of color balance error against either well balanced quality rulers depicting the same scene (thereby providing a critical reference), or against well balanced quality rulers depicting a different scene. These results are pertinent in setting tolerances for systems that produce duplicates of images (such as print copying stations), and also have implications regarding the relative importance of color and tone defects that are consistent within a customer order (e.g., biases caused by calibration error) compared to those that are variable (e.g., from image-specific rendering errors).

3. The final example, described in Sect. 21.4, shows how to combine the predictions of individual color and tone attributes to make predictions of overall color quality. This is accomplished using the multivariate formalism of Ch. 11, which provides a mathematical framework for combining quality changes arising from perceptually independent image attributes to predict overall image quality. Given the three-dimensional nature of color and tone reproduction and the existence of complicating effects such as adaptation, the definition of perceptually independent color and tone attributes can sometimes be challenging. In Sect. 11.5, the results of an experiment co-varying color balance and contrast were shown to be explained by the multivariate formalism. In the present chapter, another example, involving a greater number of color and tone reproduction attributes, is described.

21.2 Impact versus Compositional Importance

Conflicting requirements for scene selection arise in studies of attributes with significant susceptibility variations. For example, in an investigation of memory color reproduction, it would be desirable for the memory color to constitute an important part of each scene to obtain good signal-to-noise in the psychometric experiment. However, it is critical that the results of the experiment can be used to make realistic predictions for practical imaging systems, without overemphasizing the importance of the reproduction of the memory color. A common compromise is to select a variety of scenes and observers, which may be partitioned into subsets and analyzed separately, thereby allowing customization of the results to fit a particular application. In some cases, however, a much wider applicability of the results can be achieved by explicitly accounting for scene-specific effects either in the objective metric or in a perceptual transform. An example of such a scene-specific objective metric extension was the inclusion of the angular subtense of the human pupil image in the redeye metric (Sect. 19.3).

In an initial investigation of the impact of skin-tone reproduction on image quality, only scenes with prominent faces were included, leading to robust psychometric data. The clean signals obtained allowed each of the effects investigated to be accurately characterized, at least in the case where people constitute the principle subject of the image and the attention of the viewer is centered upon them. The set of scenes employed in this study sampled only a portion of the distribution of consumer images, which contains many images with smaller faces or no faces at all. It would be reasonable to assume that images containing no people would be relatively unaffected by skin-tone reproduction (except in a minor way, by virtue of affecting the reproduction of fortuitously similar colors). Furthermore, it should be straightforward to adjust predictions to account for the frequency of such images in a particular application; however, additional information is required to estimate the impact of skin-tone reproduction on intermediate images that have relatively smaller faces.

Consequently, a second study was designed to determine the relationship between subject face size in an image and the impact of skin-tone reproduction on overall quality. In the case of redeye, the size of the red pupil in an image was expressed in terms of its angular subtense at viewing, a measure that, at constant pupil color, strongly correlates with the visibility and obviousness of the artifact. Because the impact of skin-tone reproduction on quality was expected to depend more on the importance of the human subject in the composition, an image-normalized measure of face size was used instead. The measure chosen was the ratio of head size, from the chin to the top of head, to the geometric mean of the image dimensions. A probability density function (PDF) of head size ratio in consumer photography in the mid-1990s is shown in Fig. 21.1. This distribution is in the process of shifting towards the right as telephoto and zoom cameras comprise a greater segment of the market.

The vertical line in Fig. 21.1 shows the smallest head size ratio represented in the first experiment. In the second study, scenes were chosen to span the range of head ratio sizes encountered in consumer images. The selected scenes were similar in terms of the color distribution of skin-tones, depicting average Caucasian subjects. The scenes also were similar in subject matter, representing casual pictures of one or two people. The intent was that the only factor significantly affecting the impact of skin-tone reproduction on quality would be

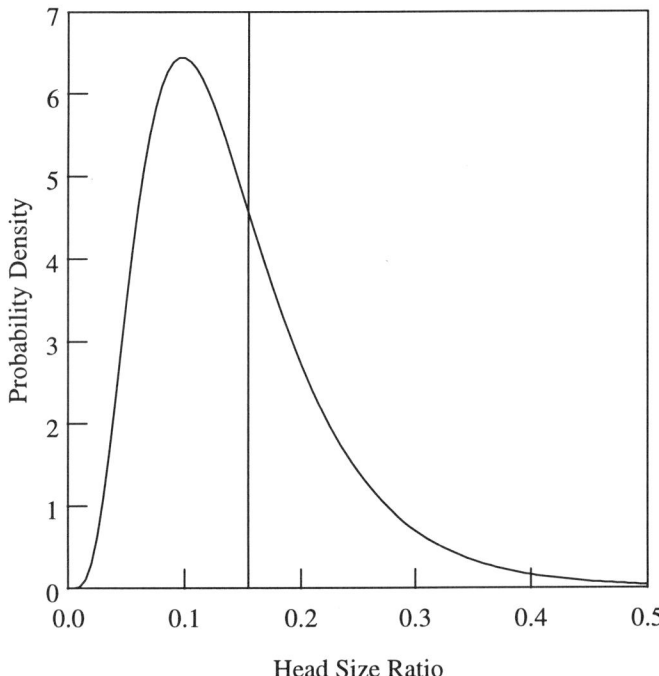

Fig. 21.1 Distribution of head size ratios in consumer photography in the mid-1990s. The head size ratio is the height of the head in an image divided by the geometric mean of the image dimensions.

Quantifying Color/Tone Effects in Perceptual Space

head size. The experimental levels represented a subset of fifteen manipulations of lightness, hue, and chroma from the first study. One scene from the original study was included to check that the two experiments were consistent with one another.

The results of this study are shown in Fig. 21.2. The x-axis is head size ratio, and the y-axis is the quality fraction, a measure of the relative impact of skin-tone reproduction on quality. More precisely, as shown in Eq. 21.1 (next page), the quality fraction ϕ_h is defined as the ratio of quality loss ΔQ arising from less preferred skin-tone reproduction at a head size ratio ρ_h, to the maximum possible quality loss, ΔQ_{max}, which occurs at the largest head size ratios.

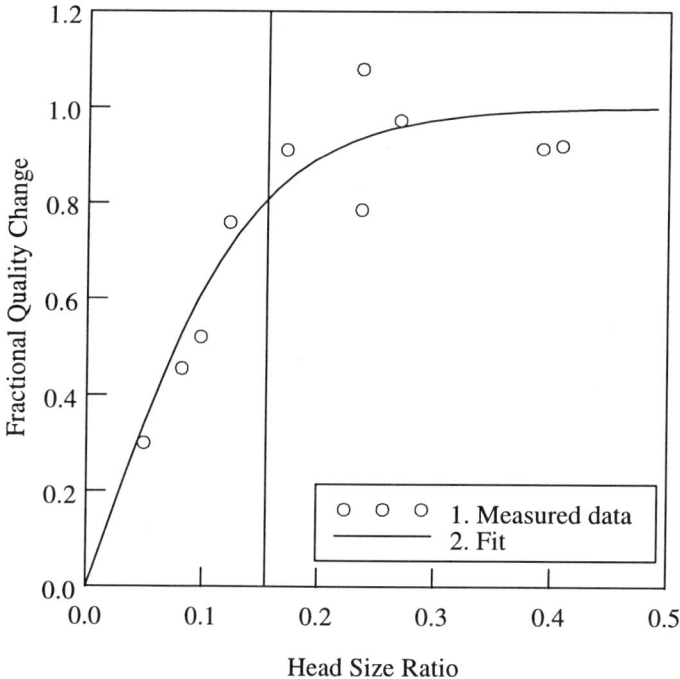

Fig. 21.2 Fraction of the maximum quality loss from a less preferred Caucasian skin-tone reproduction, as a function of head size. The impact of skin-tone reproduction on quality is diminished when the head height is less than 25% of the geometric mean image dimension.

$$\phi_h(\rho_h) = \frac{\Delta Q(\rho_h)}{\Delta Q_{\max}} \qquad (21.1)$$

As seen in Fig. 21.2, at small head size ratio the quality is approximately proportional to the head size ratio. At larger head size ratios, the quality fraction levels off, approaching its upper limit of one. The vertical reference line in Fig. 21.2 again shows the smallest head size ratio from the first experiment, from which it is evident that the scenes in that experiment fell mostly in the asymptotic regime. The equation of the regression fit to the data in Fig. 21.2 is given by Eq. 21.2.

$$\phi_h(\rho_h) = \tanh(7.155 \cdot \rho_h) \qquad (21.2)$$

Knowledge of the quality ratio function given in Eq. 21.2 permits predictions of the effect of skin-tone reproduction on image quality to be tailored to specific applications, such as consumer photography or studio portraiture. Given an application-specific distribution of head size ratios $h_h(\rho_h)$, such as that shown in Fig. 21.1, the mean quality loss arising from a less preferred skin-tone reproduction would be given by Eq. 21.3.

$$\overline{\Delta Q} = \frac{\Delta Q_{\max} \cdot \int h_h(\rho_h) \cdot \phi_h(\rho_h) \cdot d\rho_h}{\int h_h(\rho_h) \cdot d\rho_h} \qquad (21.3)$$

In a Monte Carlo simulation of the quality distribution produced by an imaging system (Ch. 28), rather than simply computing the mean quality loss as in Eq. 21.3, samples would be randomly drawn from the head size ratio distribution to build up a representative distribution of quality losses arising from skin-tone reproduction.

A principal advantage of separating the effects of head size ratio distribution and skin-tone reproduction is that predictions of image quality may still be made, without additional psychometric experimentation, if the head size ratio distribution changes over time. In consumer photography, such a change is occurring now because of greater penetration of zoom lenses into the compact camera market, leading to an increased frequency of higher head size ratios. In general, it is advantageous to independently characterize effects such as these, which are not inextricably linked, both to increase the level of understanding of image quality, and to enhance predictive capabilities.

21.3 Reference Images and Discrimination

In certain cases, the impact of a color and tone reproduction attribute on quality can depend on the type of reference image (if any) that is available during the assessment. One example of such an attribute is color balance, which relates to any consistent color shift of the entire image away from the preferred balance, creating the appearance of an overall color cast. The assessment of this attribute is potentially complicated by the adaptation of the human visual system to the colorimetric character of the light detected. If a sufficient fraction of the light received by the visual system arises from the image being viewed, either because the image subtends a significant fraction of the visual field, or because the image is much brighter than the surrounding portion of the visual field (as in slide projection), adaptation to the image can reduce the ability of the observer to ascertain the absolute color balance of the image. If two images with different color balance are directly compared under such circumstances, they will commonly be perceived to have complementary color balance (e.g., blue and yellow) because the adaptation is to a colorimetric position in between that of the two images. The experiments described herein were carried out with images of modest subtense viewed against a neutral surround having a luminance comparable to that of a rendered mid-tone in the images. Consequently, adaptation should have been largely controlled by the surround, and therefore should have been essentially constant during the experiments, and thus not have been a factor contributing to the results described below.

Three types of psychometric assessments having different practical analogues and yielding potentially different results might be envisioned.

1. Assessments may be made in a single stimulus mode, in which case no reference image is present. Category sort experiments may be conducted in this fashion by providing the observers one sample at a time and preventing them from seeing the samples they have already rated (e.g., by having them place rated samples in covered boxes with slots). In practice, this type of assessment might correspond to the case in which a viewer casually considers an isolated print and forms an impression of its quality.

2. Assessments may be made in a dual stimulus mode, where the test images are compared to one or more reference images depicting different scenes. This situation is common when evaluating samples generated by means other than controlled image simulation, e.g., in trade trial or customer intercept studies. The reference images in such cases are likely to correspond to physical standards against which a

scale of image quality is calibrated. Quality ruler experiments can be carried out in this fashion, as is the case in the experiment described below. In practice, this type of assessment is analogous to consumers viewing their photofinishing orders, which typically will contain multiple images from each of several separate events. Under such circumstances, a consistent bias across the entire order (e.g., a slight reddish color cast caused by an output device calibration problem) is not as noticeable as equally large color balance errors in different hue directions, as could be caused by scene-specific errors in balancing algorithms.

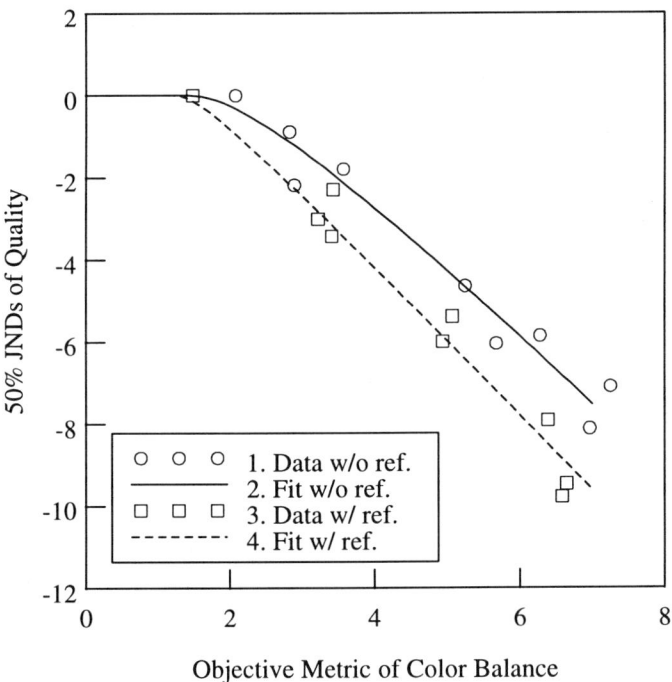

Fig. 21.3 Quality loss from color balance errors in the presence and the absence of a well-balanced reference image of the same scene. In the presence of a critical reference image, quality is assessed to be lower whenever there is quality loss arising from color balance.

3. Assessments may be made in direct comparison with other images depicting the same scene, as we commonly do in quality ruler experiments. It has already been shown in Fig. 8.10 that in the case of artifacts, the results obtained from quality rulers having matched scenes do not differ from those depicting different scenes. In practice, this type of assessment corresponds to cases in which an original image is duplicated (as at a print copying station) and, importantly, to cases in which multiple imaging systems are being compared in a controlled fashion. It is often thought that, because most customers do not themselves perform such controlled tests, assessments of this type may be unduly sensitive to differences; however, customers are strongly influenced by people who do make controlled comparisons or who are aware of the results others have obtained in such experiments.

Although we believe that most system specifications should be set based on the results of assessments of the third type (matched reference image present), because of the considerable influence of controlled tests in the marketplace, it is also valuable to understand how the results of assessments of the second type (unmatched reference image available) might be related to them, given their analogy to typical consumer viewing of an order. Both scenarios can be accommodated in image quality modeling by experimentally determining quality loss response functions in one case, and applying a transformation in perceptual space to convert to the other case. This approach is similar to that taken in accounting for the effect of head size on the impact of skin-tone reproduction quality.

Figure 21.3 shows the results of two quality ruler experiments on color balance error, one employing ruler images depicting the same scenes as in the test images, and the other utilizing unmatched reference images. As expected, greater quality loss occurs for a given degree of color balance error in the case with matched reference images, which permits particularly critical evaluation.

The data from Fig. 21.3 may be used to define the transformation from quality loss in the matched, and thus critical, case, denoted ΔQ_c, to the quality loss in the unmatched, and therefore uncritical case, denoted ΔQ_u. The transformation is conveniently quantified in terms of the fraction ϕ_c of the uncritical quality loss compared to the critical quality loss, expressed as a function of the latter. This dependence is well described by Eq. 21.4, which is plotted in Fig. 21.4.

$$\phi_c(\Delta Q_c) = \frac{\Delta Q_u}{\Delta Q_c} = \frac{\Delta Q_c}{1.094 \cdot \Delta Q_c - 1.814} \qquad (21.4)$$

It is interesting to speculate why the presence of a critical reference image increases discrimination in the case of color balance error, but does not do so in the case of an artifact such as banding. One possible explanation is that even in single stimulus viewing, it is easy to recognize an artifact because of its unnatural appearance, and then to visualize what the image would look like if the artifact were removed. Essentially, in the case of an artifact, an observer may be able to construct in his or her mind a critical reference image for virtual comparison with the test image. In contrast, as part of their everyday lives, people are exposed to illuminants having different chromatic characteristics, so a small amount of color balance error may not appear especially unnatural, even though the resulting rendition is not preferred, as determined in direct

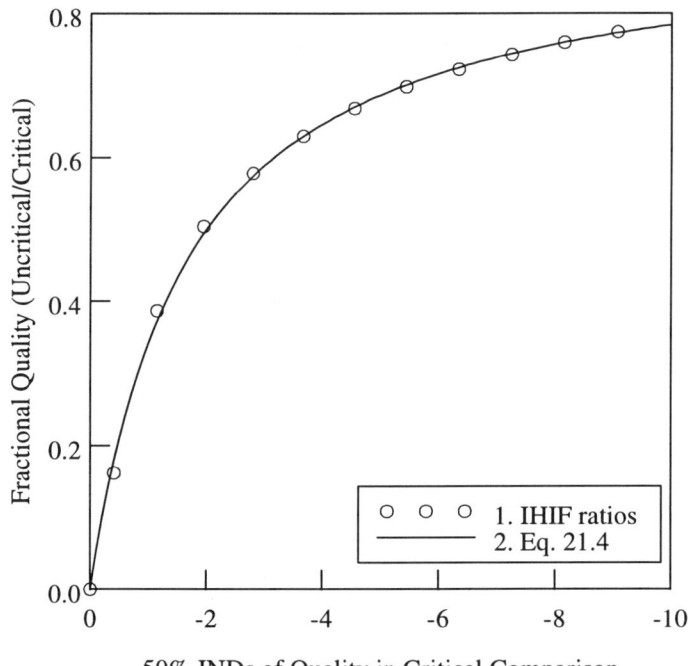

Fig. 21.4 Fraction of quality loss from color balance error in the absence of a well-balanced reference image of the same scene, compared to the case in which it is present. Smaller quality differences, which are evident in the presence of a critical reference, are almost unnoticed in its absence.

comparison with alternatives. It actually requires some degree of training to recognize subtle color balance errors and identify the direction of their biases in single stimulus viewing, as may be appreciated by individuals with practical experience in printing color negative film.

21.4 Multivariate Color/Tone Quality

The preceding chapter showed how the effect of individual color and tone reproduction attributes on image quality can be modeled, and the preceding sections of this chapter have provided examples of how the predictiveness of such individual metrics can be extended through transformations carried out in perceptual space. This section addresses the question of how to integrate the knowledge of individual attributes of color and tone reproduction into the prediction of overall image quality.

If color and tone reproduction attributes could be defined so that they were perceptually independent from one another and from other attributes such as those related to spatial phenomena, the multivariate formalism of Ch. 11 should permit prediction of overall quality from individual attribute levels. Because of the multidimensional nature of color and the complexity of perceptual effects such as adaptation, significant interactions might be anticipated among the attributes of color and tone reproduction. To what extent do such interactions actually occur?

Let us first consider the interaction of color and tone attributes with other attributes. The reader will recall from Sect. 11.2 that two attributes are said to interact if a change in the level of one attribute leads to a change in perception of the other attribute. In many cases, a simple thought experiment will provide an intuitive assessment of the magnitude of interaction. For example, it seems unlikely that color balance error would significantly affect the perception of banding, because all portions of the periodic banding pattern would be shifted similarly, and so the visibility of the artifact would scarcely be affected. But what of the case of contrast and noise? An increase in final image contrast as a result of altered rendering would certainly amplify capture noise and, if the noise were above threshold, would clearly alter its appearance in the final image. This does not constitute an attribute interaction, however, because an objective measurement of RMS granularity in the final image would properly reflect the impact of the increased rendering contrast. Stated another way, the increase in contrast did not change the perception of a fixed amount of noise, but rather changed the actual amount of noise present. Had the amount of capture noise been reduced to exactly compensate for the increased rendering contrast, such that the same RMS granularity measurement resulted at both contrast positions, the perceived quality loss arising from noise in each of the final images would be expected to match.

Considerations of this type lead to the conclusion that very few color and tone attributes significantly interact with other types of attributes. Perhaps the most notable exception is the interaction of contrast, as measured by a gradient metric, and sharpness, as quantified by a metric based only on system MTF. Improved MTF at constant gradient leads to the perception of slightly higher contrast, and, under some conditions, increased gradient at constant MTF results in slightly higher perceived sharpness, although the magnitude of both of these effects is quite small. This interaction may be handled in a fashion analogous to that described for streaking and noise in Sect 17.2. Inclusion of a small gradient-related term in the objective metric of sharpness, and addition of a minor MTF-related term in the objective measure of contrast, leads to two more nearly orthogonal objective metrics that more closely correspond to the perceived attributes of contrast and sharpness.

Let us next consider interactions among attributes of color and tone reproduction. Examples in the literature suggest that tone scale attributes, e.g., mid-tone contrast and highlight and shadow detail, are perceptually independent from colorfulness or color saturation attributes (Yendrikhovskij et al., 2000). As discussed in connection with Fig. 11.7, contrast and color balance also behave independently. In that experiment, univariate stimuli depicting one of four contrast modifications or one of four color balance variations, and multivariate stimuli containing both a contrast modification and a color balance variation in combination, were assessed against quality rulers. As shown in Fig. 11.7, the multivariate color and tone reproduction quality of the levels containing combinations of contrast and color balance treatments were quantitatively predicted by the same multivariate relationship that described various combinations of sharpness, noise, clipping, and streaking. These results provide definitive evidence that contrast and color balance behave as independent attributes.

Other color and tone reproduction attributes may not behave independently, depending on how they are defined. For example, if the reproduction of memory colors (e.g., skin-tones, sky, and foliage) is defined relative to absolute colorimetric positions of preference, then memory color reproduction appears to interact with color balance. A shift of skin-tones in the green direction might be very objectionable in an image that is otherwise well balanced; however, if the image had an overall green color cast, resulting in the same absolute colorimetry of skin-tones, the image would be perceived to have less quality loss. It remains open for discussion whether this effect is caused by partial adaptation to the color bias or whether the skin-tone reproduction simply appears more consistent with that of the remaining scene elements, and therefore less unnatural. By generalizing the preferred colorimetric positions of memory colors so that they depend upon the color balance of an image (i.e., expressing the positions relative to that of a neutral tone), more orthogonal behavior between memory color reproduction and color balance may be achieved.

To test whether the overall color and tone reproduction quality of images could be successfully predicted when many attributes were simultaneously varied, an extensive verification experiment was performed. Lightness and chroma mappings were varied in a 3 × 3 factorial design centered on a practical color and tone reproduction position, and an additional improved position was included for reference, making 10 levels in total. Hue reproduction was constant and conformed closely to the average preferred position. The resulting images varied in the attributes of contrast, colorfulness, detail in saturated colors, and memory color reproduction of skin-tones, sky, and foliage.

Eight scenes were assessed in reflection prints by 104 consumers using magnitude estimation. In a second session, the same images were displayed on a monitor and were assessed by eighteen observers using the softcopy quality ruler. The mean logarithms of the magnitude estimates for each level were linearly regressed against the mean JND differences from the quality ruler to allow transformation of the magnitude estimates into JNDs. The data from the two sessions were then combined, yielding 8·(104 + 18) = 976 assessments for each of the 10 levels.

Predictions of the expected quality were made as described below.

1. The values of the objective metrics derived in earlier studies of each attribute in isolation were calculated for each experimental level.

2. Each of the eight scenes was classified in terms of anticipated susceptibility to each attribute.

3. The corresponding IHIF regression parameter values were used to predict JNDs of quality change arising from each attribute in each scene. If a scene did not contain a memory color, the quality change associated with that memory color was set to zero.

4. The predicted quality changes arising from each attribute were then combined using the multivariate formalism.

The predicted results and the measurements, with 95% confidence intervals, are compared in Fig. 21.5. The 45° line is a direct prediction, not a fit; this figure depicts an absolute comparison. The predictions of the overall color and tone reproduction quality average ≈0.3 JNDs high; the largest discrepancy is ≈1.3 JNDs. This level of agreement is excellent given the complexity of the stimuli variations, the number of perceptual attributes involved, and the preferential nature of those attributes. Particularly considering that the objective metrics and regression parameters were defined in independent experiments using different scenes, observers, and image simulation techniques, the comparison of Fig. 21.5 provides a compelling confirmation of the ability to predict perceived quality even for complex, preferential attributes.

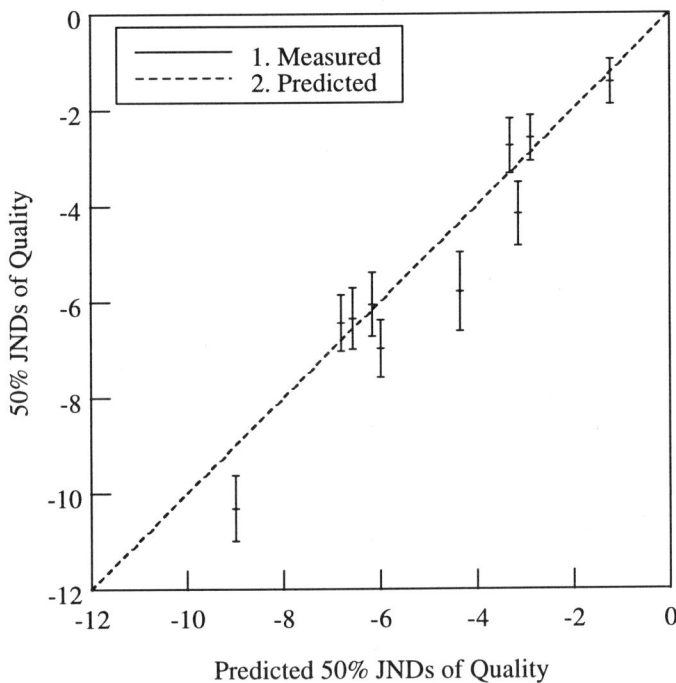

Fig. 21.5 Comparison of measured and predicted overall color and tone reproduction quality for samples varying in colorfulness, contrast, memory color reproduction, and detail in saturated colors. The predictions of the multivariate formalism using results of earlier studies of single attributes are on average only ≈0.3 JNDs in error.

21.5 Summary

Although it is not always apparent whether image quality effects are better modeled in objective space, through the extension of objective metrics, or in perceptual space, by operating directly on JNDs of quality, a few guidelines may be suggested. If attributes appear to be perceptually dependent, revision of their definition and corresponding modification of their associated objective metrics is recommended. Lower level perceptual effects such as masking, which affect the visibility of an attribute, should usually be handled in this fashion. Higher level cognitive effects, such as those that influence the importance attached to different compositional elements or attributes of an image, are probably better treated in perceptual space.

One example of a transform in perceptual space pertains to the effect that human subject prominence in an image has on the impact of skin-tone reproduction on the quality of that image. At small head sizes, the impact is approximately proportional to head size, but an asymptotic limit is approached as the head height in the image exceeds roughly 25% of the geometric mean image dimension. A significant advantage of characterizing the effect of skin-tone reproduction on quality as an explicit function of head size (rather than simply determining the relationship for a single head size distribution) is that new predictions may be made for different or evolving head size distributions.

A second example of a perceptual space transform involves the impact that the presence of a critical reference image can have on the perceived quality of a less preferred color and tone reproduction position. Subtle deviations from a preferred position that might otherwise not be noticed may become evident when the image is directly compared with an image depicting the same scene at the preferred position. This is in contrast to the response to artifactual attributes, which is independent of the nature of the available reference images. Perhaps a reference image is not needed in the latter case because artifacts are unnatural in appearance, so that an observer may readily visualize how an image would appear in their absence.

The final example of this chapter relates to the application of the multivariate formalism to color and tone reproduction quality. Color attributes, tonal attributes, and other attributes (such as those associated with spatial artifacts) usually appear to be approximately mutually orthogonal when defined in an intuitive fashion; however, delineation of independent attributes within the classification of color quality or tonal quality may be more challenging. Nonetheless, the multivariate formalism has been applied successfully, substantiating the validity of this approach in treating preferential attributes.

Part III

Modeling System Quality

Part III first discusses mathematical models used to propagate tone, color, signal fidelity, and noise through imaging systems (Ch. 22). Definition of measurement protocols, and use of parametric models for estimation of measurement results from engineering specifications, are considered next (Chs. 23–24). Following the description of a software implementation of an image quality model (Ch. 25), a number of examples of capability (peak quality) analyses, used to set component specifications, are provided (Ch. 26). Providing a transition towards full performance assessment, yield metrics based on photospace coverage are described next (Ch. 27). Following an explanation of the use of Monte Carlo modeling to generate quality distributions (Ch. 28), concepts useful in interpreting such distributions are developed (Ch. 29). Finally, after presentation of a number of examples of performance analyses based on this type of modeling (Ch. 30), the accuracy of such predictions is demonstrated (Ch. 31).

22

Propagating Key Measures through Imaging Systems

22.1 Introduction

Part III of this book is concerned with the construction and use of system image quality models. The goal of such models is to make predictions of the final image quality produced by an imaging system that is described in terms of component or subsystem properties. These models are valuable for purposes such as: (1) setting product specifications; (2) forecasting technological trends for strategic planning; (3) developing novel imaging systems; and (4) benchmarking products and technologies.

The term objective measure, defined in Sect. 12.4, refers to quantities such as the modulation transfer function (MTF) and the noise power spectrum (NPS), which are functions of one or more variables such as spatial frequency, exposure, image density, etc. Because objective metrics are calculated from objective measures of the final viewed image, mathematical methods for propagating key objective measures through imaging systems are needed to predict final image quality from component and subsystem properties. Such methods are the subject of the present chapter, which is organized as follows. Section 22.2 defines the terms system, subsystem, and component in preparation for describing propagation models. Section 22.3 describes the nature of four key objective measures: (1) mean channel signal; (2) image wavelength spectrum; (3) modulation transfer function; and (4) noise power spectrum. Sections 22.4–22.7, respectively, discuss the propagation of these four quantities through imaging systems.

22.2 Systems, Subsystems, and Components

Before discussing propagation of objective measures through an imaging system, the use of the terms imaging system, subsystem, and component must be clarified. An imaging system refers to a complete imaging chain, usually initiated by the photographic capture of an image and generally terminating with the human observation of a final image. The term component refers to a contiguous portion of an imaging system that has input and output corresponding to analog or digital images, or to the original scene.

A simple example of a system would be a digital still camera that is used to capture a scene, and a video monitor that displays the image. The camera and monitor are considered components. The input to the camera is the scene itself; its output is a digital representation of the image of the scene. The latter in turn is the input to the monitor, the output of which is an analog image. The camera can also be considered to be comprised of several components, such as the lens, which forms a real image of the scene at the sensor plane, and the sensor and readout electronics, which convert the analog image to a digital representation thereof. Thus, parsing an imaging system into components may be somewhat arbitrary and depends, in part, on the application. We will usually equate a component with what would be regarded as one piece of equipment, or a particular medium such as film, or one mathematical image processing algorithm.

The term subsystem refers to a set of contiguous components that are grouped together for some purpose, usually a measurement process. For example, a digital output device and its output medium, which might be paper or transparency material, may be treated as distinct components, but usually their combined MTF is measured by microdensitometry of a target written by the device onto the medium. We refer to such a determination as a composite or subsystem measurement. In the discussion that follows, whenever the term component is used, the phrase "and/or subsystem" is implicitly appended.

In general, a photographic imaging system may be regarded as converting the analog stimulus of a scene to an analog response displayed in the final image. In contrast, in synthetic imaging systems, such as those used for animation and special effects, there may be no original scene. In either case, the final image must be in analog form (i.e., rendered in light) to be viewed by an observer. Therefore, a photographic system having one or more digital components will contain at least one transformation from an analog signal to a digital signal, and an equal number of stages where the digital signal is converted back to an analog signal.

22.3 Key Objective Measures in Imaging Systems

Although a large number of artifactual attributes affect imaging system quality, many of these artifacts arise from single components in the imaging system and:

1. are relatively unaffected by other component properties, and so require no propagation model; or

2. are primarily geometrically magnified by passage through the imaging system, and so are propagated according to elementary optical relationships.

An example of the first type of artifact is the degree of tilt in the framing of a photograph, which if excessive, can lead to noticeably slanted horizons. Other than the possibility of digital correction through a rotation procedure, no system component subsequent to the camera affects this degree of tilt and so no propagation model is needed. Misregistration is an example of an artifact of the second type, which is affected primarily by magnification.

Nonetheless, many artifactual and preferential attributes are characterized by objective metrics deriving from one of four key objective measures, each of which typically is modified by many of an imaging system's components, and therefore requires a comprehensive propagation model. These four measures are:

1. mean channel signal (MCS), which characterizes the macroscopic signal in each color channel as a function of original scene exposure;

2. image wavelength spectrum (IWS), which characterizes the spectral properties of an analog image at particular MCS values;

3. MTF, which characterizes the signal's spatial frequency response; and

4. NPS, which characterizes various types of noise (Sect. 13.2).

A summary of selected properties of these four objective measures is given in Table 22.1 (next page). These quantities are referred to as key measures because so many objective metrics may be calculated from them, as shown in the fifth column of the table. Each key measure is discussed in turn following the table.

Key Objective Measure	Acronym and Symbol	Units	Property Described	Objective Metrics Predicted	Targets Measured	Dependencies
mean channel signal	MCS $[\mu]$	various (code value, luminance, density, etc.)	signal in each color channel	tonal clipping, contouring, contrast, density balance	color/neutral patches	IWS; N-channel exposure at scene capture
image wavelength spectrum	IWS $\phi(\lambda)$	transmittance, reflectance, or radiometry vs. wavelength (nm)	radiometric spectrum of analog image	color balance, colorfulness, memory color reproduction, detail in saturated colors	color/neutral patches	MCS, colorant or emitter wavelength spectrum
modulation transfer function	MTF $[M(v)]$	modulation transfer vs. spatial frequency (cycles/mm)	spatial frequency response of signal	unsharpness, oversharpening, reconstruction infidelity, aliasing, CFA artifacts	slanted edge, sine wave, square wave	MCS, field position, orientation
noise power spectrum	NPS $[W(v)]$	spectral density (mm^2) vs. spatial frequency (cycles/mm)	spatial frequency content of noise	isotropic noise, streaking, banding	large, uniform fields	MCS, field position, orientation

Table 22.1 Summary of selected properties of key objective measures.

Propagating Key Measures through Imaging Systems 327

The mean channel response (MCS) characterizes the macroscopic signal level in each of N channels that results from a camera exposure of some stimulus such as a neutral or color patch. Because human vision is trichromatic, most color imaging systems have three channels, usually corresponding roughly to red, green, and blue colors; however, rotation into luminance-chrominance color spaces is advantageous for certain image processing operations (see Sect. 24.3), and hardcopy output devices often use four or more colorants. The signal in each channel may be specified in any of a number of units, including density, code value, or luminance (for an emissive display such as a video monitor).

To fully characterize the behavior of a system, the MCS must be known for a sufficient number of stimuli that the signal resulting from all possible initial N-channel scene exposures can be estimated by interpolation. This is usually achieved by measuring the response to an extensive series of neutral and color patches, captured at various camera exposures to span the full dynamic range of the system. The MCS arising from one N-channel exposure stimulus is denoted by $[\mu]$, and is a column vector of dimension $N \times 1$. To avoid cumbersome notation, the dependence of MCS on N-channel scene exposure is not explicitly shown in the equations of this chapter, but this and other implicit dependencies are listed in the rightmost column of Table 22.1. The MCS derived from a series of neutral scene exposures provides the basis for many objective metrics related to tone reproduction.

At an analog stage of a system, there is an image wavelength spectrum (IWS) associated with each MCS vector. The IWS may be measured from the neutral and color patches used in MCS characterization or it may be calculated from properties of a system component. It is generally possible to predict the MCS from the IWS and the IWS may usually be reconstructed given the MCS and a knowledge of basic spectral properties of an analog component. As an example of the former (IWS $\rightarrow$ MCS), if the reflectance spectrum of a print is known, its ISO Status A RGB densities may easily be computed. As an example of the latter (MCS $\rightarrow$ IWS), if the phosphor wavelength spectra and the luminance produced by each channel of a video monitor are known, the radiometry of the image may be predicted. As discussed in Sect. 22.4, the IWS, denoted by $\phi(\lambda)$, determines or affects the radiometric stimulus to the next system component or, in the case of a final image, to the observer's visual system. The IWS of the final image forms the basis for many objective metrics reflecting color reproduction quality.

MTF is a measurement of response to spatially varying stimuli and characterizes fidelity of a signal as a function of frequency. Classically, sinusoids of different frequencies are imaged and the modulation transfer at the frequency of each sinusoid is calculated by taking the ratio of the peak-to-peak output modulation to the peak-to-peak input modulation, expressed in the same units. Many other target patterns may alternatively be used; slanted edges (ISO 12233, 1998) are often particularly convenient in digital systems. MTF is a normalized measurement in that a ratio of output to input modulations is taken, and for most spatial processes the modulation transfer approaches one as the frequency approaches zero. At zero frequency, the signal transfer characteristics of a system component are usually determined primarily by the propagation of MCS transfer (which is based on uniform, spatially extensive stimuli and so does not reflect spatial fidelity), whereas at higher frequencies, both MCS propagation and MTF affect the transfer. The MTF at a given frequency, MCS, field position, and orientation is denoted $[M(\nu)]$ and is an N × 1 column vector. Image attributes discussed previously that are largely predictable based on MTF include sharpness, oversharpening, aliasing, and reconstruction infidelity.

NPS is a spatial frequency-dependent measurement of the variation in signal found in an initially uniform field, which, in the absence of noise, would have constant signal. If the neutral and color patch targets used to characterize MCS are large enough and sufficiently uniform on a microscopic level, they may be used to measure NPS. NPS units are inverse spatial frequency squared because the NPS quantifies the unitless spectral density per area of the two-dimensional spatial frequency plane. If spatial frequency is measured in cycles/mm, the NPS units will therefore be mm^2. Image quality attributes that can be characterized via NPS include isotropic noise, streaking, and banding. In addition, aliasing and reconstruction infidelity, treated as signal distortions and mathematically described in terms of MTF in Ch. 18, can alternatively be represented as types of noise, and quantified using NPS.

The NPS at a given frequency, MCS, field position, and orientation is denoted $[W(\nu)]$ and is a square N × N matrix. The diagonal elements of this matrix are the familiar single-channel NPS. The off-diagonal elements are cross-spectra that measure the power of noise that is correlated between two channels (Bendat and Piersol, 1980). The matrix element in the ith row and j^{th} column, denoted $[W(\nu)]_{ij}$, quantifies the correlated noise between the ith and j^{th} channels. Unless the channels are misregistered, $[W(\nu)]_{ij} = [W(\nu)]_{ji}$ and so $[W(\nu)]$ is a symmetric matrix. Often, the noise is initially uncorrelated at its source, and so $[W(\nu)]$ is diagonal; however, the noise may become partially correlated as a result of subsequent channel interaction in the imaging system, which may arise from image processing operations (such as color rotations) or from optical exposures.

Propagation of these four key measures is described in detail in the following four sections.

22.4 Propagation of Mean Channel Signal

The propagation of MCS through an imaging system may be treated as a series of transformation functions $f(\mu)$ that map the MCS values from one component to the next, as shown in Eq. 22.1.

$$[\mu_i] = f([\mu_{i-1}]) \qquad (22.1)$$

This equation states that the MCS of the ith component is a transformation of the MCS of the previous component. The nature of the transformation depends in part upon whether the $(i-1)^{th}$ and i^{th} components are analog or digital in nature. Examples of transformation functions or principal constituents thereof, for each of the four possible combinations of digital and analog components, are given in Table 22.2 and are discussed below.

Transformation	Example	Input	Output
digital to digital	lookup table	code value	code value
analog to digital	opto-electronic conversion function	exposure	code value
digital to analog	video grayscale characteristic function	voltage ($\propto$ code value)	luminance
analog to analog	sensitometric H&D curve	$\log_{10}$ (exposure)	density

Table 22.2 Examples of transformation functions (f in Eq. 22.1) of mean channel signal from one component to another.

An example of a digital to digital transformation is a three-dimensional lookup table, which is a tabulation of three color channel input code values and their corresponding three output code values. The lookup table usually contains enough mappings to allow satisfactory interpolation between the listed input-output pairs. An example of an analog to digital transform is the opto-electronic conversion function (OECF; ISO 14524, 1997), which relates output code value to input exposure in a digital still camera. A similar function based on scanned image density can be used to characterize the analog to digital transform of a scanner. The grayscale characteristic function of a video monitor, giving output luminance as a function of input voltage (which is proportional to code value), is an example of a digital to analog transform. This relationship often takes the form of a power function, the exponent of which is the monitor gamma. Finally, an analog-to-analog transformation usually involves an optical exposure. In a black-and-white (single-channel) system, the results of this process can be predicted from the sensitometry (H&D, D log E, or D log H curve) of the receiving material. In color systems, additional factors must be accounted for, as discussed in the next section.

Signal transformations affect both the MTF and the NPS of the system and so their effect must be included in the propagation models of these two key measures; however, a simple mathematical shift in the numerical scale of the MCS has no impact on MTF or NPS. MTF is based on modulation ratios, and modulations are signal differences, so shifts are cancelled. Similarly, noise is a deviation of signal from a mean, and the deviations are differences and so are unaffected by a signal shift. The effect on signal differences, and thus on MTF and NPS, arising from the transformation to the ith component, may be quantified by an $N \times N$ matrix $[J_i]$ having elements in the jth row and k^{th} column that are the partial derivatives of the output MCS with respect to the input MCS.

$$[J_i]_{jk} = \frac{\partial [\mu_i]_j}{\partial [\mu_{i-1}]_k} \qquad (22.2)$$

This matrix is the Jacobian matrix of the inverse transformation, $f^{-1}(\mu)$ that would map from the ith component back to the preceding $(i-1)^{th}$ component (Searle, 1982). Accordingly, $[J_i]$ will be referred to as the inverse Jacobian matrix. If a column vector of small MCS differences $[\delta \mu]$ in the $(i-1)^{th}$ component were left-multiplied by this matrix, a column vector of the corresponding MCS differences in the ith component would result.

$$[\delta \mu_i] = [J_i][\delta \mu_{i-1}] \qquad (22.3)$$

This result may be confirmed by expanding the matrix multiplication; e.g., the first row (channel) would yield the following expansion.

$$[\delta \mu_i]_1 = \frac{\partial [\mu_i]_1}{\partial [\mu_{i-1}]_1} \cdot [\delta \mu_{i-1}]_1 + \frac{\partial [\mu_i]_1}{\partial [\mu_{i-1}]_2} \cdot [\delta \mu_{i-1}]_2 + \frac{\partial [\mu_i]_1}{\partial [\mu_{i-1}]_3} \cdot [\delta \mu_{i-1}]_3 \qquad (22.4)$$

Eq. 22.4 is the anticipated result; the overall change is simply the sum of changes arising from each channel dependency.

As one example, suppose a simple digital tone scale stretching operation were performed to increase the contrast of an image. The transformation might consist of subtracting the mean code value in each image channel from the individual pixel values in that channel, scaling these differences by 1.3× to amplify contrast, and adding the mean code value back to each channel. In this case, the inverse Jacobian matrix would have diagonal elements of 1.3 and off-diagonal elements of zero. The matrix would be diagonal because no channel interaction occurs in the transformation.

As a second example, suppose instead that a crude digital color boosting operation were performed in which, at each pixel, the average of the three MCS code values were subtracted from each code value, the difference were scaled by 1.3× to amplify color differences from a neutral, and the average code value were added back in. At a given pixel, the red output signal $[\mu_i]_R$ would depend on the RGB input signals $[\mu_{i-1}]_R$, $[\mu_{i-1}]_G$, and $[\mu_{i-1}]_B$ as shown in Eq. 22.5.

$$[\mu_i]_R = 1.3 \cdot \left\{[\mu_{i-1}]_R - \left(\frac{[\mu_{i-1}]_R + [\mu_{i-1}]_G + [\mu_{i-1}]_B}{3}\right)\right\} + \left(\frac{[\mu_{i-1}]_R + [\mu_{i-1}]_G + [\mu_{i-1}]_B}{3}\right) \quad (22.5)$$

Taking the derivative of $[\mu_i]_R$ with respect to $[\mu_{i-1}]_R$ yields 1.2, whereas the derivative of $[\mu_i]_R$ with respect to $[\mu_{i-1}]_G$ or $[\mu_{i-1}]_B$ is −0.1. Similar results are obtained in the other channels because of symmetry, giving the following inverse Jacobian matrix.

$$[J] = \begin{bmatrix} 1.2 & -0.1 & -0.1 \\ -0.1 & 1.2 & -0.1 \\ -0.1 & -0.1 & 1.2 \end{bmatrix} \quad (22.6)$$

The interaction indicated by the non-zero off-diagonal elements is caused by the use of one channel in computing the transformed signal of another channel.

The inverse Jacobian matrix will be used in the MTF and NPS propagation models described in Sects. 22.6–22.7.

22.5 Propagation of Image Wavelength Spectra

The IWS of the final image forms the basis for most objective metrics of color reproduction attributes. With the viewing illuminant's wavelength spectrum and

CIE color matching functions, the IWS may be used to calculate tristimulus values and the many quantities derived from them, such as CIELAB values. The IWS associated with earlier components in an imaging system are also important, because they control the N-channel exposures to subsequent components, and, therefore, are a critical determinant of the MCS transformations from an analog component to the next component, whether it is analog or digital.

In this section, the MCS transformation associated with an optical exposure and particularly its relationship to the IWS will be considered in some detail, to introduce several useful concepts and to clarify further the nature of the propagation of both MCS and IWS through imaging systems. To make the discussion a bit less abstract, the specific case of color motion picture film printing will be considered.

The motion picture film that is projected in a theater is referred to as a release print, even though it is a transparency, not a reflection print. A typical pathway by which such a release film is made is as follows. The original scene is captured on a camera negative film, which is contact-printed or optically printed at unit magnification onto the first stage of intermediate film. Intermediate film is a negative material having unit gradient, so a second stage of printing onto the same intermediate film stock produces a near duplicate of the camera negative. If special effects are to be incorporated into the images, two more stages of intermediate film printing may occur, during which images from multiple sources may be combined. The duplicate negative on intermediate film is finally printed onto a negative print film having a high gradient to produce the release print for projection in a dark surround. Each stage of printing offers flexibility and allows the number of copies to increase exponentially without excessive handling of any given film, especially the valuable original negative.

The printing stage we will consider is the final one, in which a duplicate negative on intermediate film (hereafter, the negative) is printed onto print film (hereafter, the print). The MCS of the negative is usually specified in terms of red, green, and blue (RGB) ISO Status M densities, which are intended for use with materials that are printed. The MCS of the print is instead expressed as ISO Status A densities, which are intended for use with directly viewed materials. The MCS transformation $f([\mu])$ must therefore map the Status M densities on the negative to the Status A densities on the print. This transformation may be modeled by the following steps.

1. The IWS of the negative is generated from the negative densities and the wavelength spectra of the image dyes and the film base.

2. The channel-specific exposures to the print film are computed based on the negative IWS, the print film spectral sensitivity, the printing illuminant radiometry, and flare, as discussed further below.

3. The initial estimate of the resulting print densities are made by mapping the logarithms of the channel-specific exposures through the sensitometric (H&D, D log E, or D log H) curve of the print film; and

4. The initial estimates of print densities are modified to account for interlayer interimage effects in the print film, in which development in one color record affects development in other records, preferably in a fashion that amplifies color differences.

A detailed discussion of all these steps is beyond the scope of this chapter; the reader is referred to Giorgianni and Madden (1998) for more information. To demonstrate the propagation of the IWS, however, the second step, involving optical exposure, will be considered in detail.

Exposure of the print film arises from two light sources: (1) the intended exposing light, in this case from the optical or contact printer lamp, denoted by $I(\lambda)$; and (2) unintended ambient flare, e.g., from light leaks, denoted $I_a(\lambda)$. Ambient light exposure is normally negligible in printing processes but is included for completeness because it is of significance in certain viewing circumstances, and the equations developed in this section can be applied with minor modification to camera exposure and final image viewing as well as to printing processes. An example of ambient light in a viewing environment is the dim aisle and exit lighting in motion picture theatres, provided to permit safe movement during projection. Both light sources are characterized by the energy they deliver per unit wavelength and per unit area on the print film during the exposure, and so have units of ergs/(nm·cm^2) or similar quantities.

Light arising from the intended source and exposing the print film may be categorized into three types.

1. Light that passes through the negative film as intended will be referred to as imaging light. It is spectrally modified by the negative film IWS, denoted $\phi(\lambda)$, which is its fractional transmittance.

2. Light that passes through the negative film, but is scattered, so that its spatial position is randomized will be called veiling flare (this term is sometimes defined more broadly to mean any light from any source that produces a uniform, undesired exposure). Such scattering events

usually originate at surfaces where the index of refraction changes, e.g., at air-glass or air-film interfaces; after reflection off a few surfaces, the spatial position becomes nearly random. The fraction of veiling flare light is denoted by ϕ_v. Because the light from all parts of the image is spatially randomized, the veiling flare is proportional to the average transmittance of the entire image.

3. Light that does not pass through the negative film but does reach the print film will be referred to as stray light (again, this term is often defined more broadly). In many instances, this light is negligible in amount, but under certain viewing conditions it may be of significance. For example, the light that reflects off the front surface of a reflection print, and therefore does not penetrate into the layers containing the colorant, is a type of stray light that significantly reduces the perceived visual density of dark areas in the print. The fraction of stray light is denoted by ϕ_s.

Combining all the effects mentioned above, the total exposing light radiance $I_e(\lambda)$ in ergs/(nm·cm^2) may be expressed as:

$$\begin{aligned} I_e(\lambda) &= I_a(\lambda) + I(\lambda) \cdot \left\{ \phi_s + \phi_v \cdot \overline{\phi(\lambda)} + (1-\phi_v) \cdot \phi(\lambda) \right\} \\ &= I_a(\lambda) + I(\lambda) \cdot \phi'(\lambda) \end{aligned} \quad (22.7)$$

where $\phi(\lambda)$ is the negative film IWS (fractional transmittance), $\overline{\phi(\lambda)}$ is the mean IWS value over the entire image, $I(\lambda)$ is the intended exposing light, $I_a(\lambda)$ is the ambient flare, ϕ_s is the fraction of stray light, ϕ_v is the fraction of veiling flare, and $\phi'(\lambda)$ is defined to be the effective IWS of the negative, accounting for the effects of veiling flare and stray light.

The exposure in lux-seconds to the jth channel of the ith component (the print film), denoted $[E_i]_j$, is now given by:

$$[E_i]_j = \frac{E_0 \cdot \int (I_a(\lambda) + I(\lambda) \cdot \phi'_{i-1}(\lambda)) \cdot \Psi_{i,j}(\lambda) \cdot d\lambda}{\int I(\lambda) \cdot \Psi_{i,j}(\lambda) \cdot d\lambda} \quad (22.8)$$

where $\phi'_{i-1}(\lambda)$ is the effective IWS of the preceding component (the negative), $\Psi_{i,j}(\lambda)$ is the spectral sensitivity of the jth channel of the print film in cm^2/erg, and E_0 is exposure in lux·s that would have occurred if there were no flare and if the negative did not attenuate the light passing through it at all (i.e., had zero density). The spectral sensitivity of a component is measured by determining how much energy per unit area is required in a narrow-band exposure to produce a specified response, such as a certain amount of density above the minimum density of the material.

Equations 22.7 and 22.8 characterize the process of an optical exposure and demonstrate the influence of the IWS on MCS propagation. With minor modifications, these equations may also be used to understand the results of camera exposures, instrumental measurements, and the viewing of images. For example, consider the case in which the RGB MTFs of a final viewed image are quite different from one another, perhaps because one system component suffered from significant chromatic aberration. These MTFs must be weighted in some fashion to produce a single objective metric value of unsharpness, which can then be converted to JNDs of quality through an IHIF regression.

This weighting may be accomplished as follows. In Eq. 22.8, the spectral sensitivities of the exposed medium are replaced by CIE color matching functions, and the exposing light radiometry is replaced by the viewing illuminant radiometry. The channel "exposures" $[E_i]_j$ are now proportional to tristimulus values, and because only their relative magnitudes will be used in this analysis, the value of E_0 is arbitrary. The inverse Jacobian matrix mapping the Status A densities of the print to the visual channel exposures is numerically determined via Eq. 22.2 for a neutral scene mid-tone and other scene exposures if desired. If the color matching functions were numbered from long to short peak wavelength, the second function is the photopic curve, and the second signal in the MCS vector is proportional to luminance. Consequently, the middle row of the inverse Jacobian matrix may be said to relate to the achromatic channel of the human visual system.

The perception of image structure is largely determined by the achromatic channel response. One indication of this behavior was discussed in Sect. 15.5, where it was shown that the visibility of noise as a function of image density could be approximately predicted from the CIE lightness equation. A demonstration of analogous behavior in the spectral domain will now be provided. The relative magnitudes of the three elements of the middle row of the inverse Jacobian indicate the influence of each channel of the image on the achromatic visual channel, and so, if normalized to sum to unity, might be taken to be visual weights of the three image channels. This argument is made more generally in the next section, where the propagation of MTF is shown to depend on the row-sum-normalized inverse Jacobian matrix.

Figure 22.1 demonstrates how well these visual weights predict perceived sharpness. Observers assessed images produced by digital simulation processes having dramatically different MTFs in the three color channels. The measured Status A RGB MTFs were combined into a visual MTF using the calculated visual weights, and the visual MTF was used to compute an objective metric of unsharpness. In Fig. 22.1, the observed quality changes are shown as open circles, and the values predicted from the visual weights are shown as a straight line. No discrepancies greater than one-quarter JND are observed, substantiating that the perception of sharpness is largely determined by the achromatic visual channel. We have obtained similar results with perceived noise, and so this conclusion probably applies to other image structure attributes as well.

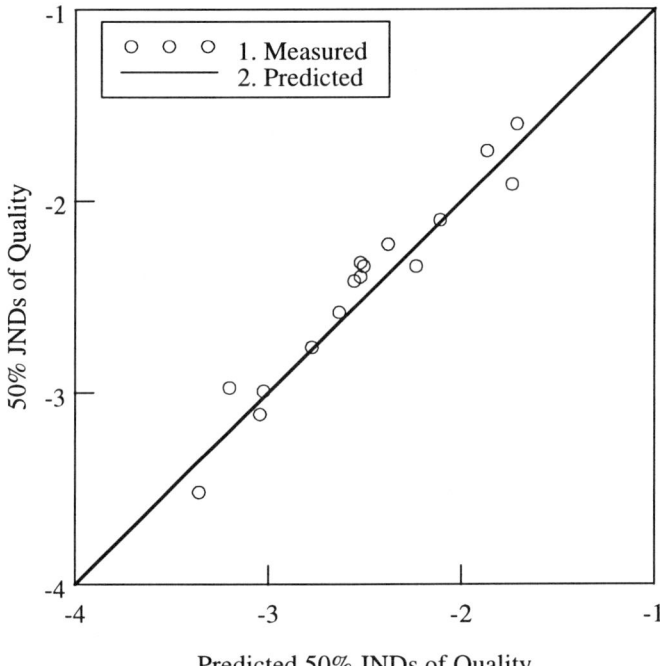

Fig. 22.1 Predicted versus assessed quality arising from sharpness in samples having very different red, green, and blue MTFs. The agreement is excellent, supporting the assumption that the achromatic (luminance) channel of the human visual system largely determines perception of image structure attributes.

22.6 Propagation of Modulation Transfer Functions

In a linear system, the system MTF can be quantitatively predicted by cascading the component MTFs, i.e., multiplying them together frequency by frequency (Dainty and Shaw, 1974). Although imaging systems often contain nonlinear processes, such as image compression, they are nearly always approximated as being linear systems for modeling purposes. This approach has been very successful in the prediction of image quality, although it is inadequate to explain precise image appearance.

Before MTFs can be cascaded, they must be mapped to the same frequency plane. For example, if the MTF of a digital camera is measured in cycles/mm in the sensor plane, and the MTF of a digital writer is measured in cycles/mm at the print plane, the two MTFs cannot be cascaded directly unless the print is the same size as the sensor, which would be very unlikely (most sensors are much smaller than even wallet-sized prints). The ratio of frequencies in two planes depends upon the printing magnification between the planes. For example, if a 4 × 6-inch print (the standard 4R print size in conventional photofinishing) were made from a sensor 0.5 × 0.75 inches in size, the printing magnification would be approximately 8×. Thus, a spatial frequency of 40 cycles/mm in the sensor would map to 5 cycles/mm in the print, because the printing magnification leads to fewer cycles fitting in a given distance. In general, frequencies v following a printing step with magnification m_p are related to frequencies v_{i-1} prior to the printing step by Eq. 22.9.

$$v_{i-1} = v \cdot m_p \qquad (22.9)$$

The MTF of digital image processing operations, such as spatial filtering, are usually expressed as a function of frequency in cycles/sample units. Care must be taken to account for all interpolations and down-sampling in converting these frequency units to a common frequency plane. Interpolation increases the number of samples per cycle and thus decreases cycles per sample, so that the frequencies before and after a 2× interpolation are related in the same way as if a 2× printing magnification were applied.

Just as MTFs must be mapped to the same frequency plane prior to cascading, they must also be transformed into the same MCS space before they are combined if the interceding transformation involves channel interaction, as discussed in Sect. 22.4. Equation 22.3 showed that the effect of a signal transformation on a column vector of small differences in MCS could be calculated by matrix multiplying the inverse Jacobian matrix by the MCS difference vector. Small modulations should transfer in a similar fashion, but when their ratio is taken, any overall scaling of the signal will be canceled out. This effect may be accounted for by normalizing the sum of each of the rows of the inverse Jacobian matrix to equal unity. In this way, the transformed MTF $[M'(v)]$ will have unit response at zero frequency, as normally desired. Thus:

$$[M'(v)] = [J'][M(v)] \qquad (22.10)$$

where $[J']$ is a row-sum-normalized inverse Jacobian matrix. In certain cases involving image differences, a row of $[J]$ will sum to zero and such a normalization will not be possible, but a subsequent transformation will generally compensate and restore proper behavior.

In a linear system, the cumulative system MTF through a given component, $[M_{s,i}(v)]$, is equal to the cumulative system MTF through the previous component, $[M_{s,i-1}(v)]$ (suitably transformed with respect to frequency and MCS space), cascaded with the current component MTF, $[M_{c,i}(v)]$. Combining this relationship and Eqs. 22.9 and 22.10, this result may be compactly written as:

$$[M_{s,i}(v)] = [\tau'_i(v)][M_{s,i-1}(v \cdot m_p)] \qquad (22.11)$$

where the elements of the normalized transfer matrix, $[\tau'_i(v)]$ are defined by Eq. 22.12.

$$[\tau'_i(v)]_{jk} = \frac{[J_i]_{jk} \cdot [M_{c,i}(v)]_j}{\sum_k [J_i]_{jk}} \qquad (22.12)$$

The inverse Jacobian matrix $[J]$ is given by Eq. 22.2. The sum in the denominator of Eq. 22.12 row-sum-normalizes the elements of $[J]$. Equations 22.2, 22.11, and 22.12 constitute a propagation model for MTF through an N-channel, linear system.

To predict perceived attributes in viewed images, the final image MTF (and NPS) are usually converted to cycles/degree at the eye for comparison with the contrast sensitivity function of the human visual system. Consequently, in our system modeling, we initially convert all MTFs (and NPS) to cycles/degree at the eye so that they may be cascaded without further magnification bookkeeping. This has the added advantage that frequencies higher than those discerned by the human visual system are easily excluded from the calculations.

22.7 Propagation of Noise Power Spectra

Noise propagation through imaging systems is affected by:

1. component MTFs;

2. printing magnification; and

3. mean channel signal transformations.

These effects will each now be described.

If a given amount of noise at some stage in a system is passed through a component that introduces blur, the noise pattern will be blurred just as the image signal is, and the amount of noise will be reduced. Specifically, because MTF is a modulation or amplitude ratio, the noise power at a given frequency, being an amplitude squared, will be shaped by the square of the component MTF at that frequency (Dainty and Shaw, 1978).

Next, consider the effect of printing magnification on NPS propagation. First, the frequency axis of the NPS is scaled just like that of the MTF (Eq. 22.9), and for the same reasons. Second, because the NPS has units of power per unit frequency squared (i.e., the power in a small area of two-dimensional frequency space), the noise power scales with the square of the printing magnification. For example, if the printing magnification were 8×, the power originally in the range 8–16 cycles/mm would be compressed into the new frequency range of 1–2 cycles/mm. This compression reduces the two-dimensional frequency space area by a factor of $(16 - 8)^2/(2 - 1)^2 = 64×$. Because the same power has been compressed into a smaller frequency area, the power per unit frequency area increases accordingly, i.e., by a factor of m_p^2.

Finally, consider the effect of MCS transformations on NPS propagation. The effect here is very analogous to that with MTF except that, whereas MTF is a ratio of modulations and so does not change with a rescaling of signal, NPS are based on squared deviations from a mean signal, and so are rescaled in proportion to the square of the signal. Therefore, the row-sum normalization of the inverse Jacobian matrix that was required in the case of MTF should not be included when propagating NPS. Anticipating its usefulness by analogy with Eq. 22.12, an unnormalized transfer matrix $[\tau_i(v)]$ is defined in Eq. 22.13.

$$[\tau_i(v)]_{jk} = [J_i]_{jk} \cdot [M_{c,i}(v)]_j \qquad (22.13)$$

Before formulating the N-channel propagation model, consider the case of a black-and-white (single-channel) analog system in which a film is being printed onto a paper. The inverse Jacobian matrix would have a single element corresponding to the derivative of reflection density on the paper with respect to the transmission density of the negative. Because the logarithm of exposure to the paper is equal to the logarithm of exposure that would occur if the negative had zero density, minus the density of the negative, the aforementioned derivative is simply the negative of the slope of the sensitometric (H&D, D log E, or D log H) curve. This slope is called the sensitometric gradient of the paper, and is denoted by γ. Combining the effects discussed above (gradient, MTF, and printing magnification) into a single equation, and assuming that any new noise power added by a component in a system is independent of noise already propagating through the system, yields:

$$W_{s,i}(v) = W_{c,i}(v) + m_p^2 \cdot \gamma_i^2 \cdot M_{c,i}^2(v) \cdot W_{s,i-1}(v \cdot m_p) \qquad (22.14)$$

where $W_{s,i}(v)$ is the cumulative system NPS through the current (i^{th}) component (the paper in this example), $W_{c,i}(v)$ is the current component NPS, $W_{s,i-1}(v)$ is the cumulative system NPS through the previous component (the negative, in this example), m_p is the printing magnification from the previous to current component (negative to print), $M_{c,i}(v)$ is the current component MTF, and γ_i is the sensitometric gradient of the current component (the paper gamma in this case). This equation was first published by Doerner (1962) and is known as the Doerner equation.

Extension of the Doerner equation to N-channel systems is complicated because of correlation effects. The task of keeping track of noise correlations arising from multiple noise sources, modifying them by appropriate MTFs, and accounting for channel interactions, seems quite daunting. Fortunately, a remarkably simple matrix formulation (Bendat and Piersol, 1980) makes the bookkeeping quite feasible. Using the transfer matrix of Eq. 22.13 and the form of Eq. 22.14 allows the relationship to be written very compactly:

$$[W_{s,i}(v)] = [W_{c,i}(v)] + m_p^2 \cdot [\tau_i(v)][W_{s,i-1}(v \cdot m_p)][\tau_i(v)]^T \qquad (22.15)$$

where $[\ ]^T$ denotes a matrix transpose. Equations 22.2, 22.13, and 22.15 constitute a propagation model for NPS through an N-channel, linear system.

Because NPS propagation depends directly on MTF and MCS propagation, and because the latter depends on IWS propagation, validation of the NPS propagation model also tests the accuracy of the other three propagation models. Figure 22.2 shows such a verification; the solid line represents the predicted values of print grain index (Kodak Publication E-58, 1994; PIMA IT2.37, 2000) and the circles are measured values. Print grain index is a linear function of the visual RMS granularity of the final viewed image, which in turn is computed from the image NPS. The experimental levels span the range from near-threshold noise to the levels of noise found in enlargements made from very high-speed film. A 50% JND of quality averages about five print grain index units, so the largest discrepancies are under one quality JND. This result substantiates the accuracy of all four propagation models discussed herein.

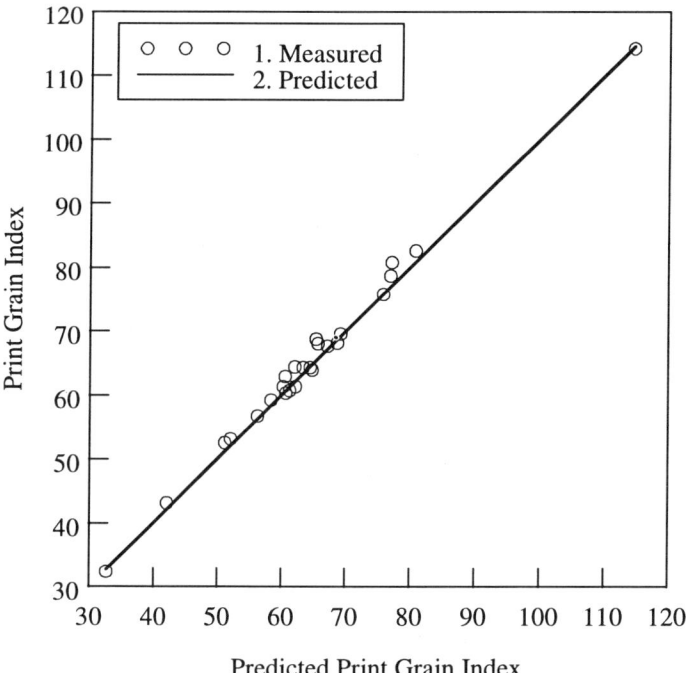

Fig. 22.2 Predicted versus measured print grain index. Because the NPS propagation depends on MTF, MCS, and IWS propagation as well, the excellent agreement (maximum error <0.8 JNDs) validates all four propagation models employed.

It is occasionally useful to propagate targets (such as uniform patches, slanted edges, etc.) through real imaging systems, and analyze the key measures of the system as a whole. However, the applicability of this technique is very limited, as the full system must have been built, and determination of the change in quality with possible substitutions of components requires tedious factorial measurements. In contrast, with propagation models of the type discussed in this chapter, it is possible to combine a modest number of standardized component measurements to make predictions of the image quality of a large number of possible imaging systems, a very powerful capability.

22.8 Summary

To make predictions of final image quality based on component and subsystem properties, the fashion in which objective measures propagate through imaging systems must be understood. Many objective metrics correlating with image quality attributes can be computed from four key measures, some properties of which are summarized in Table 22.1.

The mean channel signal (MCS) is propagated by various types of transformations, examples of which are given in Table 22.2. Objective metrics of tone reproduction are often based on properties of the final image's MCS. The transformation of small signal differences, which is of importance in image structure propagation, is described by the inverse Jacobian matrix (Eq. 22.2).

Color reproduction metrics are generally based on the final image wavelength spectrum (IWS). The propagation of the IWS of analog components is intertwined with that of the MCS. Equations 22.7 and 22.8 quantify the role of the IWS in an optical exposure, but the equations can be generalized to describe viewing phenomena as well.

Modulation transfer function (MTF) is a measure of the spatial fidelity of signal transfer as a function of frequency and is useful in predicting attributes such as unsharpness, oversharpening, aliasing and reconstruction infidelity. In linear systems, its propagation is described by Eqs. 22.11 and 22.12.

The noise power spectrum (NPS) is a measure of the variability of response to uniform stimuli as a function of spatial frequency. It provides a means of predicting attributes such as isotropic noisiness, streaking, and banding. Its propagation in linear systems is described by Eqs. 22.13 and 22.15.

23

Parametric Estimation of Key Measures

23.1 Introduction

In the preceding chapter, the propagation of key measures through imaging systems was reviewed. In that discussion it was implicitly assumed that the key measures were known for each of the system components. Determining such key measures is not necessarily a straightforward task, and even if the task is well understood, it may be costly in terms of required resources. In this chapter and the next we describe strategies for predicting or experimentally determining key measures in an efficient manner. The present chapter focuses on the use of parametric estimations of key measures, in lieu of experimental measurements.

A parametric estimation is a method for predicting the anticipated result of a measurement based on the specification of a small number of engineering parameters. For example, the standard formula for the MTF of a diffraction-limited lens based on its aperture (f-number) and the wavelength of light constitutes a parametric estimation. If a lens has minimal aberrations, the diffraction-limited formula will provide a good estimate of the actual MTF that would be measured in the plane of best focus, along the optical axis.

This chapter is organized as follows. Section 23.2 considers the interchanging roles of measurements and parametric estimation during a product development cycle. Section 23.3 reviews three examples of the application of parametric estimation to practical problems related to: (1) digital still camera noise; (2) video monitor MTF; and (3) camera lens MTF.

23.2 Evolution of Modeling During a Product Cycle

System image quality modeling is useful in the development of a product, regardless of whether that product is a component, a subsystem, or a full imaging system. In the case of a component product, image quality modeling is especially helpful in understanding to what extent changes in the component will affect the final system output. In some cases, other components may limit certain aspects of final image quality, and the customer may best be served by a balancing of the capabilities of the component under consideration with the rest of the system, so that a decrease in cost may be realized. In other cases, properties of a demanding system will challenge the component product. Design of a component in the absence of systems considerations can lead to a product that is inadequate to the task or is excessively costly given the final system output quality. In the case of products that are subsystems or full systems, and therefore include multiple components, image quality modeling is of particular value in co-optimizing the characteristics of the various components so that the overall performance is maximized, relative to its total cost.

The process of developing a new product may involve a number of stages during which system image quality modeling contributes in different ways. A typical product development cycle might consist of stages such as these:

1. selecting a technical approach;
2. writing specifications;
3. prototyping;
4. manufacturing; and
5. verifying performance.

In the first stage, a variety of proposed inventions and technologies may be under consideration. Even the general nature of the product may be open to discussion, with only its strategic intent specified. For example, the intent of a product might be to increase the number of photographs taken by people on vacation. Products that might meet this criterion include a very small camera for increased portability on hikes and other excursions; a very inexpensive camera for use in potentially adverse conditions; or a generic waterproof housing for compact cameras to enable underwater photography. The product would not even have to be a piece of hardware; it might instead be a service or promotion, such as rental of cameras by tour companies or production of finished vacation albums using digital photofinishing.

When choosing among an array of potential product designs, image quality considerations are one important factor, complementing other manufacturing, business, and financial factors. At this stage, what is commonly needed is an assessment of whether the image quality associated with a novel proposal could potentially meet certain minimum requirements. In such analyses, there is usually missing information, often because physical measurements cannot be made of devices that have not yet been built. Nonetheless, through the use of parametric estimations and measurements made from similar, existing products, semi-quantitative estimates are usually possible. Because of the diversity of proposals likely to be under consideration and the difficulty of direct experimental work, image quality modeling frequently contributes heavily to decisions regarding the technology or approach adopted in a product program.

The second stage of the product cycle involves specification of the required characteristics of the product, including both aim positions and tolerances about the aims. Many of these characteristics are related to image quality and may be quantified in terms of engineering parameters, such as those in parametric estimations. Alternatively, the requirements may be expressed in terms of objective measures such as MTF. A common approach is to relate the properties of the proposed product to that of existing products. Frequently, the specification will require that the quality or an attribute thereof equal or exceed that of some currently popular device. It is often useful to anticipate desired advertising claims and the evidence that will be needed to substantiate them. If a claim is to be made of superior image quality compared to some reference, the required performance might be specified to exceed that of the reference by, conservatively, two 50% JNDs (Sect. 3.4). From this requirement, expressed in terms of JNDs, the aim engineering parameters or measurement results may be determined by image quality modeling. Even if the specification is to be in terms of a measurement result, it is usually easiest to generate aims using parametric estimations. For example, the MTF of a reference product might be fit by a parametric estimation formula, so that likely improvements in the parameters, based on engineering knowledge, could be used to generate a new aim MTF.

If a product includes multiple components or distinct subcomponents, image quality modeling can be very valuable in determining how best to balance the properties across the entire product. For example, a certain amount of total product noise that is deemed permissible may be spread among several components in a variety of ways. Obviously, the designers of each component or subcomponent would like the entire noise budget to make their task easier; identifying an optimal partitioning is difficult, but critical to the success of a product. Image quality modeling is very valuable in this context because many combinations may be tested and the best balance between image quality and factors such as manufacturability and cost may be quantified. Consequently, image quality modeling frequently plays an important role in this stage, and as was the case in the first stage, parametric estimations are of particular utility.

The third stage in the product development cycle involves the production of prototype devices. This is often the first stage in which it is possible to make truly relevant physical measurements, although in earlier stages they can sometimes be made on individual components or modular subcomponents. Typically, modeling in this stage consists primarily of comparing the predicted image quality from prototype measurements with that required by the specifications. Much of the value provided by the modeling is in assessing the significance of any differences found. Parametric estimations are largely replaced by prototype measurements in this stage.

The fourth stage, manufacturing, is less dependent on image quality modeling than earlier stages. The primary usefulness of modeling at this stage is in setting meaningful pass/fail test criteria for assembly line quality control. Essentially all analyses are based on measurements by this stage.

The fifth stage, verification of performance, involves empirical assessment of the results produced by usage of the product in the hands of the customer. Image quality modeling is sometimes useful here in clarifying the origin of discrepancies between the specified and observed performance. The results of the verification process may also be useful in improving the predictive capabilities of the image quality models.

In summary, image quality modeling usually has greater impact in the earlier stages of a product cycle but continues to provide value throughout most of the product cycle. Parametric estimations often play a larger role earlier in the modeling process, with measurements assuming a dominant role later.

23.3 Examples of Parametric Estimation Applications

In this section, several examples of the application of parametric estimations are presented. In the first example, noise measurements from a digital still camera at one exposure index, in one color, are fit by a two-parameter model, from which the noise in all three colors at all exposure indices may be predicted. In the second example, the expected MTF of a video monitor is estimated within certain bounds by a Gaussian spot model, in combination with an engineering rule of thumb regarding the anticipated spot size relative to the line spacing. In the third example, the range of conditions over which accurate predictions are obtained from three models of lens MTF are shown to increase with greater model sophistication.

Figure 23.1 shows a very good two-parameter fit to measured digital still camera noise in the green channel at the base exposure index (the lowest recommended speed setting). The x-axis is logarithmic exposure, and the y-axis is logarithmic noise in RMS electrons. RMS electrons are equivalent to the square root of the volume under the NPS in the spatial frequency plane and are used here to collapse the spatial frequency dimension and thereby simplify the example. One

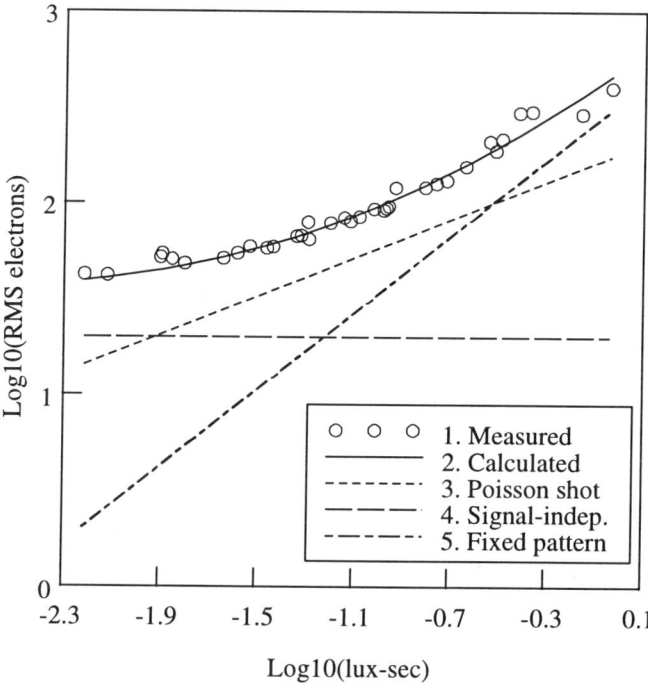

Fig. 23.1 Fit of two-parameter model to digital still camera noise data in the green channel at the base exposure index. From this one fit, using available design information, the model permits prediction of noise in all three colors at all exposure indices, substantially reducing the number of measurements needed to characterize the device.

model parameter characterizes the level of signal-independent isotropic noise arising from sources such as thermal dark current. This noise source has a constant value, independent of signal level (although it does depend on ambient temperature and integration and readout time). A second parameter quantifies the amount of signal-dependent fixed-pattern (deterministic) noise arising from gain variations between pixels and other sources. The magnitude of fixed pattern noise is a constant percentage of the signal, so on a log-log plot like Fig. 23.1, it has a slope of one. A final contributor is photon shot noise, which, because of the nature of the Poisson distribution, varies with the square root of the signal, and therefore has a slope of one-half on a log-log plot.

The total noise is the RMS sum of the contributing noise sources after modification by channel-dependent gain, color matrixing, and color filter array (CFA) interpolation (Sect. 24.3). Figure 23.1 shows each contributing noise source and the final sum, accounting for these factors. The two fit parameters (signal-independent isotropic noise and percentage fixed pattern noise) have been optimized to best fit the measured data. The fit is very satisfactory, so from these two values, and normally available design information, it is possible to predict the noise dependence of all three colors at all practical exposure indices (which are typically 2×, 4×, and 8× the base index).

In this example, the use of a parametric model dramatically reduces the number of measurements that need to be made to characterize a device, in a manner analogous to the measurement interpolation algorithms that will be described in the next chapter. A more common application of parametric models involves the estimation of model parameter values to make predictions that can be substituted for (rather than extending) experimental measurements. Such applications are referred to as parametric estimations. Thus, in our first example, if independent estimates of dark current and fixed pattern noise were available, as well as the other information noted above, the noise versus signal relationship of a digital still camera could be predicted even before the first prototype device was assembled.

Our second example pertains to the MTFs of high-quality video monitors. In digital output devices, the size of the writing spot compared to the line spacing affects both MTF and banding. If the spot size is considerably smaller than the line spacing, then adjacent lines will not overlap sufficiently to yield output with a uniform appearance in a region with constant input signal, producing a banding pattern at the raster frequency. Because the spot size determines the MTF, excessively large spot sizes lead to reduced sharpness. Maintaining a good balance between sharpness and banding more or less constrains the range of practical spot size to line spacing ratios.

Given an engineering rule of thumb regarding the optimum range of values of this ratio, and a model of the shape of the spot, the MTF of a digital output device may be estimated from its pixel pitch within certain tolerances. Figure 23.2 shows a measured MTF and a fit to the measured data using a Gaussian spot model, which evidently describes the spot profile quite accurately. Also shown are the highest and lowest MTFs that would be anticipated based on an engineering rule of thumb stating that the preferred values for the ratio of the spot profile standard deviation to the line spacing fall in the range 0.45–0.70. The measured MTF falls close to the middle of the anticipated range, which is reasonably narrow given that only a single engineering parameter, the video monitor line spacing, was required to produce the estimation.

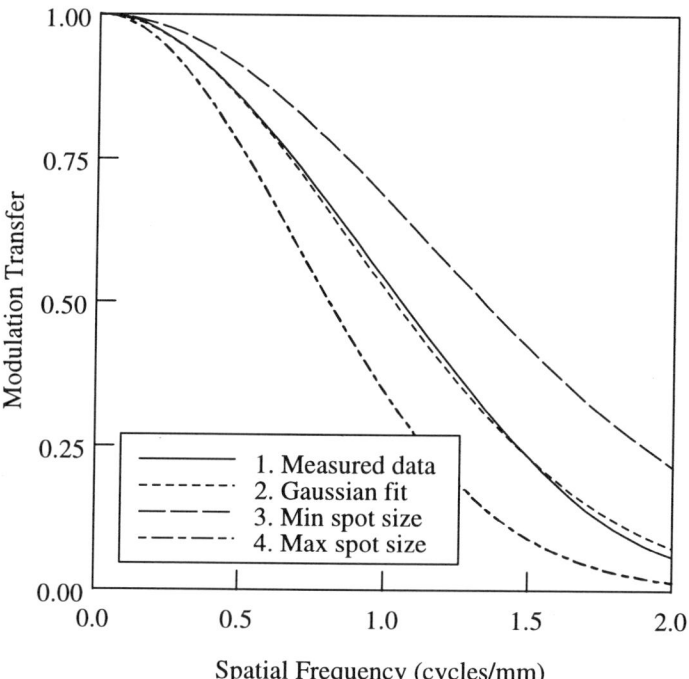

Fig. 23.2 Parametric modeling of video monitor MTF by Gaussian spot model. A measured device is fit well by the model using a spot size to line spacing ratio intermediate between the extremes expected from engineering rules of thumb.

Our third example involves the prediction of camera lens MTF. Camera lenses are often used at a variety of f-numbers (apertures) to help control exposure, so their MTF must be characterized as a function of f-number. Furthermore, the behavior of camera lens MTF with varying amounts of defocus must also be quantified because many images are not taken at the best focus position. Defocus may occur for a variety reasons, including: (1) the camera is a fixed focus model; (2) there are only a small number of zones of autofocus; (3) the autofocus ranging system inaccurately measures the distance to the subject; or (4) the aim separation between the lens and film/sensor plane is not achieved in manufacturing.

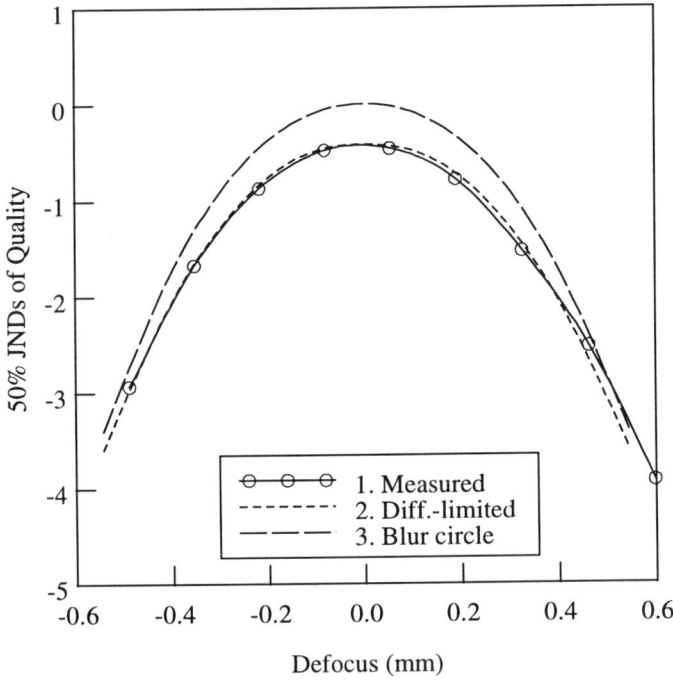

Fig. 23.3 Through-focus behavior of a prime lens at f/8. The blur circle (geometrical optics) prediction is poor but the diffraction-limited model quantitatively matches measured on-axis data.

Figure 23.3 compares measured and predicted lens through-focus behavior of a prime lens at f/8 for two parametric models of differing sophistication. To simplify interpretation of results, rather than depict MTF, quality changes arising from the MTF variations in a representative system are shown instead. The y-axis is JNDs of quality and the x-axis is defocus in mm, with zero defocus corresponding to the best focus position. The simplest model of lens MTF as a function of defocus is the blur circle (geometrical optics) approximation, in which the point-spread function of the lens is assumed to be a uniform disk, the diameter of which is proportional to the defocus (and inversely proportional to the lens f-number). This approximation is fairly good for predicting depth of field (the range of subject distances rendered sharply) but, as seen in Fig. 23.3, it is progressively less accurate as optimum focus is approached, because it predicts that the point-spread function at best focus vanishes, corresponding to a unit MTF. This is physically unrealizable, because the point-spread function of even an ideal lens is limited by the diffraction of light. Consequently, although the predictions of the blur circle parametric model approach the measured values at large defocus, the MTF is notably overestimated near optimum focus.

A better parametric model is the diffraction-limited MTF, in which the behavior of the lens is assumed to be ideal in the sense that it is limited only by the fundamental wave nature of light. A convenient method for calculating the diffraction-limited MTF as a function of defocus is given by Stokseth (1969). Deviations from ideal behavior are termed aberrations and are often a strong function of the lens aperture, with aberrations often decreasing markedly at higher f-numbers. As seen in Fig. 23.3, in the case of a high-quality lens at an aperture of f/8, the diffraction-limited model quantitatively predicts the through-focus measurement data. Similar predictiveness would be expected at higher f-numbers as well, where aberrations would be even further reduced.

In contrast, at lower f-numbers and in less expensive lenses, more extensive aberrations are expected to lead to deviations from diffraction-limited behavior. In such a case, the diffraction-limited model is expected to overestimate the MTF except far from optimal focus, where the defocus dominates the MTF. Fortunately, a much more sophisticated approach than either the diffraction-limited or blur circle models is available, namely, the use of commercially available lens design software. In such programs, which constitute extremely complex parametric models, the shapes and positions of lens surfaces and the limiting aperture, and the indices of refraction of the lens materials, are used to predict lens MTF based on fundamental laws of wave propagation.

The predictive capabilities of lens design software and the diffraction-limited model are compared in Fig. 23.4. The data presented pertain to the f/3.5 behavior of a point-and-shoot compact camera lens, costing perhaps two orders of magnitude less than the prime lens of the previous figure. As expected at this wider aperture and with this less sophisticated lens design, the diffraction-limited model progressively overestimates the MTF as best focus is approached because the effects of aberrations are neglected. This behavior is analogous to that of the blur circle approximation in the previous figure. In contrast, the predictions of the lens design software agree very well with the measurements except at high levels of defocus, where a slight systematic underestimation of MTF apparently occurs.

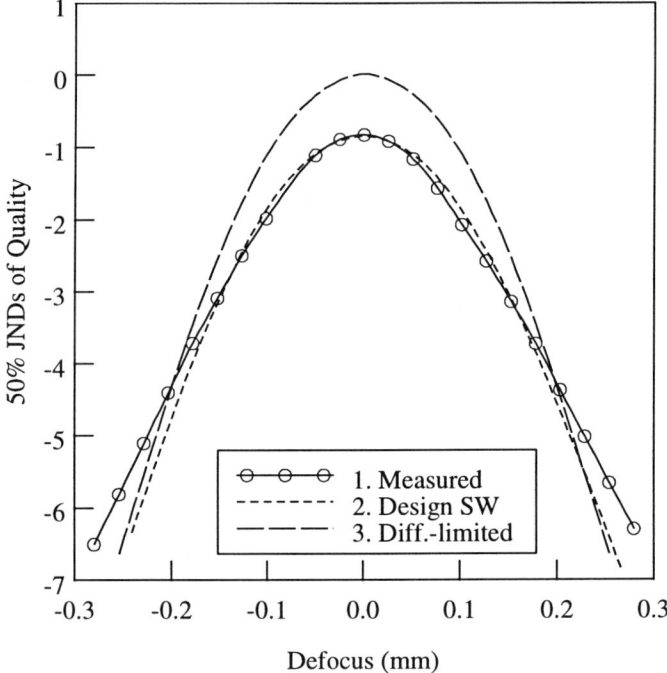

Fig. 23.4 Through-focus on-axis behavior of a compact camera lens at f/3.5. Because of the presence of aberrations, the diffraction-limited approximation is inaccurate, but lens design software predictions agree very well with measured data.

Parametric Estimation of Key Measures 353

Figure 23.5 further demonstrates the predictive capabilities of lens design software. The figure compares lens design software and diffraction-limited MTF predictions with measurements made at 550 nm wavelength for an f/2.8 lens that is used in a non-critical image viewing application. The large difference between the diffraction-limited curve and the measured curve implies the presence of extensive aberrations, yet the lens design program quantitatively predicts the exact MTF shape.

In this example, we have seen an increase in the range of conditions for which valid predictions could be obtained as parametric model sophistication increased. The blur circle approximation required only the defocus and f-number as input, but failed to accurately predict quality except at large defocus. The

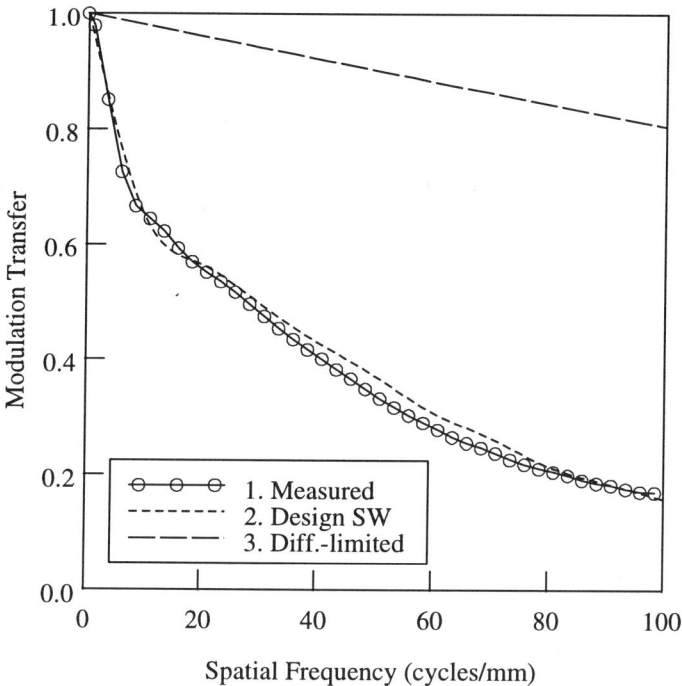

Fig. 23.5 MTF of a highly aberrated lens at f/2.8. The lens design software predictions match measurements superbly, despite the extensive aberrations that are evidenced by the large departure from diffraction-limited behavior.

diffraction-limited model required the additional input of wavelength, and was predictive except at lower lens f-numbers. Finally, the lens design software required many additional inputs (surface shapes, refractive indices, etc.), but it made quantitative predictions under all conditions tested. Development of advanced parametric models usually requires considerable effort, but the advantages accrued are likely to justify the resources expended if significant numbers of measurements may be avoided and/or if reliable design decisions are enabled earlier in product programs.

Although parametric models are very valuable, especially in the early stages of the product development cycle, it is nearly always necessary to make measurements later in the cycle to confirm the actual behavior of the manufactured components. The next chapter considers how such measurements can be made in an efficient and reliable manner.

23.4 Summary

Image quality modeling makes key contributions to product definition, selection of technology, specification of aims and tolerances, assessment of prototype performance, development of manufacturing pass/fail tests, and identification of sources of performance loss. Although image quality modeling is useful throughout the product development cycle, it is particularly valuable earlier in the process, where parametric estimation is frequently used. Parametric estimation is the process of predicting what the result of a measurement would be, based on a small number of readily available engineering parameters. Later in the product development cycle, measurements of prototype and ultimately manufactured devices largely replaces use of parametric estimation.

Parametric estimation is valuable in several contexts. A parametric model may be used to fit a limited set of experimental data so that a complete description of the behavior may be estimated. Alternatively, engineering rules of thumb may be used to estimate the parameter values so that expected measurement results may be predicted within certain tolerances in the absence of any experimental data. In the most advanced applications, the values of the parameters are known from a detailed understanding of the product design, and quantitative predictions are possible over a range of conditions, the breadth of which depend upon the sophistication of the parametric model.

24

Development of Measurement Protocols

24.1 Introduction

Some measurements are made repeatedly during the development cycles of a variety of products. For example, any lens under consideration for inclusion in a product invariably will have its MTF measured. Similarly, the noise of a sensor that might be used in a digital camera or scanner will certainly be characterized. In such cases, there is considerable value in standardizing the measurement protocols used. The primary advantage of standardization is the improved consistency of measurements, increasing the reliability of comparison between results from different devices. A secondary benefit is that parametric estimations may be optimized for maximum consistency with measurements made according to the protocol, thereby minimizing modeling discrepancies when a transition is made from parametric models to measured data. In Sect. 24.2, various factors that should be considered when defining measurement protocols are discussed. Examples are provided regarding lens MTF interpolation over defocus, and film NPS estimation for color patches from neutral exposures.

In some instances, it is desirable to try to extend the use of a measurement protocol outside the range of conditions for which it was originally intended. Such extensions may violate one or more of the assumptions in the theory underlying the measurement, or they may compromise the robustness of the results because of signal-to-noise degradation or other practical considerations. Section 24.3 reviews an example of such a verification process, involving the MTF of color filter array interpolation schemes.

24.2 Definition of Measurement Protocols

In some cases, standards organizations have published measurement protocols that merely need to be adopted, perhaps augmented, and implemented, but in other instances customized procedures must be developed. Often the consistency with which a method is applied is more important than the details of the method. When developing a measurement protocol, the modeling calculations that it will support should be identified and the protocol should be designed to generate exactly the data needed with a minimum use of resources. If there are multiple applications of the same measured data, which is a common situation, the protocol should either support all applications or should require that the requestor specify the intended application.

The terms capability and performance will be carefully defined in Sect. 26.1, but in brief, the capability of an imaging system is the peak quality it produces, whereas the performance is the full distribution of quality resulting from its use in the hands of the customer. In many cases, performance modeling requires more extensive measurements than does capability modeling. For example, a capability calculation might require only the best focus MTF for a film scanner lens, whereas a performance calculation might require through-focus data to determine the impact of lens, film and sensor plane positioning variability.

A protocol for a through-focus lens MTF measurement would have to specify factors such as the number of planes of focus, the separation between the planes, etc. Figure 24.1 shows the sort of data and analysis that might be used to define such a protocol. This figure depicts measured and estimated lens MTFs as a function of defocus. The lens is a high-quality 35-mm format macro lens and is measured in the blue at an f/2.8 aperture. Lens MTFs were measured in-focus (0) and out of focus by ±1, ±2, and ±3 arbitrary units (one unit being 0.07 mm). The 0, +1, and +3 data were used to predict the +2 data based on a 3-parameter Gaussian fit at each frequency (and similarly for negative defocus). The agreement between the Gaussian interpolated data and the measured data at the +2 position is very good, providing an indication of the type of spacing between defocus planes that can be supported by the interpolation algorithm. Based on similar tests with many lenses over a wide range of conditions, a measurement protocol was defined based on five equally spaced planes of defocus, with spacing proportional to f-number, and interpolation via a Gaussian function passing through the three nearest measured points. The protocol also included rules for determining how many and which apertures need to be measured to provide sufficiently accurate interpolation over the full aperture range. In this case, the development of a robust interpolation scheme reduced the amount of data that needed to be collected by as much as a factor of four.

Our second example also involves a case in which the number of potentially required measurements was greatly reduced, albeit at a greater loss of accuracy. A complete description of noise in an imaging system might involve specification of the NPS of a number of color patches in the final image. If this number of patches were large, NPS measurements of the various system components at each of the color patch positions would become unwieldy. In principle, in the case of a photographic film, it should be possible to approximately predict the NPS of color patches from: (1) the NPS of a small number of neutral patches; (2) the RMS granularity (a simpler measurement) of a modest number of neutral patches; and (3) the film interlayer interimage effect matrix (Giorgianni and Madden, 1998), which is likely to have been measured to characterize color reproduction.

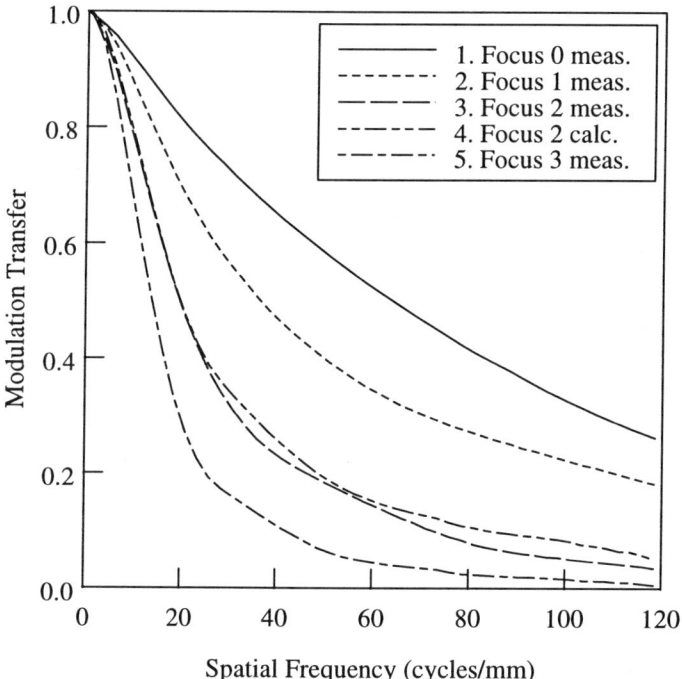

Fig. 24.1 Measured and interpolated lens MTFs as a function of defocus. Data such as this are useful in developing measurement protocols that involve the fewest measurements possible to adequately characterize a system component.

Figure 24.2 compares the results of such predictions to measured red, green, and blue NPS of a Caucasian skin-tone patch in a reversal (slide) film. The agreement is only fair, with neither the magnitudes nor the shapes of the NPS consistently predicted accurately. These discrepancies might arise from at least two sources. First, RMS granularity and NPS shapes were interpolated linearly between neutral patches based on ISO Status A film density, whereas the change might have been quite nonlinear. This is especially likely in reversal films, which often have stronger noise versus signal dependencies. Second, the interimage matrix is determined from large-area sensitometric measurements, and so may not reflect amplification of high-frequency noise by chemical effects. The lesser accuracy of these predictions might be acceptable given the

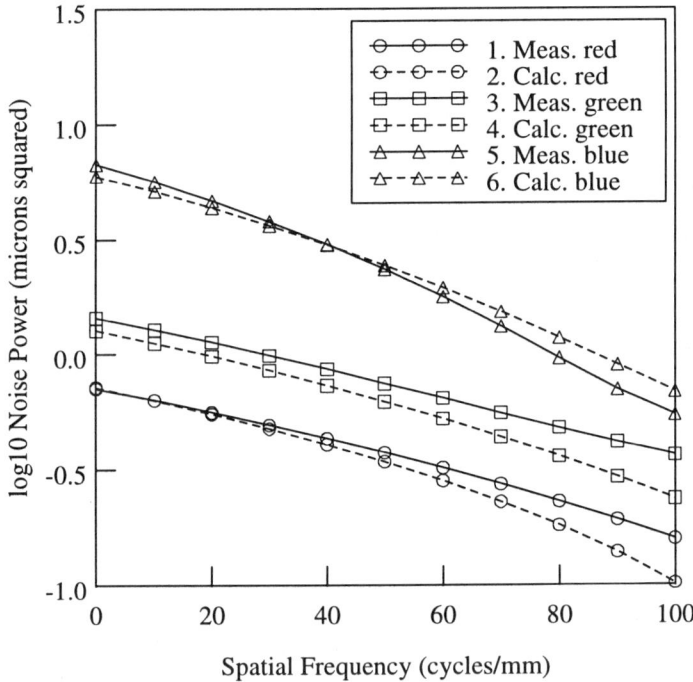

Fig. 24.2 Measured and predicted NPS of a Caucasian skin-tone patch, which exhibit fair agreement. If the NPS of many color patches were required, and this level of accuracy were acceptable, such an estimation from neutral patch determinations would drastically reduce the number of measurements needed.

Development of Measurement Protocols 359

substantial reduction in measurement resources required, particularly if some sort of weighted average of results from a number of color patches were to be taken. In such a case, it is likely that errors in complementary color patches would be correlated and would tend to partially cancel one another, so that the accuracy of the weighted average result might be substantially improved.

24.3 Verification of Measurement Protocols

In some cases, it may be desirable to test whether a measurement protocol yields results that produce accurate predictions of image quality attributes. Such cases may involve a potential violation of the assumptions of the theory underlying the measurement, or practical limitations of signal-to-noise or other factors might raise a concern regarding the robustness of the determination. This section describes an example of a verification study related to the measurement of the MTF of mosaic color filter array (CFA) interpolation algorithms.

A variety of CFA patterns are used in digital still and video cameras to permit simultaneous capture of different color channels by a single sensor. The CFA is formed during sensor manufacture and the pattern defines the color channel of each pixel in the sensor array. An interpolation algorithm is then used to estimate the value that each pixel would have had in each of the missing color channels. For example, if the CFA pattern consisted of alternating columns of red, green, and blue pixels, the green value of a blue pixel might be estimated by linearly interpolating between the nearest green pixels in that row.

Shown below are four different CFA patterns that are discussed subsequently. The "Striped" pattern is efficiently manufactured because each column (or row) is a single color; however, visually noticeable interpolation artifacts occur because continuous color stripes and their associated artifacts are aligned vertically (or horizontally), a direction in which the eye is very sensitive. The "Staggered" pattern reduces visibility of artifacts by rotating the color stripes to the 45° diagonal, a direction in which artifacts are less noticeable (halftone printing screens are usually rotated 45° for the same reason).

Striped	Staggered	Bayer	3Gs
R G B	R G B	G R G R	G G R G
R G B	B R G	B G B G	B G G G
R G B	G B R	G R G R	G G R G
		B G B G	B G G G

Recall from Sect. 22.5 that the perception of image structure is largely determined by the response of the luminance channel of the human visual system. In a CFA having red, green, and blue channels, the green spectral responsivity is closest to that of the luminance channel, so it is important to achieve as high a spatial fidelity as possible in the green channel. Both the Striped and Staggered patterns are deficient in this regard because only one-third of their pixels are green. The "Bayer" pattern assigns a larger fraction (one-half) of the pixels to the green record and maintains the diagonal orientation of the green stripes, while breaking up the blue and red stripes, thereby suppressing chromatic artifacts very successfully. Finally, the "3Gs" pattern allocates three-quarters of the pixels to the green channel. Several variants of this pattern exist, with different dispositions of the red and blue pixels.

Many CFA interpolation schemes have been proposed (Adams, Parulski, and Spaulding, 1998) and most are optimized for use with a particular type of CFA pattern. Some interpolation schemes are quite sophisticated in accounting for, and taking advantage of, the properties of pictorial images and the human visual system, and it is these advanced features that raise questions about the applicability of linear systems theory and MTF.

One interpolation-based technique for improving green fidelity involves using values from red and blue pixels in addition to those from green pixels when interpolating the green color plane. For example, in the Bayer pattern, when estimating the green value of a blue pixel, rather than simply averaging the adjacent green pixel values, the values of the nearest blue pixels can be used to estimate local signal curvature in that pixel neighborhood, allowing a higher order estimate (Adams, 1997, 1998). To the extent that spatial detail in the three channels is correlated in the scene captured, this will improve the estimate of the missing green values for red pixels and blue pixels. In many images, this correlation is quite high; for example, a set of twelve representative pictorial scenes exhibited correlation coefficients of 0.86, 0.92, and 0.80 between red and green, green and blue, and red and blue film channels, respectively (Töpfer, Adams, and Keelan, 1998). Consequently, the use of all three colors to interpolate the green plane is often quite beneficial.

Once a good estimate of the full green plane is available, it may be treated as a near-luminance, or luma, channel, and the interpolated green values of red and blue pixels may be subtracted out, leaving two sparsely populated color difference, or chroma, planes. To the extent that the spatial detail of the chroma and luma channels is correlated, this subtraction will reduce the high-frequency content of the two chroma planes, making them less susceptible to interpolation error. After interpolating the chroma planes, the red and blue channels may be regenerated by adding the luma channel back into each of the chroma channels.

A second advanced feature of some CFA interpolation schemes is the use of adaptive algorithms to decide, on a local basis, in which direction to interpolate. Interpolating across an edge blurs and/or distorts it, whereas interpolating parallel to an edge preserves it. CFA patterns that avoid solid rows or columns of a single color generally can be interpolated in either of two directions. At each pixel position, adaptive algorithms evaluate the variation within the pixel values that would be used if interpolation were done in either of the two possible orientations. The interpolation is then performed in the direction having less apparent signal variation.

The improvement resulting from the adaptive approach depends upon the angular distribution of edges in, and the noise characteristics of, the captured image. If there are many edges that are nearly horizontal or vertical in orientation, as are frequent in man-made objects, then an adaptive interpolation may yield distinctly superior results. Its effectiveness depends, however, upon its ability to identify the preferred interpolation direction, which is compromised by image noise.

To summarize, the success of chroma-luma interpolation depends on inter-channel correlation of high-frequency spatial detail in the captured scene, whereas that of adaptive interpolation depends upon capture noise and the distribution of edge orientations. There are ambiguities in the application of linear systems theory to each of these two methods, and different MTFs will be inferred depending upon the properties of the targets used in the analysis.

The use of neutral targets, which are pervasive in such determinations, will yield the best results possible from a chroma-luma process because of the complete correlation between the three captured color channels. In contrast, a determination based on a target consisting of uncorrelated isotropic noise in three color records would yield a lower MTF. An MTF can be inferred from a noise target based on the Doerner equation (Eq. 22.14). A synthetic digital target is processed in a fashion emulating sampling and CFA interpolation, and with no sensitometric gradient or magnification affecting the results, the MTF may be computed at each frequency by taking the square root of the ratio of the noise power after the processing to the noise power originally present.

Similarly, the angle of a slanted edge or the direction of modulation of a sine or square wave will affect the MTF obtained from an adaptive interpolation scheme. Usually such targets are essentially free of noise and are captured so that their axes of variation are parallel or nearly parallel to the sensor columns and rows, and so again, the highest possible MTFs are typically measured.

Figure 24.3 shows measured and calculated red/blue MTFs resulting from application of a chroma-luma interpolation with a Bayer pattern capture. The frequency axis is in cycles/sample of the native sensor, not the CFA pattern. The green channel Nyquist frequency is also 0.5 cycles/sample but the red/blue channels have Nyquist frequencies half as large because only half the columns and rows contain red/blue pixels. Neutral slanted edge, square wave, and color-correlated noise targets yielded MTFs in very close agreement with each other. The interpolation MTF can be also be quantitatively predicted from first principles of linear systems theory, as demonstrated by the excellent agreement between the calculated and measured MTFs. The boost above unit MTF at intermediate frequencies is caused by imperfect reconstruction of the luma plane, leading to systematic amplification of modulation.

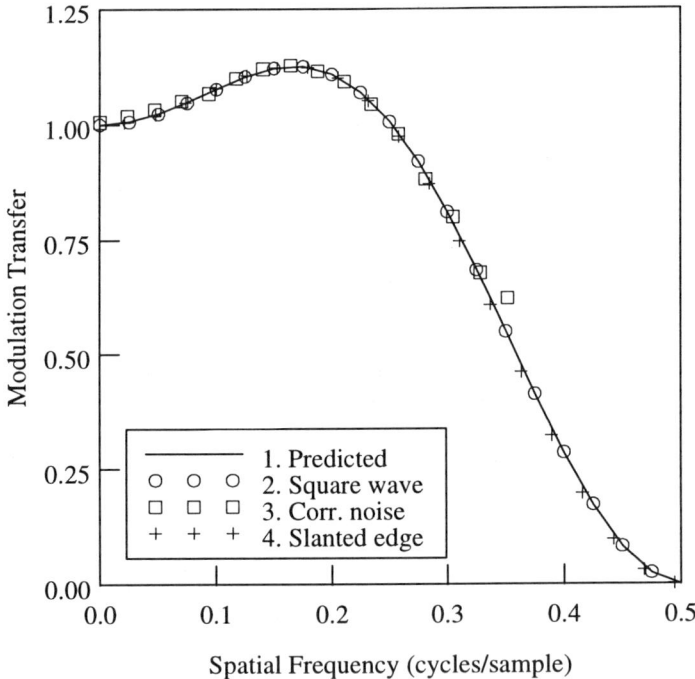

Fig. 24.3 Measured and predicted MTFs of a non-adaptive chroma-luma CFA interpolation process. Linear systems theory quantitatively predicts the MTF obtained from neutral slanted edge, square wave, and isotropic noise targets.

Figure 24.3 demonstrates that quantitative prediction of chroma-luma CFA interpolation MTF is possible if the target is neutral (i.e., the channels are perfectly correlated). Similar results are obtained in the case of adaptive interpolation if the edge is nearly vertical or horizontal and noise does not degrade algorithm performance. But are neutral, roughly x- and y-oriented targets representative of actual scenes? One pragmatic way of answering this question is to determine whether the MTFs deduced from such targets can be used to quantitatively predict perceived sharpness of pictorial images. In Fig. 24.4, the calculated quality change arising from unsharpness (solid line) is compared to assessments of representative scenes for five CFA interpolation schemes. Although the impact of the sharpness variations is expressed in JNDs

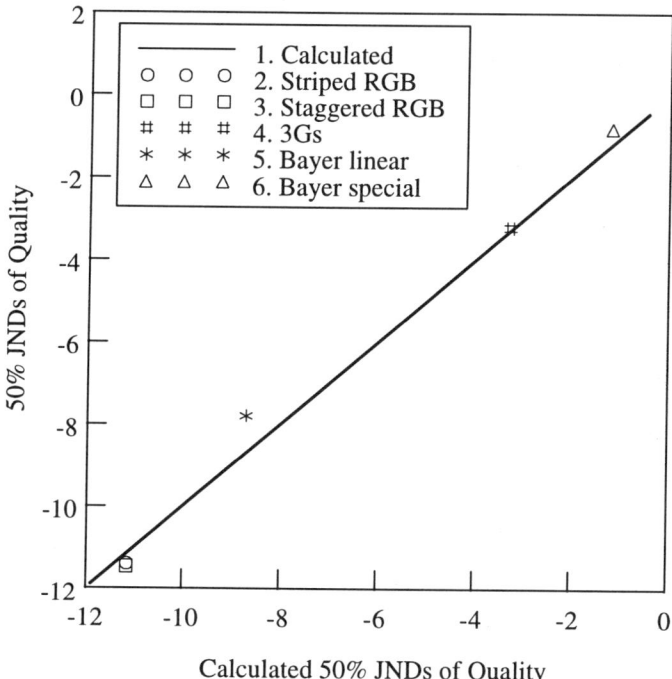

Fig. 24.4 Psychometrically measured and predicted quality changes from unsharpness for five CFA interpolation algorithms. Even the chroma-luma (3Gs) and adaptive chroma-luma (Bayer special) algorithms, which cannot be described rigorously by linear systems theory, are predicted to within one JND of quality.

of quality, the psychometric task involved assessment of sharpness via projection against a quality ruler varying in degree of blur, thereby avoiding ancillary effects caused by CFA interpolation artifacts. The Striped, Staggered, and "Bayer linear" cases are simple linear interpolations, whereas the 3Gs case involves chroma-luma processing. The "Bayer special" case incorporates adaptive interpolation direction in addition to chroma-luma processing. In all five cases, predictions of perceived sharpness based on the MTFs measured from nearly horizontal or vertical slanted edge neutral targets are accurate, the largest deviation being less than one JND of quality. Evidently, the protocols developed for MTF measurement of chroma-luma and adaptive interpolation processes produce results from which the perceived sharpness of pictorial images may be accurately predicted.

We have undertaken similar experiments to test the practicality of determining MTF by various methods in halftone imaging systems, in which nonlinear rendering processes are often employed, and which pose special experimental challenges because of the spectral consequences of their regular patterns. Very good agreement between predictions from measurements of targets and psychometric data from assessment of pictorial scenes has been found except in cases where both resolution and number of gray levels are very low. Identification of such problematical regions is valuable; where higher discrepancies are likely, more frequent verification experiments are warranted.

There are a number of instances where measurements traditionally performed on analog components may potentially be applied to digital devices and processes. Verification experiments of the type described in this section are valuable in assessing suitability of such existing protocols and in identifying modifications to extend their range of validity.

Considerations regarding the organization of measured data for greatest utility in image quality modeling software are deferred to the next chapter, which describes the integration in image quality modeling software of psychometric results, objective metrics, key measure propagation, parametric estimation, and measurement databases.

24.4 Summary

A measurement protocol is a set of instructions for characterizing one or more aspects of the performance of a component or subsystem. Standardization of protocols improves the reliability of comparisons between different measurements. Ideally, a measurement protocol should minimize expenditure of resources while providing adequate information to support image quality modeling applications. Development of algorithms for robustly interpolating between measurement positions, or accurately predicting more complex measurement results from simpler sets of measurements, may lead to considerable increases in efficiency.

Measurement protocols may be tested and sometimes extended in scope by comparing predictions based upon them with psychometric data from assessment of representative scenes. Such verification experiments are useful when assumptions of the underlying theory are violated, or when practical aspects of the measurement are compromised.

25

Integrated System Modeling Software

25.1 Introduction

The quality that would be produced by an imaging system can be predicted from the characteristics of its components, given:

1. a database of component measurements of key measures made according to standardized protocols;

2. a series of parametric models allowing estimation of component key measures when measurements are unavailable;

3. a set of propagation models permitting calculation of the key measures of viewed images as a function of key measures of system components;

4. a collection of objective metrics, derived from key measures of viewed images, and having known relationships to image quality changes; and

5. a multivariate formalism for combining the effects of different attributes to predict overall quality.

Although such assets may be applied as required on a problem-by-problem basis by writing specialized analysis programs, it is more efficient to integrate all these capabilities into a unified, general software application. When a specialized program is written to analyze a particular problem, time and resource constraints usually preclude taking the most general and mathematically rigorous approach, and it is rarely possible to test the predictions of the program as extensively as would be desirable. Furthermore, such programs are typically intended for the use of the originator, and so often have cryptic user interfaces and minimal documentation. Thus, specialized programs are usually of limited utility except to the researcher who designed them, and for the particular problem they were intended to address. In addition, such programs are prone to inaccuracy because of simplifying assumptions and/or undetected errors.

In contrast, more extensive resources justifiably can be devoted to developing unified programs, because they will be used many times. Their calculational procedures may be designed to be general in nature and to incorporate rigorous mathematical treatments, and their predictions may be tested exhaustively. Their user interfaces, including online help, may be designed to facilitate use by individuals who are not experts in image quality modeling research. Enabling use by a broader audience has many potential advantages, such as ownership of results by product development teams, rapid cycle time, and improvement of the systems understanding of the analysts. In aggregate, the investment to produce unified, general system image quality modeling software is justified because the existence of such software greatly reduces barriers to use of modeling in a variety of applications.

In this chapter, various aspects of the design of unified image quality analysis software are considered. Section 25.2 describes our approach to user interface organization. Section 25.3 addresses necessary supporting resources such as measurement databases and online help. Section 25.4 reviews the types of model output that may be of use to analysts. Finally, in Sect. 25.5, a sample modeling session is described to exemplify the use of modeling software.

25.2 User Interface Design

When describing an imaging system, it is common practice to draw a block diagram of the system so that its elements are enumerated, and their relationships to one another are exhibited in a simple pictorial fashion. Because of its intuitive appeal, we have used the block diagram as the basis for our user interface. The individual blocks in the diagram represent imaging system components, and sets of contiguous blocks may be identified as subsystems. Figure 25.1 shows the main user interface screen with a sample system block diagram. The block diagram specifies an imaging system having analog (film) capture followed by digitization through film scanning and display on a video monitor after digital image processing.

Integrated System Modeling Software

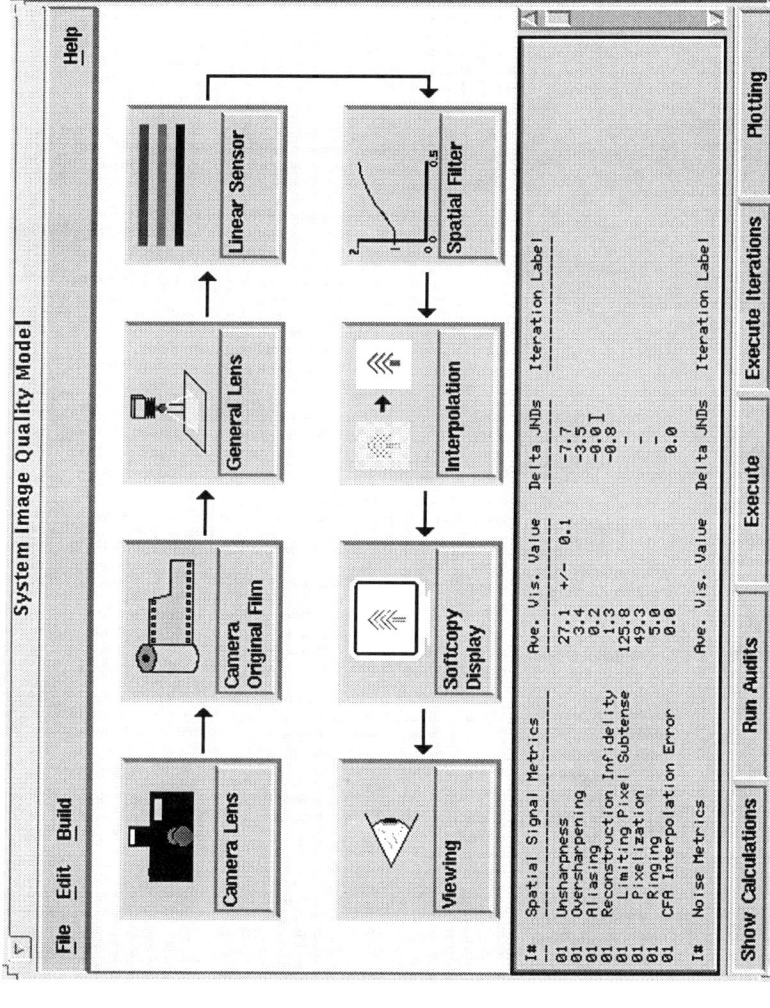

Fig. 25.1 System image quality model interface. In a menu-driven process the user defines a block diagram of the imaging system to be modelled.

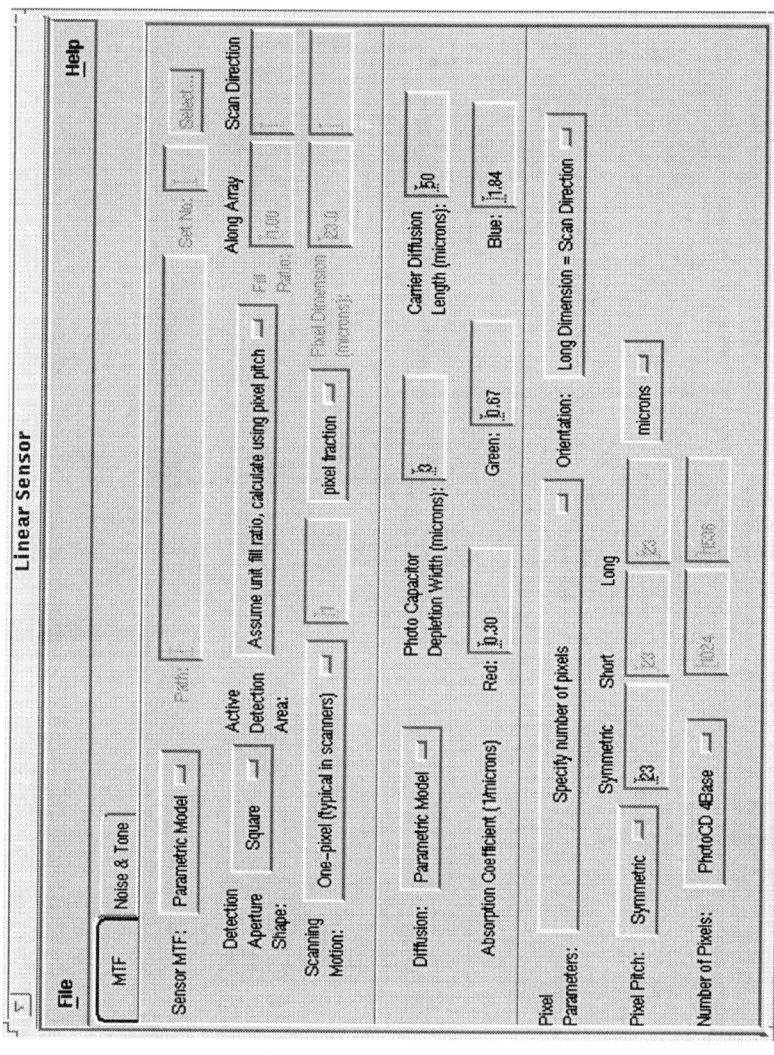

Fig. 25.2 Sample data entry screen for linear sensor MTF. The parametric MTF is based on a 23-μm square aperture, one-pixel scanning motion, and default charge carrier diffusion.

Integrated System Modeling Software 371

This diagram could have been built component by component through selection from the "Build" menu, or one of dozens of pre-defined systems could have been loaded and modified as necessary using "Edit" functions such as adding or deleting components. All components and pre-defined systems initially have pertinent default settings so that a calculation can be run as soon as a full system, beginning with image capture and ending with viewing, has been defined. The settings are modified by clicking on a component icon to bring up one or more data entry screens. When all settings have been satisfactorily specified, the user may execute a computation. A series of audits are run to ensure that the software can unambiguously interpret the information provided and that no apparent inconsistencies are detected. The information is then packaged in a file and a calculational engine is invoked. The results from this engine are stored in a second file and control is returned to the interface, which displays a high-level summary of the results. A separate plotting package may be invoked directly from the interface, with the most recent results automatically loaded into the application. This well-defined separation of user interface, calculational engine, and plotting application is desirable from the standpoint of software maintenance and flexibility.

Figure 25.2 shows the data entry screen for the MTF-related properties of the linear sensor in the film scanner. The screen contains: (1) fields; (2) buttons; and (3) pull-down options menus. Fields contain numerical values or strings that may be specified by the user. Buttons cause a particular action to occur a single time, such as bringing up a file finder. The pull-down options menus serve a number of functions, including setting, activating, and deactivating other options menus, fields, and buttons. The first pull-down options menu, "Sensor MTF", allows the user to choose between use of a parametric model, measured data, or a unit MTF. Until a user clicks on the menu, only the selected option, in this case "Parametric Model", is shown, thereby avoiding clutter on the screen. A unit MTF might be selected when a very general block diagram is being used to represent multiple systems, and the particular system being modeled does not contain a linear sensor. A unit MTF is also used when a composite measurement has been made that includes more than one component. For example, if the combined MTF of a digital writer and an imaging medium has been measured, both components may be included in the block diagram, but to avoid double counting, the MTF of one of the two components would be set to unity.

With "Parametric Model" selected, the field for the path to measured data and the button for invoking a file finder are deactivated, whereas options menus associated with the parametric model are activated (e.g., those relating to aperture size and shape, scanning motion during integration, and charge carrier diffusion). The opposite situation would occur if use of measured data had been selected instead. By inferring intent from the options menu settings, the software leads the user to enter only the minimum amount of information required to fully specify a calculation. If the user changes the setting of the "Sensor MTF" pull-down options menu, the configuration of activated and deactivated elements on the data entry screen is changed immediately, to reflect the newly signaled intent.

25.3 Supporting Resources

In Fig. 25.2, if the user chooses to load measured MTF data, the "Path" field is activated, and a default path is displayed in the field. To modify the path, the "Select" button is pressed, which invokes a file finder. Information is passed to the file finder regarding the type of file needed and the location of database files of this type. Users may switch between the public database and their private data repositories. Only files of the proper type are displayed in the file finder for selection. The public database contains data measured according to standard, documented protocols. Although the assignment of the default data may be somewhat arbitrary, it is desirable to choose a measurement that is representative of current technology. In many analyses, attention is focused on one component, but it is important to analyze the component in the context of a representative imaging system. In such cases, the analyst, who may be considerably more familiar with the component under consideration than with other elements of imaging systems, can partly rely on the provided defaults to define a reasonable reference position. Well-chosen defaults help to enable users without full systems expertise to perform reliable and useful analyses efficiently.

The predictions of parametric models should be compared to representative measured data to ensure suitability and to reduce the probability of discrepancies when switching between parametric estimations and measurements. This comparison, combined with engineering knowledge, can help establish the range of values that are typical for each parameter in the model. Default values for parametric model fields should be chosen to yield predictions that resemble the default measured data.

Although parametric models are often simple, most require at least some explanation for effective use by users other than specialists. Information on parametric model usage and recommended input values, as well as instructions for use of data entry screens and discussions of measurement caveats, are included in online help screens. As an example, the help screen associated with the data entry screen of Fig. 25.2 is included in Appendix 4. Also included in the online help is a step-by-step tutorial example demonstrating the application of the software to solve a product design problem. New users can run this tutorial

Integrated System Modeling Software

to learn how to apply the software effectively. Our tutorial takes about 90 minutes to complete, but the time expended is usually recovered during the first major analysis undertaken by the user.

25.4 Program Output

The image quality modeling program output consists of:

1. key measures data;

2. objective metrics;

3. associated quality changes caused by each attribute;

4. overall multivariate quality; and

5. summary performance statistics.

Key measures data includes both component and cumulative system measures. For example, in the system of Fig. 25.1, there would be MTF, NPS, and MCS curves for each of the eight components and cumulative system curves showing the composite MTF, NPS, and MCS after each component in the system. The cumulative system curves at a given component generally reflect the properties of that and all preceding components, according to the propagation models of Ch. 22. Summary performance statistics include cumulative distribution functions of overall quality, individual perceptual attributes, objective metrics, and diagnostic quantities such as camera exposure. These statistics are based on Monte Carlo calculations of performance, which are discussed in Ch. 28.

The software output is written to one of several files. Tabulations of average or capability values of objective metrics, attribute levels, and total quality are also shown in the messages area at the bottom of the main screen, as depicted in Fig. 25.1. Key measure data and summary performance statistics are usually examined in an associated plotting package, which may be invoked from the main screen, and which is customized to facilitate analysis of the data produced.

25.5 A Sample Modeling Session

Suppose that an analyst were setting specifications for a proposed film scanner. To predict the effect that scanner hardware characteristics might have on the imaging system quality, it is necessary to specify certain aspects of the image processing that will be done prior to the output of a digital image file. For example, the scanned image would probably be sharpened by a spatial filtering operation to partially compensate for unavoidable loss of high-frequency response in the scanning process. In addition to boosting the MTF, the spatial filtering operation will amplify existing artifacts such as noise, and may create new oversharpening artifacts (Sect. 16.5). Therefore, the existence of an optimal position of compromise, which best balances these effects, may be anticipated. To provide an example of the practical use of system image quality modeling software, in this section we follow the steps taken by the analyst to optimize a spatial filtering procedure.

After building the block diagram in Fig. 25.1, the analyst reviews the default settings in each data entry screen. The camera lens affects system MTF but not its NPS, because no noise has yet been introduced to the system. The default MTF setting might entail use of measured data from an average quality 35-mm format compact camera lens at an intermediate aperture and at its best focus position. The analyst, whose expertise is scanner hardware, temporarily accepts this default, but makes a note to try different camera lens MTFs later to see how robust the derived spatial filtering operation is to such variations. The camera film affects both the system MTF and NPS, with higher speed films typically introducing higher noise levels. The default MTF and NPS are measurements from a current color negative film, of a popular ISO speed, receiving a normal ISO exposure. Wanting to adopt a conservative approach with regard to artifact amplification, the analyst searches the database and replaces the default measurements with those from a higher speed film having higher noise, which will shift the identified optimum to a less aggressive boost level. Essentially, the analyst is choosing to optimize the spatial filtering process for more critical, rather than average, conditions, which is a common approach.

The next four components are all part of the film scanner, although the image processing steps of spatial filtering and interpolation could also be done externally via independent software. This analysis is being carried out at the product inception stage, so no measurements are available for the actual scanner. Consequently, parametric calculations are used to estimate key measures of the film scanner elements. The lens that images the film onto the linear sensor affects system MTF and system NPS, the latter because the sensor MTF shapes the system NPS, in accordance with the Doerner equation (Eq. 22.14). At this stage, detailed lens design data is not yet available, so a diffraction-limited parametric model of the lens MTF is chosen (Section 23.3). The analyst specifies an f-number slightly higher than that likely to be used, to degrade the MTF slightly (through greater diffraction) and thereby crudely emulate the MTF loss arising from aberrations in the real lens. A pull-down options menu is set to

Integrated System Modeling Software 375

automatically calculate the lens magnification based on the film format and sensor dimensions.

In the linear sensor MTF data entry screen of Fig. 25.2, a parametric MTF model is selected. The pixel pitch (spacing between pixel centers) is specified and through an options menu the program is instructed to assume a unit fill factor, implying that the entire pixel area is sensitive to light. Even though parts of the pixel are of necessity insensitive because of circuitry and/or other limitations, this condition can be closely approached by using lenslets that collect one pixel area of light and concentrate it on the active part of the pixel. One-pixel scanning motion is specified, meaning that the film moves continuously past the sensor (or vice versa) at such a rate that the translation distance is one pixel during the integration plus readout time of the sensor. Some blur is introduced by this motion, but the mechanical advantages of continuous motion compared to stepped motion, make this the most common situation. The analyst accepts the default values for charge carrier diffusion parameters after reading the online help, which supplies information indicating that the provided values are likely to be approximately correct given the sensor technology. Finally, the analyst identifies the direction of the scanning motion relative to the film format and indicates the total number of pixels collected. This information, with the pixel pitch, allows the scanner lens magnification to be calculated automatically. On a separate data entry screen the parametric noise model values, which include those described in connection with Fig. 23.1, are entered.

Under the spatial filter component, a database of useful convolution kernels are listed. Given a knowledge of processing speed requirements, a small set of candidate kernels of appropriate size may be identified for testing in an unsharp masking procedure (Sect. 16.3). Initially, one candidate is selected and, with the cursor in the unsharp masking gain field, an "Iterate" function is invoked from a pull-down options menu. The "Iterate" function permits specification of a series of values in one or more fields. For example, the analyst may enter "0 to 5 by 0.5" to obtain calculations at eleven values of gain that are likely to bracket the best value. Once the approximate value of the best gain is known, a second program execution can be made over a narrower range of values to improve precision. The calculated MTF of the unsharp masking procedure will affect both system MTF and NPS.

Next, an interpolation factor and interpolation method are specified. This component is not needed if the number of scanned pixels equals the number of displayed pixels, after accounting for any image cropping to adjust aspect ratio. The analyst chooses to estimate the MTF of the video monitor using a parametric model like that described in connection with Fig. 23.2, because that will allow particularly convenient testing of the impact of monitor addressability (resolution) on the recommended unsharp masking gain. Finally, characteristics of the viewing environment, notably viewing distance, are specified under the viewing component. This completes the initial system specification. Because spatial filtering significantly affects only image structure attributes, additional information regarding other aspects of image quality would probably be accepted at default values at this stage of analysis.

The analyst now executes calculations and locates the unsharp masking gain that best balances the effects of sharpness, noise, oversharpening, and other artifacts on quality. This optimum may be identified by reference to the overall image quality in JNDs based on those attributes included in the analysis and the multivariate formalism (Ch. 11). An example of the type of results obtained from such an analysis are given in the next chapter. The multivariate sum may also be used to select which of several potential kernels produces the highest quality at its best gain level. The analyst then returns to particular data entry screens to test the robustness of the result against variations of factors that are outside the control of the film scanner design (such as camera lens focus) or are not precisely known at present (e.g., scanner lens MTF). With the tentative spatial filtering process identified, more complete analyses of the expected system image quality become possible.

In addition to the benefits listed earlier, constructing general software for system image quality modeling organizes available knowledge in a particularly clear and accessible fashion, enabling a broader understanding of image quality by a wider audience. In addition, by periodically reviewing the analyses that have been done successfully with the software, and even more particularly by identifying needs that could not be met because of scientific limitations, it is possible to prioritize future research efforts in a relevant and objective manner.

25.6 Summary

The construction of a unified software package for system image quality modeling has a number of advantages. Because such an application will be reused many times, it may incorporate more general and rigorous calculations, and be tested more extensively, than specialized programs. Greater resources can be expended on the interface and online help, including tutorials, making the software usable by a larger number of people with a wider range of skills and training. The latter has a number of benefits, such as ownership of results by product development teams, rapid cycle time, and improvement of the systems understanding of the analysts. Furthermore, construction of general software

Integrated System Modeling Software 377

organizes the available knowledge of image quality effects, which has many beneficial aspects, including facilitation of research project prioritization.

The generalized software should incorporate parametric models and provide direct access to databases of measurements made according to documented protocols. Carefully chosen default settings help to enable users without full systems expertise to perform reliable and useful analyses efficiently. Default measurements should be chosen to represent current technology. Default values in parametric models should be chosen to approximately reproduce the default measured data. Online help screens should include information such as instructions for use of data entry screens, recommended values for parametric models, and discussions of measurement considerations.

26

Examples of Capability Analyses

with Richard B. Wheeler
Eastman Kodak Company
Rochester, New York

26.1 Introduction

As will be discussed in greater detail in the next three chapters, the performance of a system is characterized by the distribution of quality (or quality attributes) produced by the system in the hands of actual customers, subject to practical levels of variability in manufacturing and usage. Although many applications of image quality modeling require complete performance analyses, in some instances simpler calculations are sufficient and may be preferable because of the rapidity and ease with which they are carried out and interpreted.

The simpler analyses are usually based on evaluation of quality under one or a few selected conditions, which nearly always include the conditions producing the peak quality of which the system is capable, i.e., its capability. We therefore will refer to these as capability analyses, even though conditions corresponding to other points on the quality distribution may also be assessed. One application in which capability analyses are often adequate is the specification of aims and tolerances for certain properties of individual components. Because component specification is such a common application of image quality modeling, this chapter is devoted to presentation of a variety of brief examples of this type. Sections 26.2–26.7 each describe a different application of capability analysis to component specification.

26.2 Unsharp Masking Gain Selection

In the previous chapter, optimization of the unsharp masking gain in a digital boost operation exemplified the use of image quality modeling software. Figure 26.1 shows the results of such an analysis. The x-axis is unsharp masking gain, with a gain of zero corresponding to no boosting. The four curves show quality changes associated with: (1) unsharpness; (2) noise, which is amplified by the boost operation; (3) edge artifacts from oversharpening; and (4) the multivariate sum of these three contributors. As the gain is increased, sharpness improves but the noise becomes worse and, above a gain of ≈2, oversharpening begins to degrade quality noticeably. The multivariate sum is optimized at a gain of ≈2.6.

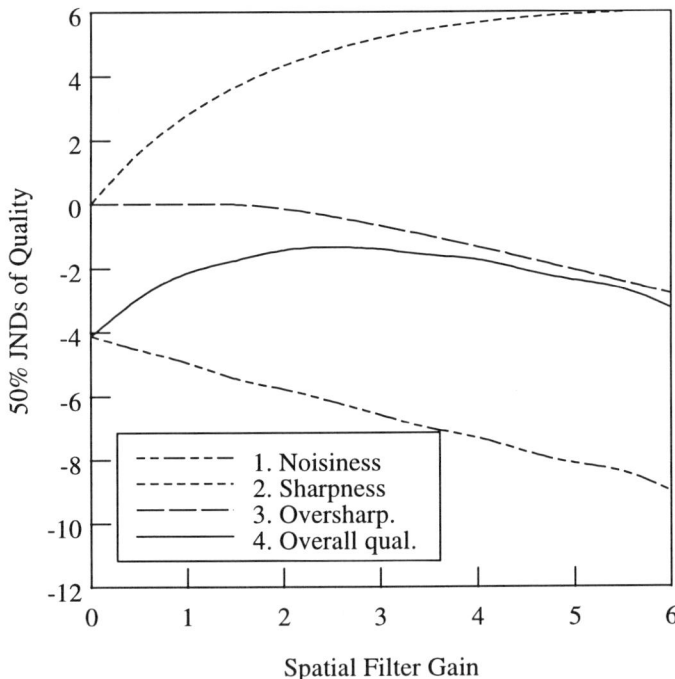

Fig. 26.1 Quality versus unsharp masking gain. As the spatial filter boost is increased, sharpness improves but noise is amplified and edge artifacts (oversharpening) become more noticeable, leading to the existence of an optimum gain value.

Examples of Capability Analyses

Fig. 26.2 Original, unsharp image that has not been boosted, serving as the first of three images in an unsharp masking gain series.

Fig. 26.3 Image boosted to a reasonable balance between unsharpness, edge artifacts (rims along the outlines of profiles), and amplified noise.

Examples of Capability Analyses

Fig. 26.4 Oversharpened image, exhibiting edge artifacts, amplified isotropic noise, and a harsh appearance in the faces.

If the noise in the system were substantially lower, so that it was initially subthreshold, the optimum gain value would have been greater because there would have been no quality loss from noise until the noise reached suprathreshold levels.

The changes occurring as unsharp masking gain is increased, are demonstrated using pictorial images in Figs. 26.2–26.4. Figure 26.2 depicts the original, unsharp image. Figure 26.3 shows an approximately optimal degree of sharpening for the average observer. Some noise is evident in the uniform background, and there are dark rims along the outlines of the people against the light background. This edge artifact is caused by oversharpening, and its appearance might be compared to that in Fig. 16.1, which has white rims instead of dark ones, because the boosting was done in a different signal space (exposure vs. density). Figure 26.4 shows excessive boost, leading to obvious noise in the background, severe edge artifacts, and a harsh appearance in the faces of the subjects.

26.3 Scanner Optical Positioning Tolerances

In our second example, we consider positioning tolerances in a film scanner and their dependence on film format and scanner lens aperture. The film frame being scanned is held in a mechanical gate so that the portion of the frame that is actively being scanned is approximately flat and lies within a specified plane. The spacing between the film and the scanner lens that images the film onto a sensor must be controlled to very tight tolerances to maintain acceptable focus. Deviation of this spacing from the best focus distance leads to reduced MTF.

Figure 26.5 shows quality loss arising from unsharpness as a function of film position error in mm. The results are shown for several combinations of film formats and scanner lens apertures. The reference system, Curve #1, pertains to a 35-mm format scan. Because the Advanced Photo System format is smaller than 35-mm format, it requires about 1.4× higher printing magnification to reach the same final print size. As can be seen by comparing Curves #1 and #2, at fixed lens aperture this causes a given position error and resulting lens MTF degradation to produce a greater sharpness and quality loss in the case of the smaller format. Comparing Curve #3 with Curve #1 shows what happens when the scanner lens is stopped down two stops (f-number doubled) at fixed printing magnification. The 2× increase in depth of focus (see Eq. A5.8, Appendix 5) substantially decreases the rate of quality loss with positioning error.

Examples of Capability Analyses

If a maximum quality loss in JNDs arising from positioning error is specified, a corresponding positioning tolerance may be inferred graphically as shown in Fig. 26.5, where a one JND limit has been adopted. This criterion yields maximum permissible film positioning errors of 0.15, 0.11, and 0.35 mm for Curves #1 through #3, respectively. The ratio of the first two tolerances is nearly identical to the ratio of the print magnifications of the two formats, as would be expected from the blur circle approximation, which was mentioned previously in connection with parametric estimation of defocused lens MTFs (Sect. 23.3) and is developed in greater detail in the next example. Similarly, the ratio of the first and third tolerances (0.35/0.15 ≈ 2.3) is fairly close to the 2× ratio of f-numbers, also as expected from the blur circle approximation.

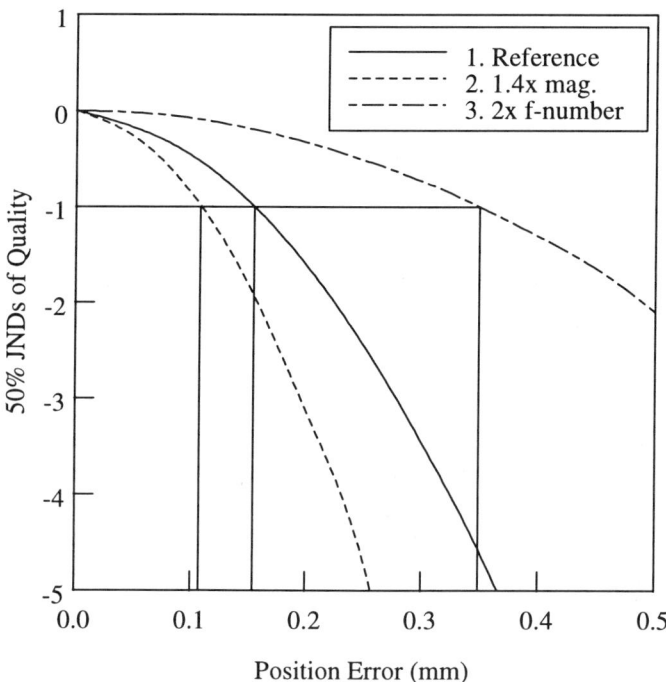

Fig. 26.5 Film positioning tolerance in scanners as a function of scanning lens aperture and film format. At an equal quality loss from defocus, smaller film formats (and thus higher magnifications) and lower f-number scanner lenses (which support greater throughput) require tighter positioning tolerances.

Although increasing the lens f-number relaxes positioning tolerances, it decreases scanner throughput by requiring longer sensor integration times for proper exposure. This might be compensated by a brighter illumination system, but increased cost, decreased lifetime, greater power consumption, and/or greater cooling requirements would be likely to result. This type of interaction between subcomponent properties is quite common, and image quality modeling is very helpful in identifying the best compromise among competing factors.

26.4 Autofocus Ranging and Lens Aperture Specifications

The third example demonstrates another subcomponent interaction, between maximum autofocus ranging distance and minimum camera lens f-number. It will be convenient to describe the behavior of the autofocus system in terms of the amount of blur produced by a particular amount of defocus (i.e., the distance from the best focus position to the actual image plane). In the geometrical optics approach, a defocused point is approximated as a solid disk, which is called the blur circle. As shown in Fig. 23.3, although the blur circle approximation does not accurately predict lens behavior near best focus, it becomes more accurate at greater defocus, and so can reasonably be used to set defocus tolerances. The maximum tolerable value of the blur circle diameter (without unacceptable loss of sharpness) is usually called the circle of confusion. Image quality modeling such as that in the preceding example may be used to establish a criterion for the allowable circle of confusion based on the permitted loss of quality from unsharpness.

Defocus may be shown to be proportional to the difference of the reciprocal distances between the best focus position and the actual object plane (Appendix 5, Eq. A5.6). For example, if a camera lens were focussed at a point ten meters away, the amount of defocus, and the diameter of the blur circle, would be equal for points five meters away and infinitely far away, because the reciprocal object distance differences would be equal, i.e., $5^{-1} - 10^{-1} = 10^{-1} - \infty^{-1}$. Furthermore, as shown in Eq. A5.3, the blur circle diameter is equal to the amount of defocus divided by the lens f-number. Consequently, a plot of blur circle diameter against reciprocal object distance is a straight line, and the absolute value of its slope is inversely proportional to lens f-number (Eq. A5.7).

Figure 26.6 presents blur circle diameter versus inverse object distance plots for four different autofocus subcomponent designs. In each of the four plots, zero blur circle diameter is indicated as a fine dashed line and the allowable circle of confusion is depicted as a coarse dashed line. Consider the first (uppermost) of the four plots. A five-zone autofocus equally partitions the reciprocal object distance space from infinite distance, at the far left, to the minimum camera focus distance at the far right. As the reciprocal object distance deviates from a position corresponding to a lens set point (perfect focus), the blur circle diameter increases linearly. When the switch to the next set point occurs, the blur circle diameter trend reverses and it decreases linearly with further change in reciprocal object distance until perfect focus occurs and the cycle starts over. In this case the blur circle is equal to the circle of confusion at the switch point, so the autofocus design is adequate but has no margin for error. The maximum distance that can be measured by the ranging system is shown by an asterisk; object distances between this value and infinity are not distinguishable from one another, and so a single set point must be chosen to span this range.

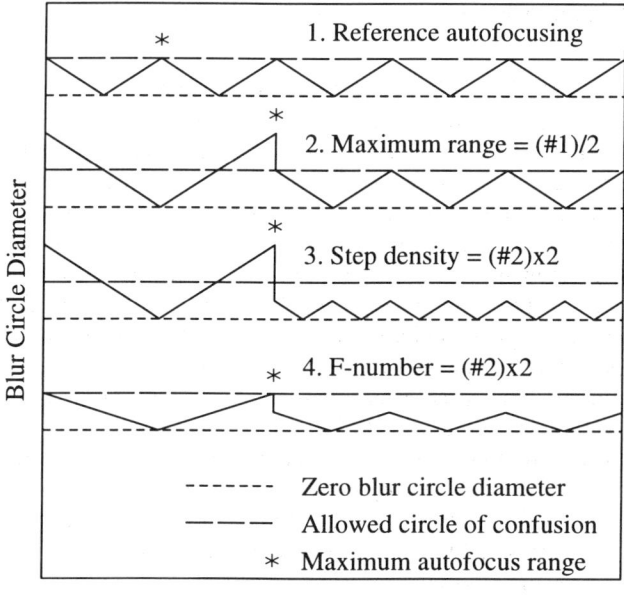

Fig. 26.6 Defocus versus reciprocal object distance for four combinations of minimum camera lens f-number, maximum autofocusing range, and number of autofocus zones (see text). An inadequate maximum autofocus range leads to the requirement for a higher minimum camera lens f-number, reducing low-light capability.

In the second plot, a maximum ranging distance half as large as in the first plot leads to unacceptable blur circle diameters at some distances because adequate ranging information in not available. Doubling the density of autofocus zones as in the third plot reduces blur circle diameters below the maximum ranging distance, but has no affect above it, again because the information needed to divide up the zone deterministically is lacking. In the fourth plot, the lens f-number has been doubled instead, thereby halving the rate of change of blur circle diameter with defocus. This brings the blur circle diameters back within tolerances at all distances, but at the expense of light-gathering capacity, which is reduced by two stops (4×). Thus, the required maximum autofocus ranging distance and the supported minimum lens f-number are interdependent, and neither should be specified in the absence of consideration of the other.

26.5 Spot Separation in Optical Anti-Aliasing Filters

The fourth example involves an analysis of the interaction of sharpness and color filter array (CFA) interpolation artifacts arising principally from aliasing. As discussed in Sect. 18.2, aliasing can be suppressed by band-limiting the signal prior to a sampling operation. The bandwidth of the signal that is sampled in a digital still camera is affected by the frequency content of the original scene, the camera lens MTF, and the sensor MTF. The sensor MTF decreases as the size of the light-sensitive (active) area of each pixel increases. The ratio of active pixel size to the pixel spacing (pitch), which is called the fill factor, has a strong impact on sharpness and aliasing. The fill factor can closely approach one in full-frame devices, but is often closer to one-half in interline devices, although lenslet arrays on the sensor may be used to concentrate light into the active pixel area, thereby effectively increasing the fill factor. At constant pixel pitch, lower fill factors correspond to smaller active areas and therefore higher MTFs, which increase both sharpness and aliasing. In a monochrome sensor (no CFA), if the fill factor is close to one, the active area MTF band-limits the signal to a degree that is reasonable for the sampling frequency (reciprocal of the pixel pitch), and aliasing is rarely serious.

When a CFA is present, the active area MTF no longer sufficiently band-limits the signal, relative to the sampling frequency in the individual color channels, so significant color-dependent (chromatic) aliasing may occur. Additional optical anti-aliasing filtration may then be desirable, depending upon the frequency content of the original scene and the camera lens MTF. If the sensor fill factor is significantly lower than one, the importance of such anti-aliasing filtration is even greater. One practical method for optically band-limiting a signal in a well-controlled manner is to use a birefringent filter (Hecht and Zajac, 1976). If a

Examples of Capability Analyses

beam of monochromatic light passes through a birefringent material, two beams each with half the flux emerge. The separation between the two spots can be controlled by material thickness, and will be wavelength-dependent if the material is dispersive (Greivenkamp, 1990). By cementing two such filters together at right angles to one another, a diamond-shaped pattern of four identical spots may be produced. By varying spot separation, a precisely controlled amount of two-dimensional blur may be introduced into an image.

Figure 26.7 shows an example of the quality losses arising from unsharpness (coarse dashes), aliasing (fine dashes), and their multivariate sum (solid lines) versus birefringent filter spot separation for two fill factors, 0.5 (circles) and 1.0 (squares). The spot separation is expressed as a fraction of the pixel pitch, which

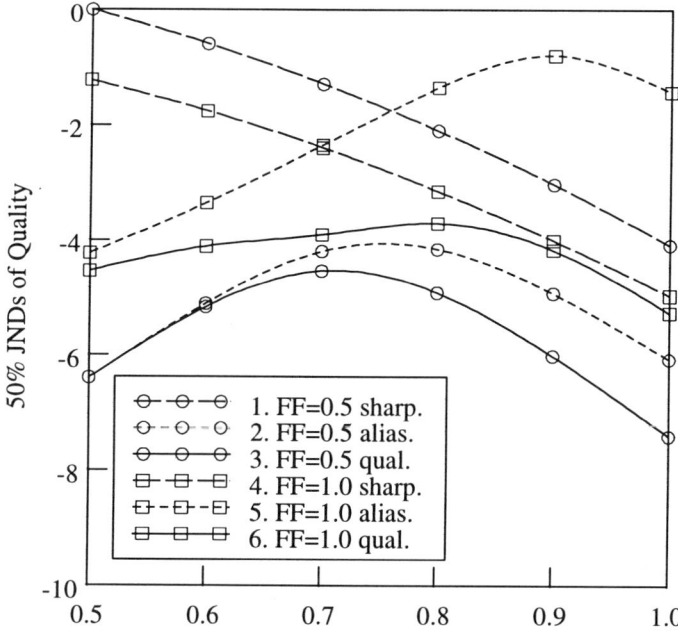

Birefringent Filter Spot Separation (fraction of pixel pitch)

Fig. 26.7 Quality changes versus birefringent anti-aliasing filter spot separation. Optical anti-aliasing filters trade quality loss arising from unsharpness with that caused by aliasing; the optimum spot separation depends upon the sensor fill factor (FF).

is constant in this example. As expected, the higher fill factor leads to lower MTF because the signal is integrated over a larger active spatial area. This reduced bandpass relative to the sampling frequency decreases the amount of aliasing, however, leading to the exchange of sharpness for aliasing as fill factor is increased. As the spot separation increases, the MTF decreases, and, initially, the aliasing decreases also, as anticipated. However, the poor stop-band characteristics of the birefringent filter lead to periodic variation in the amount of aliasing as a function of spot separation, so the amount of aliasing actually starts to increase beyond a certain separation.

The position of the peak in the multivariate sum may be identified as the best compromise between sharpness and aliasing, which occurs near spot separations of ≈0.71× and ≈0.80× the pixel pitch at fill factors of 0.5 and 1.0, respectively. The peak quality of the 1.0 fill factor exceeds that of the 0.5 fill factor by about one JND in this case. Apparently increasing fill factor provides a more favorable exchange of sharpness and aliasing than does increasing spot separation, probably because of the poor stop-band characteristics of the birefringent filter.

26.6 Capture and Display Resolution

Our fifth and sixth examples pertain to the interaction of capture and display resolution. The number of pixels needed in a digital still camera for different final image sizes and electronic zoom cropping factors is investigated in the fifth example. The sixth example involves the prediction of the impact of output resolution on quality, as a function of capture resolution. These examples assume an imaging system having: (1) a digital still camera with a high quality lens, a birefringent anti-aliasing filter, and an adaptively interpolated Bayer CFA pattern; (2) an image processing pathway including an optimized unsharp masking procedure and a cubic convolution interpolation to the requisite number of output pixels; and (3) digital writer output onto a high quality reflection medium with a well chosen spot size to pixel pitch ratio.

Figure 26.8 shows quality at 300 pixels per inch output versus camera megapixels for three display conditions: (1) a standard 4 × 6-inch (4R) print; (2) an 8 × 10-inch print; and (3) a 4 × 6-inch print that has been cropped by 2×, as might be done using an electronic zoom feature in the camera. The 8 × 10-inch print is more demanding of system characteristics than a 4 × 6-inch print because the printing magnification is 2× greater. The 2× cropped 4 × 6-inch print has the same printing magnification as the 8 × 10-inch print, but is even more demanding because hand-held viewing distances are shorter for smaller prints. Typical hand-held viewing distance d_v (mm) depends on d_D, the diagonal of the viewed image (mm), as shown in Eq. 26.1 (Wheeler and Keelan, 1994).

$$d_v = 288 \cdot \log_{10}(d_D) - 312 \qquad (26.1)$$

As can be seen in Fig. 26.8, in the uncropped 4 × 6-inch case, the quality quickly saturates, with only about one JND of quality to be gained above 2 megapixels. In contrast, in the 8 × 10-inch case, the asymptotic limit is approached more gradually, and a higher number of capture pixels is warranted. In the now common case of 2× electronic zoom in standard 4R prints, quality is still improving noticeably with increasing capture resolution at 6 megapixels. This

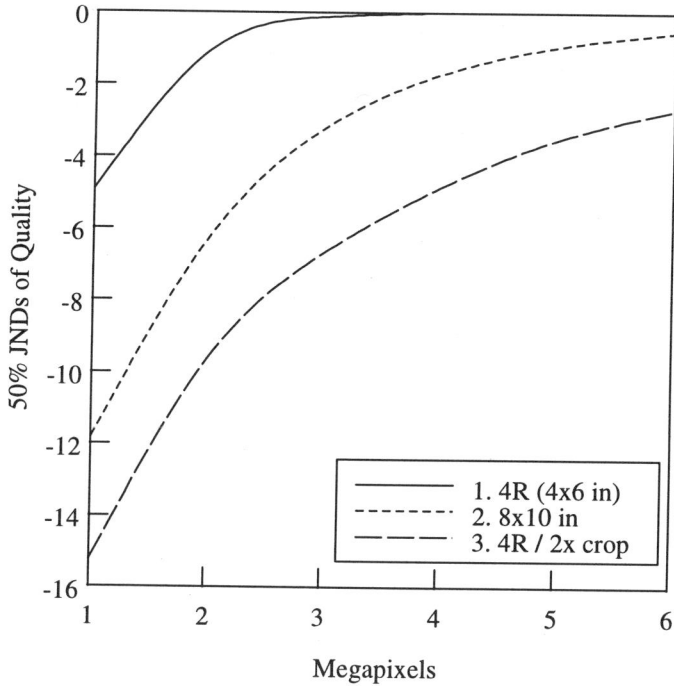

Fig. 26.8 Quality versus number of capture pixels for three types of output. With uncropped 4 × 6-inch (4R) prints, there is diminishing quality improvement above 2 megapixels, but larger or cropped prints continue to improve noticeably in quality at higher capture resolutions.

result is not surprising because only $6/2^2 = 1.5$ megapixels actually contribute to the final cropped image.

Figure 26.9 depicts the dependence of quality on the number of output pixels per inch at three capture resolutions, 0.4, 1.6, and 6.4 megapixels. We quantify the output in terms of addressable pixels per inch rather than dots per inch because we desire a quantity that relates to spatial resolution. In halftone output, there may be many dots per pixel so that intermediate gray levels, rather than just black and white levels, may be represented at a given pixel. In the 0.4 megapixel case, the capture resolution severely limits the possible quality of the system, and quality falls off slowly with decreasing output resolution even below 256

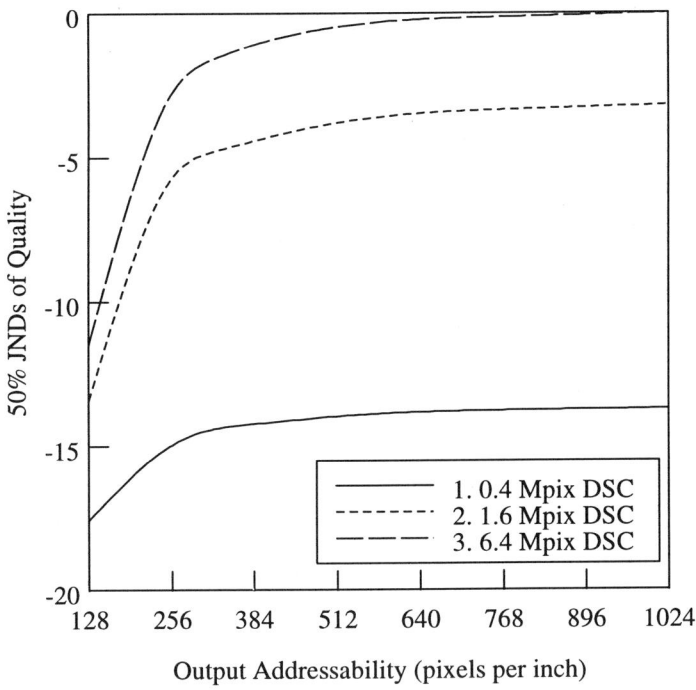

Fig. 26.9 Quality versus output addressability for several digital still camera (DSC) resolutions (megapixels). As capture resolution increases, a correspondingly greater demand is placed on output resolution, causing it to affect quality more strongly and to approach asymptotic behavior more slowly.

pixels per inch. In the other two cases the capture resolution is not limiting, so low output resolutions significantly degrade final image quality. At these adequate capture resolutions there is about one-half JND of improvement possible above 512 pixels per inch.

When an output device has selectable output resolution, quality may be exchanged for printing speed. For example, in one inkjet printer measured, a 4× increase in output resolution and use of unidirectional printing increased printing time by about 6× but also led to a roughly three JND improvement in quality.

26.7 Digital Image Compression and File Size

Our seventh and final example pertains to the compression of digital images to reduce file size, thereby saving storage space and increasing transmission speed. The information contained in a digital image consists of the code values in each color channel at each pixel location. The simplest way of representing this information is to store the code value for each pixel and channel at a fixed quantization level, frequently chosen to be 8 bits ($2^8 = 256$ levels). Although straightforward, this approach is inefficient. There is some correlation between the code values of pixels that are near one another spatially, because they are likely to represent similar scene tones and so have similar values. Furthermore, as discussed in Sect. 24.3, there is also correlation between code values of different channels. These correlations lead to a certain amount of numerical redundancy, which can be exploited by a compression algorithm to reduce the total number of bits required to represent the data.

For example, to reduce color channel redundancy, the green code values at each pixel may be subtracted from the red and blue code values, so that only chromatic differences remain. This is analogous to the formation of chroma channels in color filter array interpolation (Sect. 24.3). The differences of code values typically are smaller numbers than the code values themselves, and so can potentially be encoded using symbols requiring fewer bits. Similarly, if the code values are ordered by channel, and within a channel are ordered in a spatial progression, spatial correlation and redundancy can be reduced by encoding the difference between the current code value and the previous code value, rather than the current code value itself. The code value difference between adjacent pixels in a given channel will usually be small unless an edge is traversed, and these smaller values can be encoded using fewer bits, thereby saving space.

Efficient numerical encoding of digital image information often reduces the required electronic file size by a factor of 2–3×. Compression of this type is called lossless, because the original code values in each channel at each pixel can be exactly reconstructed by inverting the encoding scheme. In many applications, even higher compression ratios are needed to reduce file size further. In such instances, a lossy compression process may be employed, at the expense of errors in the reconstructed image, which can lead to quality loss.

There are a number of different lossy compression techniques. In many of them, the information in each color channel is decomposed into spatial frequency components (Rabbani and Jones, 1991). This frequency decomposition allows the information in each channel and frequency band to be encoded separately. Because the human visual system is less sensitive to higher frequencies, especially in chroma channels, larger losses of information can be tolerated in these cases. Information reduction is achieved by more coarsely quantizing the signal to produce fewer possible numerical values, which can be encoded more compactly. More aggressive quantization will lead to greater loss and greater resulting compression. By quantizing the higher frequency signals and chroma signals in a fashion that reflects the discrimination of the human visual system, greater compression ratios may be achieved at a given level of quality loss.

Two schemes have been standardized for compression of still images by the Joint Photographic Experts Group (JPEG) committee: the JPEG (Pennebaker and Mitchell, 1992) and JPEG 2000 (Taubman and Marcellin, 2001) methods. There are many feature differences between JPEG and JPEG 2000, with the latter offering significantly greater flexibility. The most notable difference from an image quality standpoint is the type of frequency representation employed. The JPEG method uses a discrete cosine transform (DCT) of 8×8 blocks of pixels, whereas the JPEG 2000 method uses a discrete wavelet transform (DWT) of the entire image. The frequency representation affects the nature of artifacts produced when the signal is quantized during lossy compression, and the balance between loss of sharpness and artifacts. For example, because 8×8 blocks of pixels are treated independently in the JPEG method, the individual blocks may become evident at high compression ratios, producing what is called a blocking artifact. In contrast, because JPEG 2000 transforms the entire image, rather than small blocks within the image, it does not normally produce blocking artifacts. Another difference between the standards is that the JPEG 2000 method uses an adaptive arithmetic encoding scheme that is often more efficient than the Huffman encoding employed in the JPEG DCT method.

Figure 26.10 shows quality loss caused by compression as a function of compressed file size for JPEG DCT and JPEG 2000 methods. These results pertain to the mean observer and scene and are applicable only for specific viewing conditions and particular quantization strategies, but they provide a

Examples of Capability Analyses

fairly representative comparison of the two methods. The quality loss values include the effects of artifacts and the loss of sharpness that accompanies coarser quantization of higher spatial frequencies. This quality loss function is a perceptually calibrated example of what is called a rate-distortion (R-D) curve in the compression literature. As shown in Fig. 26.10, if the permissible quality loss from compression were specified to be two JNDs, the achievable file sizes for the two methods would be ≈123 kilobytes for JPEG and ≈107 kilobytes (13% smaller) for JPEG 2000. The 13% smaller file size at equal quality with JPEG 2000 results from both the reduction of blocking artifacts and from the more efficient encoding scheme. As seen in Fig. 26.10, the reduction in file size at equal quality with JPEG 2000 becomes even more pronounced at higher compression ratios

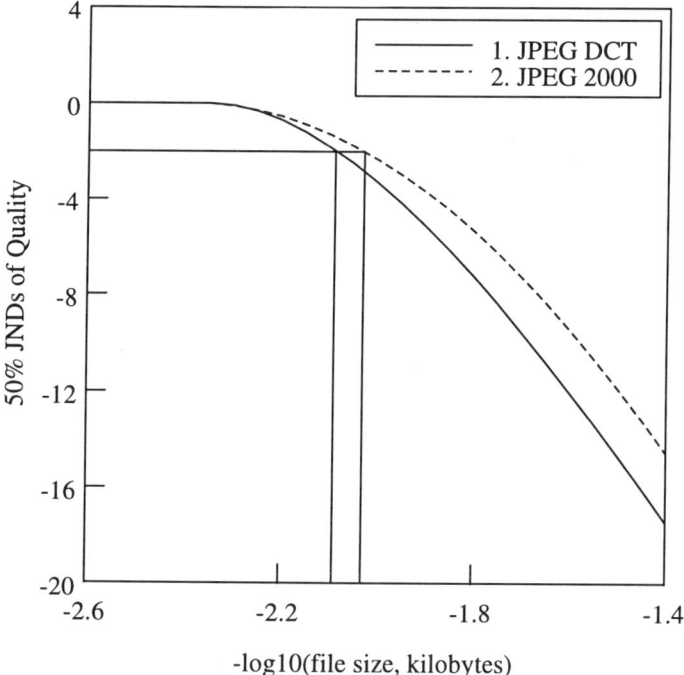

Fig. 26.10 Quality loss arising from artifacts and unsharpness caused by two standard lossy compression methods. The JPEG 2000 method is superior under these test conditions, producing smaller file sizes at equal quality loss.

The examples described in this chapter demonstrate a variety of uses of capability modeling to set component specifications. The next three chapters will: (1) explain the interpretation of image quality distributions; (2) describe Monte Carlo modeling of performance; and (3) provide a number of examples of analyses based on performance.

26.8 Summary

Capability analyses involve the investigation of system quality under one or a few selected conditions, which nearly always include the condition yielding the peak quality that can be delivered by the system. In contrast, performance analyses simulate all pertinent conditions through Monte Carlo modeling to generate the full image quality distribution expected in the hands of customers. The advantage of capability analyses lies in their relative simplicity compared to the more rigorous and powerful performance analogues. A pervasive application in which capability analyses are often suitable is the setting of component specifications, a number of examples of which were presented in this chapter. These examples showed how the characteristics of one component or subcomponent could greatly affect the requirements of other parts of the system, and how more or less demanding imaging applications can strongly influence the impact that changes in a component will have on final image quality.

27

Photospace Coverage Metrics

27.1 Introduction

The distinction between capability and performance has already been made several times in passing, and a number of examples of capability analyses were presented in the preceding chapter. This chapter provides a transition from a capability viewpoint to a performance viewpoint, the latter of which is emphasized in the remainder of the book. The transition is affected by consideration of an intermediate approach, namely, the development of metrics correlating with the yield of a photographic system. A system's yield under specified conditions is defined as the fraction of images it produces that meet or exceed some minimum quality criterion. Although yield metrics will ultimately prove to be an incomplete substitute for full quality distributions, the intuition gained in their elucidation will be of considerable value in understanding factors affecting system performance.

This chapter is organized as follows. Section 27.2 introduces the concept of a photospace distribution, which is the probability density function (PDF) of the light levels and distances at which photographs are taken. Section 27.3 discusses requirements that a system must meet to cover photospace adequately, i.e., to capture images with a high success rate at the light levels and distances most frequently encountered. System design constraints related to depth of field and exposure are considered in Sects. 27.4 and 27.5, respectively. The results of these sections are used to quantify the photospace coverage of a simple camera in Sect. 27.6. Finally, the limitations of such photospace coverage metrics are discussed in Sect. 27.7, further motivating the value of complete performance characterization of photographic systems. The derivations of a number of the equations used in Sects. 27.4 and 27.5 are collected in Appendix 5, which the interested reader may wish to consult before or after reading this chapter.

27.2 Photospace Distributions

Light levels and camera-subject distances at which consumers take photographs are two factors that lie largely outside the control of the system designer but that

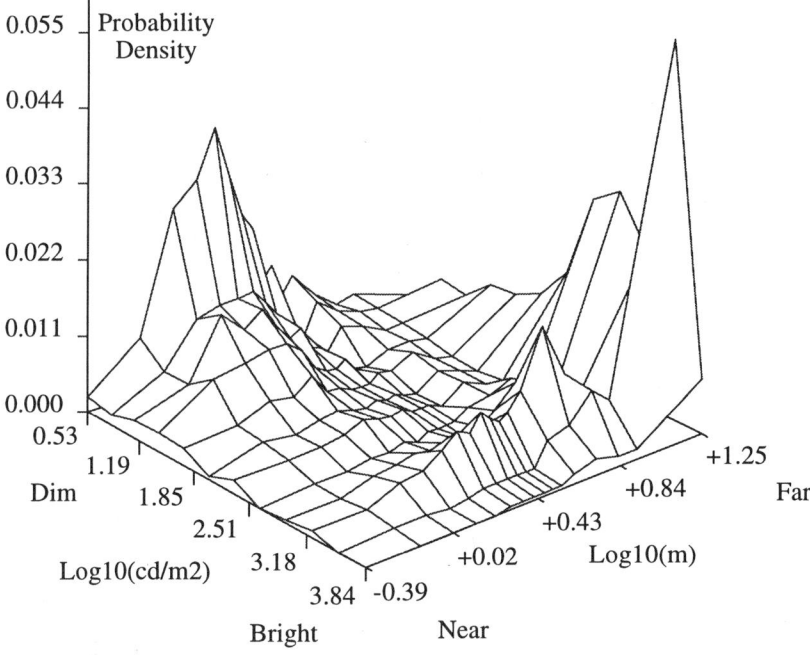

Fig. 27.1 Photospace distribution of a representative compact point-and-shoot 35-mm format camera. The peak to the right, at high light level and long distance, is produced by outdoor ambient light photographs, whereas the peak to the left, at low light level and short distance, primarily corresponds to indoor flash pictures.

substantially influence the performance of an imaging system. Imaging systems tend to perform better at high light levels, which permit use of: (1) smaller apertures for better depth of field; (2) faster shutter speeds for better freezing of subject and camera motion during exposure; and (3) lower film speeds or sensor gains, leading to lower noise and other desirable characteristics. The camera-subject distance affects many aspects of image quality, such as the level of detail visible in the primary subject, depth of field, level and uniformity of flash exposure, and severity of redeye.

We shall refer to the two-dimensional coordinate system of light level and distance as photospace. The PDF of the usage of an imaging system as a function of photospace will be called the system's photospace distribution. Finally, the distribution of situations in which photographers would like to take pictures will be defined as the photomotivation distribution (Rice and Faulkner, 1983). The photomotivation distribution corresponds to the photospace distribution that would result from a system capable of obtaining excellent images at all light levels and distances. Photospace distributions of consumer imaging systems have changed little over time, and these distributions are rather weakly correlated with the limitations of the photographic system employed. Consequently, it may be assumed that consumer system photospace distributions closely resemble the photomotivation distribution.

A photospace distribution typical of consumer imaging (for point-and-shoot cameras) is shown in Fig. 27.1. Because light meters built into cameras measure reflected rather than incident light, it is convenient to express ambient light level in terms of average scene luminance in cd/m^2 rather than illuminance in lux. Two major peaks occur in the distribution: (1) the one to the right, at high light levels and moderate to longer distances, corresponding primarily to outdoor photographs in daylight; and (2) the one to the left, at low light levels and short to moderate distance, comprised principally of indoor shots. The grid shows slices of the distribution at constant distance or light level; the shape of the slices (beyond a normalization) varies substantially in both dimensions, so the two factors are interdependent upon one another. This is expected, because being indoors places constraints upon both the maximum distances and light levels likely to be encountered, introducing a correlation between the two factors.

27.3 Photospace Coverage Requirements

The fraction of a photospace distribution that is likely to be captured in a satisfactory manner is referred to as the system yield. Objective quantities that are correlated with system yield are called photospace coverage metrics. To have good photospace coverage, a system must produce images of adequate quality with high frequency when the photospace coordinates are near either peak in the distribution of Fig. 27.1. Covering the "outdoor" peak region primarily requires that:

1. objects far from the camera are rendered sharply;

2. the exposure is constrained sufficiently to prevent quality loss from overexposure at high light levels; and

3. adequate exposure is obtained at moderate light levels to cover the considerable long-distance, low-light-level "tail" along the back right edge of Fig. 27.1.

As shall be seen in the subsequent analysis, the "indoor" peak is mostly at light levels too low to be accessible by hand-held ambient light exposures. Therefore, covering the indoor peak primarily requires that:

4. objects close to the camera are rendered sharply;

5. adequate illumination is provided by an electronic flash to reach the farther distances at which there is significant probability in the photomotivation distribution; and

6. the flash exposure is constrained sufficiently at close distances to avoid quality loss arising from overexposure.

Requirement #2, the constraint of exposure at high ambient light levels, is usually straightforward to meet, because discarding excess light is relatively easy to do, e.g., by stopping down the lens aperture or using a shorter exposure (shutter or sensor integration) time. Similarly, Requirement #6, regarding exposure at close flash range, is readily met by stopping down the lens aperture or electronically quenching (limiting) the flash output. In fact, in color negative film systems, the wide film exposure latitude may completely obviate the need for exposure control to meet Requirements #2 and #6, and thereby permit the use of fixed aperture, exposure time, and flash output. We will assume in the following discussion that Requirements #2 and #6 have been met, and will concentrate on what is required to satisfy the remaining four requirements.

27.4 Depth-of-Field Constraints

Let us next consider Requirements #1 and #4 from Sect. 27.3, which relate to coverage of the distance coordinate of photospace. The range of camera-subject distances that can be rendered sharply by a camera at a single focus point is called the depth of field. In an autofocus camera, the range of distances covered is increased because the depth of field associated with each of the multiple focus points overlap to create an extended region that can be rendered sharply, as discussed in connection with Fig. 26.6. To simplify the present analysis and allow the results to be cast in terms of easily understood equations, we will assume that the camera has a fixed focus point and a fixed aperture. These results can readily be extended to cameras with autofocus and independent shutter/aperture control. The reader may question the relevance of fixed focus and aperture cameras, but at the time of this writing, about one-half of the point-and-shoot cameras sold in the United States were of this type. The reason for their popularity is that such cameras are an excellent value, because their prices are significantly reduced by the omission of autofocus and autoexposure subsystems, while reasonable quality is still maintained.

As derived in Appendix 5, the maximum achievable depth of field occurs when the camera is focused at its hyperfocal distance d_h, at which point distances from one-half the hyperfocal distance to infinity are rendered sharply. From Eq. A5.10, the minimum depth-of-field limit is given by:

$$d_{\min} \approx \frac{d_h}{2} \approx \frac{F^2}{2 \cdot A \cdot d_c} \quad (27.1)$$

where F is camera lens focal length in mm, A is camera lens aperture (f-number), and d_c is the allowable circle of confusion (blur circle diameter in mm) in the capture plane. The minimum depth-of-field limit is the first of three photospace coverage metrics derived in this chapter; the other two are the minimum ambient light level and the maximum flash range at which ISO normal exposures may be obtained.

As capture format dimensions change, lens focal length must be changed to maintain the same field of view, and printing magnification must be changed to maintain the same final display size. The change in printing magnification further affects the allowable circle of confusion. Therefore, it is useful to recast Eq. 27.1 explicitly in terms of format size and quantities that are independent thereof, which can be accomplished using three equations from Appendix 5.

First, Eq. A5.14 relates the diagonal angular field of view θ_d to the capture format diagonal dimension d_d and the camera lens focal length F.

$$\tan(\theta_d/2) = \frac{d_d/2}{F} \Rightarrow F = \frac{d_d}{2 \cdot \tan(\theta_d/2)} \quad (27.2)$$

The second result needed from Appendix 5 concerns angular magnification, which is the ratio of the angular subtense of a feature, viewed in the final displayed image, to the angular subtense of that feature in the original scene, as viewed by the naked eye of the photographer from the camera position. A system with unit angular magnification will appear to reproduce objects at their "natural" size. Higher angular magnification is often desirable because it allows capture of images with greater detail or with similar detail, but from a more comfortable or convenient distance. Equation A5.15 relates the angular magnification m_a to the focal length F, the viewing distance of the final displayed image d_v, and the printing magnification m_p.

$$m_a = \frac{F \cdot m_p}{d_v} \quad (27.3)$$

Recall that printing magnification m_p is approximately the linear ratio of final displayed image size to capture format size. In optical printing systems it is a few percent higher than this ratio to provide overfill and permit some positioning tolerance in the printer without imaging the edge of the film frame.

The third result needed from Appendix 5 is Eq. A5.16, which relates the circle of confusion in the capture plane d_c to the final display viewing distance d_v, the printing magnification m_p, and the maximum allowable angular subtense of blur at the eye θ_{max}.

$$d_c = \frac{\theta_{max} \cdot d_v}{m_p} \quad (27.4)$$

As discussed in Appendix 5 following Eq. A5.16, a reasonable value for θ_{max} is ≈2.4 arc-minutes, which is slightly larger than the angular resolution of the eye. For example, the Snellen eye chart defines 20/20 vision based on the ability to resolve letters composed of segments subtending one arc-minute (Bartleson and Grum, 1984).

Combining Eqs. 27.1–27.4 yields an expression for the minimum depth-of-field limit as a function of format size and format-independent quantities:

$$d_{\min} = \frac{d_d \cdot m_a}{4 \cdot \tan(\theta_d/2) \cdot A \cdot \theta_{\max}} \qquad (27.5)$$

which is Eq. A5.17. Consideration of the photomotivation distribution allows identification of a desired value of $d_{\min}$. If field of view, format size, and angular magnification have been specified, Eq. 27.5 can be solved for the required aperture for adequate depth of field. From Eq. 27.1 and 27.5 it is evident that sufficient depth of field is more difficult to obtain with longer focal lengths, higher angular magnifications, and larger capture formats. This further indicates that adequate photospace coverage may be challenging to achieve in systems having zoom camera lenses, electronic zoom and crop features, and/or optional larger print sizes (e.g., panoramic images).

27.5 Exposure Constraints

Requirement #3 of Sect. 27.3, related to low ambient light capture, will be addressed next. Before deriving a general equation for the minimum light level at which a satisfactory exposure may be obtained, as a function of basic system properties, a numerical example will be worked, to help clarify several important relationships. This example, summarized in Table 27.1 on the following page, involves computation of the lens f-number required to obtain adequate depth of field for a variety of capture formats, at matched field of view and displayed image size. The combination of fixed field of view and fixed display size forces the angular magnification of all the systems to match (neglecting the small effect of overfill in the film systems), because equal scene content is mapped to equal display size, viewed from an equal distance.

We first choose a reference system having: (1) 35-mm capture format (23.8 × 35.7 mm); (2) a representative 30 mm focal length camera lens; and (3) optical enlargement to standard 4R print size (101.6 × 152.4 mm). The reference printing magnification, including 4% overfill for positioning tolerance in the printer, is m_p = 101.6·1.04/23.8 ≈ 4.44. From Eq. 27.2, the reference field of view is given by θ_d = 2·tan^{-1}[(23.8^2 + 35.7^2)$^{1/2}$/(2·30)] ≈ 71°. Point-and-shoot 35-mm systems often are designed to have depth of field from three feet (914 mm) to infinity, based on a circle of confusion of d_c ≈ 0.05 mm. These criteria imply an f-number of A = 30^2/(2·0.05·914) ≈ 9.8, according to Eq. 27.1. These values are shown in the second row of Table 27.1

Values of focal length, printing magnification, and aperture for other formats in Table 27.1 are calculated as follows. To avoid cropping issues, the long format dimension is assumed to equal 1.5× the short format dimension, even though some of the actual formats have different aspect ratios. To match field of view to the reference case, focal length is taken to be proportional to format diagonal in accordance with Eq. 27.2. Printing magnification is calculated geometrically as above, including 4% overfill in film systems. From Eq. 27.4, at constant viewing distance, the circle of confusion in the capture plane d_c is inversely proportional to printing magnification. Finally, with d_c known, Eq. 27.1 yields the aperture providing the required depth of field.

The last column in Table 27.1 gives the light-gathering potential of the systems in stops (factors of two) relative to the reference 35-mm case. This quantity is termed relative format speed, because, all else being equal, greater light gathering potential may be equated with lesser requirements for ISO capture (film or sensor) speed. As discussed below, exposure is inversely proportional to the square of the f-number, so formats requiring lower f-numbers may be capable of successful capture of images at lower ISO speeds or light levels (however, this advantage may not be realized because of associated increased lens costs and tighter positioning tolerances; see Sect. 27.7). Sensors used in consumer digital still cameras at the time of this writing have ISO speeds that are typically a couple of stops slower than those of film in analogous products, but nonetheless have similar photospace coverage because of the relative format speed advantage shown in Table 27.1.

Format	F (mm) for $\theta_d = 71°$	m_p to 4R print	A for $d_{min} = 3'$	Format speed (stops)
120 roll film	68.8	1.9	22.6	−2.4
35-mm film	30.0	4.4	9.8	0.0
6 Mpixel 9 μm sensor	23.2	5.5	7.3	+0.9
Advanced Photo System	20.4	6.5	6.7	+1.1
3 Mpixel 9 μm sensor	16.4	7.8	5.2	+1.9
110 film	16.2	8.2	5.3	+1.8
1.5 Mpixel 9 μm sensor	11.6	11.0	3.7	+2.9
Disc film	10.2	13.0	3.4	+3.1

Table 27.1 Relative speeds (in stops) of different capture formats. At equal displayed image characteristics and depth of field, smaller formats are advantaged by their greater light gathering capacity.

Photospace Coverage Metrics 405

The analysis of Table 27.1 may be generalized by deriving an equation giving the minimum ambient light level at which a satisfactory capture may be made with adequate frequency. Equation A5.20 (from ANSI/ISO 2720, 1994) gives the average scene luminance Y (cd/m^2) at which an ISO normal exposure results as a function of aperture A, ISO capture speed S, exposure (shutter or sensor integration) time T (seconds), and the reflected light metering constant K.

$$Y = \frac{K \cdot A^2}{S \cdot T} \qquad (27.6)$$

A representative value of K is 12 cd·s/m^2, which is consistent with typically assumed values of average scene reflectance and the commonly used "sunny sixteen" rule of exposure (see discussion following Eq. A5.20). The minimum ambient light level at which a normal exposure may be obtained at a given ISO speed will occur when the shutter time is maximized. As discussed in Appendix 5 in some detail, the longest practical shutter time in hand-held photography is limited by camera motion during the exposure arising from imperfect stability of the photographer. Equation A5.19 gives the maximum allowable exposure time T_{max} in terms of the camera stability ratio ρ_s (having unit value for an average camera and higher values for more stable cameras), a characteristic rate of rotation of an average camera during exposure ω_c (radians per second), angular magnification m_a, and allowable angular blur at the eye θ_{max}.

$$T_{max} = \frac{\rho_s \cdot \theta_{max}}{\omega_c \cdot m_a} \qquad (27.7)$$

As discussed in Appendix 5 following Eq. A5.19, ≈1.9 degrees per second is a reasonable value for ω_c that is consistent with the commonly used "one over focal length" rule (for 35-mm format, the exposure time in seconds should not exceed the reciprocal of the lens focal length in mm). Substituting Eqs. 27.7 and 27.5 (solved for A) into Eq. 27.6 yields Eq. A5.21 (next page), which is the desired equation relating the minimum light level allowing a satisfactory ISO normal ambient exposure (Y_{min} in cd/m^2), to two sets of complementary fundamental quantities (related by substitution of Eqs. 27.2–27.3).

$$Y_{min} = \frac{K \cdot \omega_c \cdot d_d^2 \cdot m_a^3}{16 \cdot \rho_s \cdot S \cdot d_{min}^2 \cdot \theta_{max}^3 \cdot \tan^2(\theta_d/2)}$$

$$= \frac{K \cdot \omega_c \cdot F^5 \cdot m_p^3}{4 \cdot \rho_s \cdot S \cdot d_{min}^2 \cdot \theta_{max}^3 \cdot d_v^3}$$

(27.8)

To review, K is a constant set by standards; ω_c is a measure of the stability of photographers; θ_{max} is related to the resolution of the human visual system; d_{min} is the minimum depth-of-field limit; S is ISO speed; θ_d is field of view, which affects composition; ρ_s is a measure of relative hand-held camera stability; d_d is capture format diagonal; m_a is angular magnification; m_p is printing magnification; d_v is viewing distance; and F is camera lens focal length. The first three parameters lie outside the control of a system designer. The minimum depth-of-field limit must be selected to adequately cover photospace, and so is largely determined by the photomotivation distribution. Stability of a camera depends upon its mass distribution and the smoothness of its shutter button actuation, but for a given size class of cameras (overall size being significantly correlated with format size), the variation in this parameter is relatively modest. The field of view and angular magnification are related because both depend on focal length and both are competitively constrained features; lower angular magnifications and wider fields of view are undesirable in cameras with fixed focal length lenses.

Given the above considerations, the first equality in Eq. 27.8 suggests that the greatest leverage for controlling minimum ambient light level Y_{min} in a fixed focus/aperture system lies in the format size (a squared dependence) and ISO speed. However, as seen in the second equality of Eq. 27.8, even a small decrease in focal length, which might not compromise the field of view and angular magnification values too severely, can have a large positive impact on photospace coverage. Because Y_{min} is proportional to focal length raised to the fifth power, a mere $\approx 13\%$ reduction in focal length halves Y_{min}.

More sophisticated systems, having zoom camera lenses, electronic zoom (cropping), and/or optional larger prints sizes (such as panoramic prints) provide access to higher angular magnifications and, in the first case, to longer focal lengths as well. Given the cubic dependence of Y_{min} on angular magnification and the quintic dependence on focal length, it is clear that obtaining adequate photospace coverage in advanced systems may be challenging.

Finally, we consider Requirement #5 from Sect. 27.3 concerning maximum flash exposure range. From Eq. A5.22, the product of the aperture A, and the distance d_n at which an ISO normal exposure results, is equal to the flash guide number G, which is proportional to the square root of the product of the ISO capture speed S and the flash output O.

$$G = A \cdot d_n \propto \sqrt{S \cdot O} \qquad (27.9)$$

Substituting Eq. 27.5 (solved for A) into Eq. 27.9, and setting d_n to d_{max}, the maximum distance allowing a normal or greater flash exposure, yields:

$$d_{max} = \frac{G}{A} = \frac{4 \cdot d_{min} \cdot \theta_{max} \cdot G \cdot \tan(\theta_d / 2)}{d_d \cdot m_a}$$
$$= \frac{2 \cdot d_{min} \cdot \theta_{max} \cdot G \cdot d_v}{F^2 \cdot m_p} \qquad (27.10)$$

which is Eq. A5.23 (the second form follows from Eqs. 27.2 and 27.3). As in the ambient case, photospace coverage, as affected by maximum flash range, is influenced strongly by lens focal length (varying with its square). At fixed angular magnification and field of view, flash range can be increased by increasing guide number (via increased capture ISO and/or flash power) or by decreasing format size.

27.6 Example of Photospace Coverage

The results of the last four sections can be integrated into a single graphical analysis of photospace coverage by superimposing flash exposure (d_{max}), ambient exposure (Y_{min}), and depth-of-field (d_{min}) constraints onto a photospace distribution. An example of such an analysis for a fixed focus/aperture system is provided in this section.

The assumptions of Table 27.1 for 35-mm format with regard to focal length (30 mm), f-number ($A = 9.8$), and circle of confusion at the capture plane (0.05 mm) are followed in this example, leading to a minimum depth-of-field limit of three feet (≈ 0.91 m). A value of $Y_{min} \approx 86$ cd/m^2 is estimated from Eq. 27.6 with $K = 12$ cd·s/m^2, an assumed ISO speed of 400, and an exposure time of 1/30 second, in accordance with the one over focal length rule. Finally, the maximum flash range is calculated from Eq. 27.10 with $G = 24$ meters (about the maximum practical value in a compact camera at ISO 400), yielding $d_{max} \approx 2.4$ m.

In Fig. 27.2, the photospace distribution of Fig. 27.1 is plotted as a contour diagram against the same photospace coordinates. The outdoor peak is in the upper right corner and the indoor peak is on the left. The minimum depth-of-field limit ($d_{min} \approx 0.91$ m) plots as a horizontal line, below which images are

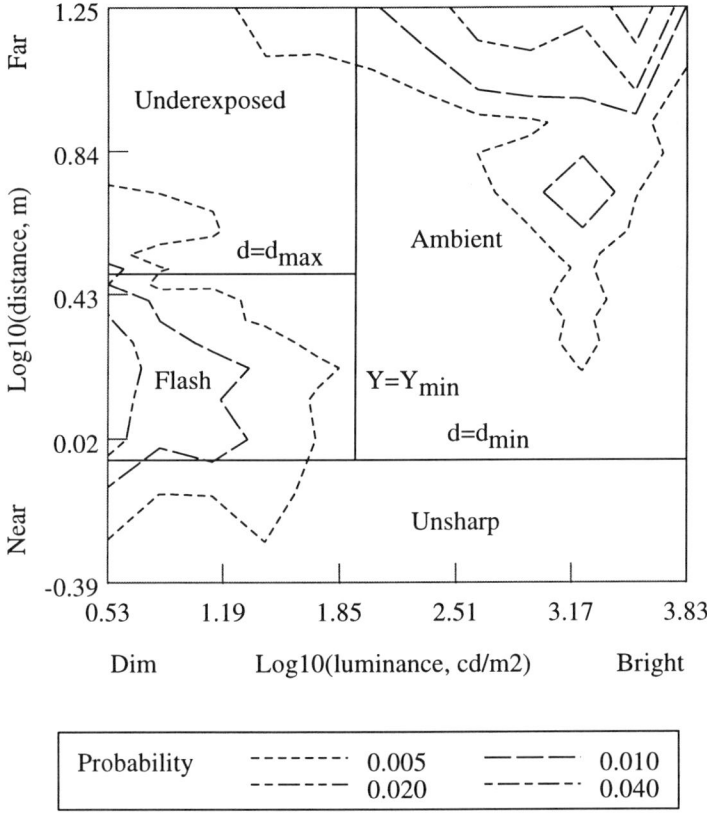

Fig. 27.2 Photospace of Fig. 27.1 shown as a contour plot, with limits of photospace coverage of a fixed focus and aperture camera. The regions labeled Ambient and Flash are successfully covered and include the peaks of the photospace distribution, although much greater coverage would be possible with autofocus and aperture control.

usually unsharp. The minimum luminance limit, $Y_{min} \approx 86$ cd/m^2, plots as a vertical line, to the right of which satisfactory ambient exposures are normally obtained. Finally, the maximum flash range, $d_{max} \approx 2.4$ m, plots as a horizontal line above which flash exposures are usually underexposed. These three lines divide photospace into four regions, labeled "Ambient", "Flash", "Unsharp", and "Underexposed" in Fig. 27.2. The photospace coverage of this system is not unreasonable. The outdoor peak and much of its tail towards low light levels and long distances is successfully covered. The summit of the indoor peak is also covered, but some of its slopes are not.

Several additional remarks regarding this figure are in order.

1. There is a small area in the bottom right corner of the underexposed region where significant exposure is provided by both ambient and flash illumination, and although neither alone can provide an ISO normal exposure, the combination of the two is sufficient. This crescent-shaped area is quite small and of limited impact and so is not shown in Fig. 27.2.

2. Depending upon the characteristics of the system and especially those of the capture medium, adequate image quality may result from exposures somewhat lower than that of an ISO normal exposure. An underexposure latitude of one stop is often assumed, which would move the Y_{min} line to the left by 0.3 units ($\log_{10}[2] = 0.3$) and the d_{max} line upward by 0.15 units (because the guide number varies with the square root of ISO speed by Eq. 27.9).

3. The positions of the three boundary lines in Fig. 27.2 may be rearranged by adjusting the aperture value. If a smaller f-number is chosen, d_{min} will increase (Eq. 27.1), so that more images are unsharp, but Y_{min} will decrease (Eq. 27.6) and d_{max} will increase (Eq. 27.10), both of which reduce the occurrence of underexposures.

4. Photospace coverage, while tolerable in this system, could be substantially increased by an autofocus facility. This would extend the depth of field by virtue of the multiple focus positions, as discussed earlier, and thereby lower the d_{min} line, and so reduce the frequency of unsharp images. Furthermore, the autofocus would permit use of smaller f-numbers for most scenes (excepting those of great depth), which effectively would raise the d_{max} line and shift the Y_{min} line to the left, both of which would reduce frequency of underexposures.

The fraction of the photospace distribution falling within the flash and ambient regions, and therefore likely to be captured successfully, may be computed numerically. This fraction may be interpreted as the yield of the photographic system under conditions of consumer usage.

27.7 Limitations of Photospace Coverage Metrics

Although the discussion of photospace coverage in the preceding section served to elucidate a number of fundamental constraints and interactions of importance in photographic system design, many simplifications were made during the course of the analysis. One example is the neglect of the impact of depth of focus, which is defined as the range of lens to capture plane separation yielding acceptable sharpness. Depth of focus is approximately symmetric about the best focus position, with total depth of focus d_f being given by Eqs. A5.8 and A5.18.

$$d_f = 2 \cdot A \cdot d_c = \frac{F^2}{d_{min}} = \frac{d_d^2}{4 \cdot d_{min} \cdot \tan^2(\theta_d/2)} \qquad (27.11)$$

The latter two equalities follow from Eqs. 27.1 and 27.2, respectively. Smaller formats, with their inherent light gathering advantage (see Table 27.1), are disadvantaged in terms of depth of focus, which is proportional to format diagonal squared. Consequently, smaller formats require tighter tolerances around the lens to capture plane separation and, if applicable, film flatness. Because this effect is neglected in the photospace coverage analysis (and simple optical analyses like that of Table 27.1), smaller formats may be given undue credit. The 35-mm format case from Table 27.1 would have a depth of focus of about 1.0 mm, whereas that of the 1.5 Mpixel sensor with 9 μm pixel spacing would be only about one-sixth as large.

Although calculation of a photospace coverage metric might initially seem to provide an attractive alternative to the analysis of a full quality distribution, in practice, photospace coverage is of very restricted utility. Its primary limitations may be stated as follows.

1. The only attributes of image quality that are considered are unsharpness and whatever aspects of image quality limit the ISO capture speed. In the case of film capture, ISO speed depends only on sensitometry, i.e., signal level, and is set to avoid tonal clipping. There are three defined ISO speeds for digital still cameras, two based upon signal-to-noise ratio and one upon saturation exposure (beyond which the signal is

clipped). In total, then, photospace coverage as formulated in this chapter reflects only unsharpness and either tonal clipping or noise (but not both), and ignores all other image quality attributes.

2. Other than final image subtense (i.e., display size and viewing distance), properties of all system components other than the camera are neglected. It is common experience that other system components, such as digital image processing operations, optical printing, digital printers, and display media have a substantial effect upon image quality. Therefore, photospace coverage reflects how quality is limited by the capture, rather than the actual quality finally realized.

3. Some key capture characteristics are intrinsically treated as being optimal and invariant. For example, the camera metering system, lens aperture, exposure time, capture speed, flash guide number, and focus point are all assumed to have no bias or variability. As discussed above, the neglect of focus point variability, which is affected by lens to capture plane separation tolerances, is particularly serious, because it varies systematically with photospace coverage. Furthermore, some camera properties are ignored completely, such as film or sensor MTF, which certainly affect sharpness.

4. The choices of values of θ_{max}, ω_c, and underexposure latitude are somewhat arbitrary and because of the empirical nature of their determination are not even guaranteed to be consistent with one another. Even if they were internally consistent, choosing a threshold value for each parameter and making binary distinctions of capture success based upon them is at best equivalent to abstracting a single point from the performance distribution of the system. Thus, photospace coverage and other yield metrics technically may be regarded as the results of specialized capability analyses, and they exhibit limitations analogous to those of typical capability metrics.

In conclusion, photospace coverage is a capture-specific yield metric that reflects only certain determinants of at most two image quality attributes. Consequently, although it is useful in understanding system constraints and interactions, photospace coverage is inadequate for use in system design optimization. Although some of its serious limitations could be mitigated by increasing the sophistication of the metric beyond that described here, the effort required would scarcely be less than that involved in undertaking a full and rigorous performance analysis. There seems to be no avoiding the necessity of predicting the complete quality distribution that is expected from an imaging system.

27.8 Summary

A photospace distribution quantifies the relative frequency of occurrence of different combinations of ambient light level and camera-subject distance in the actual usage of a photographic system. These two factors fall largely outside the control of a system designer, but profoundly influence system performance. One approach that is intermediate between capability and performance modeling is the delineation of the portions of a photospace distribution that are likely to produce images of satisfactory quality. Such photospace coverage metrics, examples of which are minimum ambient light level (Eq. 27.8), minimum depth-of-field limit (Eq. 27.1 and 27.5), and maximum flash range (Eq. 27.10), are correlated with the fraction of images produced that exceed some threshold quality, which is referred to as a yield.

At fixed final displayed image size, greater photospace coverage can be obtained through shorter camera lens focal length, lower printing magnification, higher ISO capture speed, higher flash output, and greater camera stability. Smaller formats are advantageous for photospace coverage because the effect of the shorter focal length dominates that of the higher printing magnification; however, the higher magnification necessitates tighter manufacturing tolerances. Obtaining adequate photospace coverage in systems having zoom camera lenses, electronic zoom (cropping), and/or optional larger prints sizes (such as panoramic prints) usually requires camera autofocus and aperture control features.

Although photospace coverage metrics are valuable in understanding system design constraints and interactions, they are capture-specific yield metrics that reflect only certain determinants of a few image quality attributes. Consequently, they are rarely adequate for use in system design optimization, a task that normally requires analysis of full performance modeling results.

28

Monte Carlo Simulation of System Performance

28.1 Introduction

In capability modeling, quality is predicted under just a few selected conditions, which may be chosen somewhat arbitrarily, often to simplify the analysis. In contrast, performance modeling predicts the complete distribution of quality produced by an imaging system in actual customer usage. Compared to capability analyses, performance calculations require more complete underlying models, software of considerably greater sophistication, and more extensive input data. The principal advantage of performance predictions is that they include all pertinent information, and the relative importance of different photographic situations is properly weighted. Although capability calculations are simpler to implement, they are frequently subject to misleading interpretations because the available information is incomplete.

This chapter explains the mathematical technique of Monte Carlo simulation, which forms the basis for predictions of performance. The method is described in somewhat general terms, but then examples specific to image quality modeling are provided. The process of organizing and classifying factors that influence image quality, which is needed prior to implementation of a Monte Carlo simulation, is itself valuable for understanding the complex interactions between different aspects of imaging systems and their usage.

This chapter is organized as follows. Section 28.2 discusses the intent of performance modeling in general terms. Section 28.3 describes the basic organization and mechanics of Monte Carlo modeling. Finally, an example of proper sampling of a complicated distribution is presented in Sect. 28.4.

28.2 Performance Modeling

The performance of an imaging system is quantified by the full distribution of image quality produced by the system under specified conditions. These conditions usually correspond to: (1) the production of the system components under representative manufacturing circumstances; (2) the distribution of the system components in a typical marketplace; and (3) the use of the components of the system by actual customers (defined broadly to include anyone operating components of the system or viewing images produced by the system). Consequently, system performance is affected not only by aim or mean system properties, but also by variation or tolerances about those positions. Furthermore, performance may be strongly dependent upon usage factors that are beyond the direct control of the system designer, manufacturer, sales representative, retailer, or customer. It is, therefore, important to understand in detail as many factors affecting performance as possible, so that designs may be optimized within cost, manufacturability, environmental, feature requirement, and other constraints.

There are a great number of factors that can affect the quality of an individual image, some of which are constant for all images produced by a given photographic system, and others of which are variable, having a distribution of possible values. If all such factors were constant, a system would always perform at its capability, and performance modeling would not be required. In practical photographic systems, however, most pertinent factors are variable in nature, and their distributions commonly do not correspond to convenient mathematical functions such as Gaussians. Furthermore, interactions between different factors are frequent and often complex in nature, rarely being quantifiable in terms of simple correlations. A standard mathematical approach to modeling such a complicated set of circumstances is Monte Carlo simulation. Although this technique requires that factor distributions and deterministic relationships be specified, it provides a general mechanism whereby such information may be efficiently integrated to predict the final resulting distributions.

28.3 Organization of Monte Carlo Calculations

In this section, we will briefly describe some of the fundamental aspects of the mechanics of Monte Carlo modeling. One way to organize such a model is to start by listing the final quantities that are to be predicted, i.e., the desired output. In our case, the most important output is the distribution of overall quality produced by the system. Next, the factors from which the desired output are most simply predicted are identified. These factors may not be simple to predict themselves, but rather they are the quantities needed to directly compute the output.

We shall refer to factors that are hierarchically closer to the output as being more derived, and factors that are closer to the most fundamental input quantities as being more basic. In this application, the most derived factors would be the quality changes in JNDs arising from individual attributes of image quality, from which the overall quality (the output) may be predicted using the multivariate formalism of Ch. 11.

In general, it is not possible to include every single factor influencing the desired output in a model. Prioritization of inclusion of factors should be based upon criteria such as: (1) the likely variation in a factor; (2) the sensitivity of the output to that factor; (3) the interdependence between that factor and others of importance; and (4) the amenability of that factor to quantification.

Having selected the subset of most derived factors to be included in the model (e.g., image quality attributes), the next step is much like the previous one. Factors one level less derived (i.e., one level more basic), from which the most derived factors may be predicted, are identified, prioritized, and selected for inclusion or exclusion from the model. In our case, these more basic factors logically would correspond to: (1) the objective metrics associated with the image quality attributes; (2) observer sensitivity; and (3) scene susceptibility. Given this information, and a table of regression fit parameters, the integrated hyperbolic increment function (IHIF), Eq. 9.5, could be used to predict quality change arising from an attribute.

The process of tracing back dependencies to ever more basic levels is continued until each pathway is exhausted and terminates at a factor that need not, or cannot, be further divided. Such a fundamental factor is called a primitive. The exact criteria that must be fulfilled by a primitive are described below.

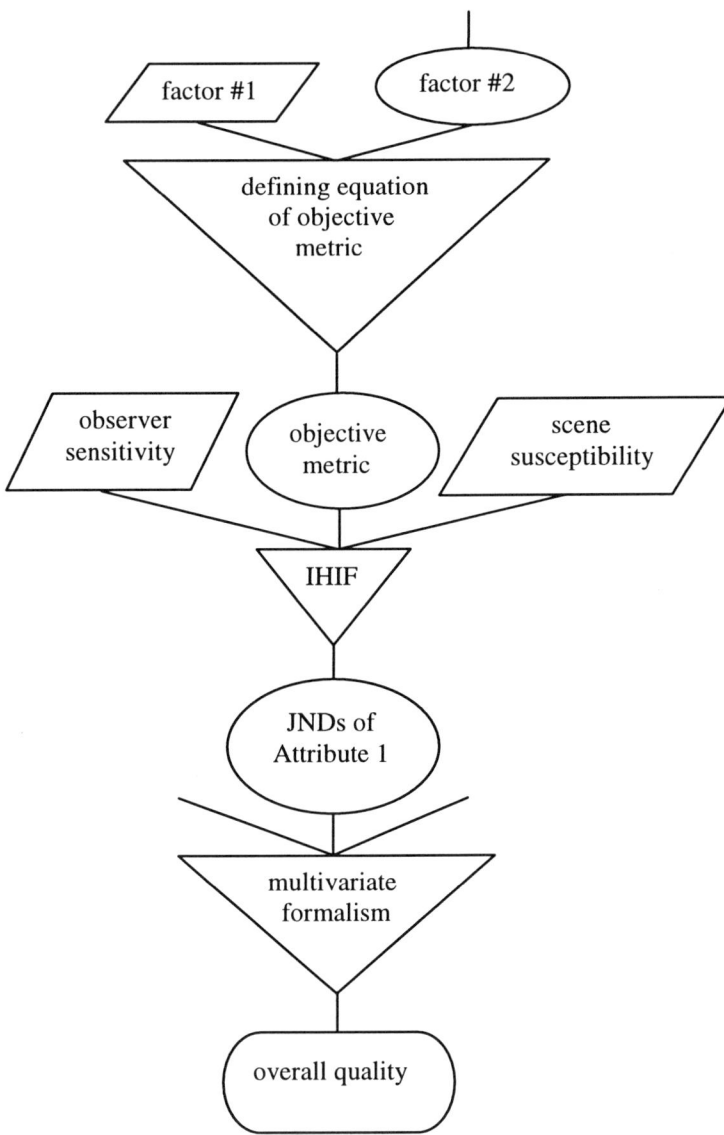

Fig. 28.1 Simplified diagram showing the organization of a Monte Carlo simulation of image quality. Symbols are as follows: rhombus = primitive; ellipse = other factor; triangle = deterministic relationship; and round-ended box = final output.

The process described above may be visualized by constructing a tree diagram, like that shown in Fig. 28.1. The trunk of the tree is the final model output (round-ended box) and the tips of the branches are primitives (rhombuses), which constitute the input to the model. As discussed below, other factors (ellipses) are deduced from deterministic relationships (triangles). Factors, including primitives, may be referred to as the nodes of the tree, and nodes that are directly related through a deterministic relationship may be said to be connected by branches. The unterminated lines in Fig. 28.1 indicate that it is only a section cut out of the complete tree.

Primitives may be: (1) simple system characteristics, such as capture format dimensions; (2) key measures, if relatively invariant, such as the NPS of a display monitor; (3) individual values in parametric models for the prediction of key measures; or (4) usage characteristics, such as a photospace distribution. Some primitives may appear at the ends of many branches, indicating the ubiquity of their impact. Similarly, derived factors may appear in duplication. For the purposes of this discussion, such redundancy is immaterial, but of course, in a software implementation of the model represented by the tree, no computation would be repeated needlessly.

Once the tree is complete, the nature of each primitive is considered. A primitive must meet the criterion that the distribution of values it assumes under the conditions of usage being modeled is uncorrelated with the values assumed by any other primitive of the tree. By definition, a factor that is invariant, or varies by so little that its impact on quality is unmodified, constitutes a primitive. For example, a fixed flash-lens separation in a given model of camera, which affects quality loss arising from redeye, would be a primitive.

Variable factors can also be primitives if they are unaffected by other primitives. For example, as discussed in the previous chapter, because depth of focus can be very shallow for smaller formats or f-numbers, even small variations in lens to capture plane separation may have a significant impact on quality. Therefore, to predict the quality distribution resulting from the usage of a camera model by a population of photographers, a random sampling from the distribution of manufactured lens to capture plane separations within the camera model might be needed. This distribution would probably be a primitive because it is unlikely to be correlated with other primitives included in the tree. It might well be correlated with excluded factors, such as the particular factory that produced the individual camera, but primitiveness is a property defined only in the context of a particular tree. A distribution such as this might be approximately normal, in which case it could be described by just a mean and variance, and standardized software routines could conveniently be used to sample randomly from the distribution.

Suppose that upon review a presumed primitive was recognized to be correlated with another putative primitive on the tree. How would this situation be resolved? One possibility is that a case of causality was overlooked; e.g., the first factor may depend upon the value of the second, but not vice versa. Then the first factor is not a primitive, and it should be branched back to the second factor, which is a true primitive. If the two factors mutually influence each other, rather than one depending upon the other but not vice versa, then the co-distribution of the two factors can be defined as a single primitive. A photospace distribution is a good example of this, being a distribution over two rather basic parameters, camera-subject distance and ambient light level, which are correlated but not causally so. Such co-distributions are likely to be complex in shape. An example of how to properly randomly sample such a distribution is given in the next section.

Once the primitives have been reviewed and modified as necessary, and their distributions specified, there remains the task of tracing causal relationships through the tree. If the tree has been properly constructed, it should be possible to write a deterministic relationship between each non-primitive factor and those connected factors residing on the next more basic level. Restated, it should be possible to predict each non-primitive factor solely from the less derived factors occupying nodes to which it directly branches. For example, if lens focal length, camera-subject distance, and the separation between the lens and capture plane are known, the amount of defocus can be computed from the thin lens equation, Eq. A5.1. Therefore, the factor defocus might reside at a node that branches to more basic nodes occupied by the three factors listed. Defocus might then feed through a branch to a more derived node occupied by the lens MTF. The latter might be generated from a deterministic relationship incorporating a diffraction-limited parametric model (in which case aperture and wavelength would need to converge upon the lens MTF node) or a method for interpolating between a collection of three-color lens MTFs measured at varying aperture and defocus values (requiring a convergence of an aperture branch).

With the completion of specification of deterministic relationships, sufficient information exists to perform a Monte Carlo simulation. Essentially, a table of random numbers or a pseudorandom number generator is used to draw values from each primitive distribution independently (which is why an absence of correlation between primitives is required). The resulting values of the primitive factors are propagated through the branches to higher level nodes, where the deterministic relationships yield the values of each derived factor. This propagation continues until the final output (in our case, JNDs of overall quality) is determined. The process is then repeated (with different random numbers) thousands of times to build up a distribution of output values. Once a specified number of iterations have been performed, or the output distribution has stabilized, the simulation is halted. In a Monte Carlo software application, it is desirable that the number of iterations be under the control of the user, permitting a smaller number to be used for speed in preliminary analyses, and a larger number to be chosen for final computations, so that the quality distributions can be made as smooth as desired.

Several practical choices must be made regarding how random numbers are handled. Tables of random numbers, drawn from actual physical events, are available, but it is much more common to use pseudorandom number generators, which are mathematical transforms that convert one number into another in a nearly uncorrelated fashion. The first input number supplied to the pseudorandom number generation process is called the seed; subsequently, the output number from one stage of the process is used as the input number for the next stage, thereby generating a series of nearly random numbers. For example, the pseudocode below generates a stream of pseudorandom numbers between zero and one.

```
integer seed=19, I1=1029, I2=221591, I3=1048576;
I0 = seed;
do forever;
    I0 = modulo((I0*I1+I2), I3);
    random = I0/I3;
end do;
```

The same seed may be used each time a program is executed, as shown above, or it may vary, usually depending on the exact time and date at which a process is initiated. In the case of a fixed seed, the output of the model is deterministic and repeatable; if instead the seed varies, sequential runs of the model yield independent estimates. The advantage of a variable seed is that the user may pool the results of sequential runs to obtain better estimates, rather than having to run a large number of iterations all at once. The disadvantage of this approach is that in software testing, two sequential runs having identical input data and calculations will yield slightly different results, so it is not possible to use simple comparisons of output files for diagnostic purposes.

28.4 Sampling Complex Distributions

The procedure for randomly sampling from a multidimensional, irregular distributions such as that of photomotivation warrants a brief discussion. To perform such a sampling, the multidimensional distribution is encoded as a one-dimensional cumulative distribution function (CDF). Consider Fig. 27.1, reproduced below, in which a photospace distribution is described in terms of the probability of occurrence in a finite set of cells (rectangular regions of luminance by distance), each centered upon the intersections of the grid lines. An arbitrary ordering of the cell locations is chosen so that the distribution can be mapped to a single (non-physical) dimension. The 12 luminance values and 16 distances yield a total of 12 × 16 = 192 grid intersections and associated cells. These cells might be ordered from 1 to 192 in raster fashion, starting at the minimum luminance and maximum distance (back corner), increasing luminance at constant distance to the maximum luminance (right corner), and then repeating the process at successively lower values of distance. From this ordering, a one-dimensional CDF could be generated by starting at zero and accumulating each cell's probability in sequence. The final value of cumulative probability would be one if the PFD were properly normalized; if not, the CDF must be rescaled for the sampling procedure to be valid.

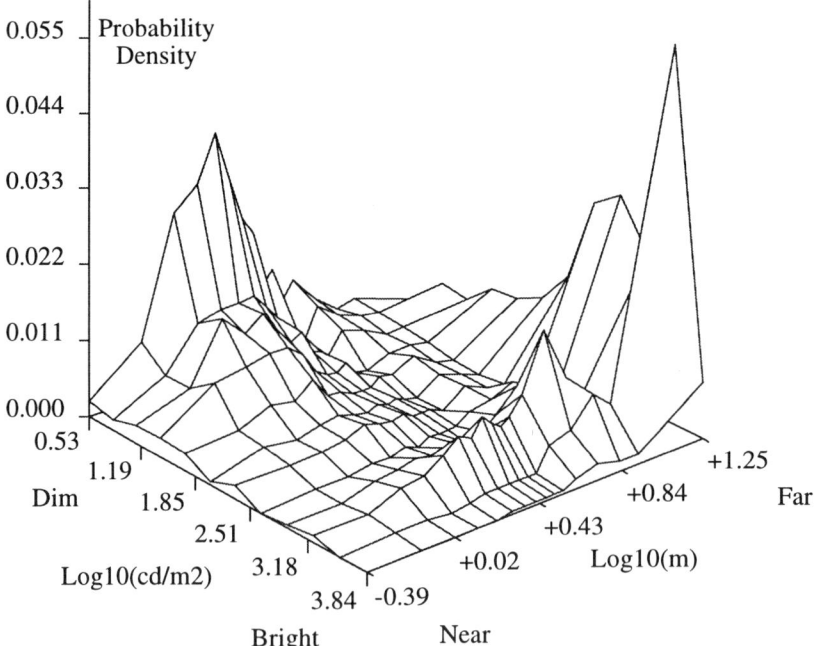

Fig. 27.1 Photospace distribution, repeated from Ch. 27 for reference.

The results of this process are shown in Fig. 28.2. The regular series of vertical lines mark the positions at which a new line of cells (new distance) is started. Suppose that the number 0.867 were drawn randomly from a uniform distribution over the interval zero to one, or were generated by a pseudorandom number generator. This number would be mapped through the CDF to yield an x-axis value of 149.6, as shown in Fig. 28.2. This value would then be rounded up to 150 to correspond with an integral cell index (rounding up is correct because the CDF was accumulated from zero cell index but the first valid cell number is one). Cell #150 corresponds to the 6th luminance (of 12) in the 13th distance series because $150 = 12 \cdot (13 - 1) + 6$. The 6th luminance and 13th distance are then used in the current Monte Carlo iteration. In this fashion, a

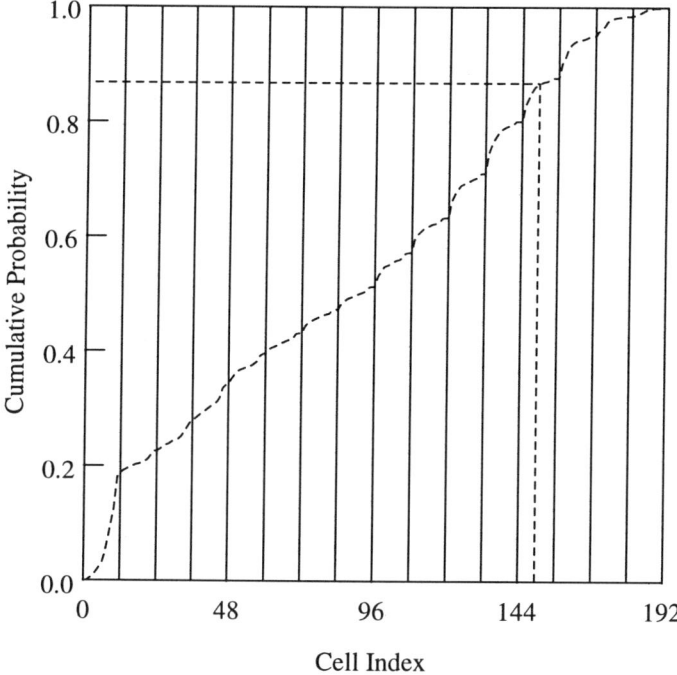

Fig. 28.2 Cumulative distribution function (CDF) derived from the photospace distribution of Fig. 27.1 by applying an ordering rule to the cells in the two-dimensional coordinate system. By mapping uniform random numbers from zero to one through this CDF, a properly correlated sample of light levels and distances may be obtained.

proper sampling of an irregularly shaped, two-dimensional distribution can be accomplished using tables of random numbers or pseudorandom number generators. If the cells are coarser than desired, the starting distribution may be interpolated prior to conversion to a CDF, and/or the fractional part of the unrounded x-value (i.e., 0.6 from 149.6) may be used to generate continuous distance and luminance values within a cell, assuming the distribution within the cell to be flat.

In conclusion, the Monte Carlo methodology provides a mechanism for the integration of a great deal of segregated information, specifically the primitive distributions and the deterministic relationships at each branch point of the tree. Although Monte Carlo calculations may seem complicated, and indeed they can be, in complex systems they still provide a vastly simpler approach than trying to construct closed form mathematical descriptions of all pertinent behavior.

28.5 Summary

The performance of an imaging system is quantified by the full distribution of image quality it produces under representative manufacturing and customer usage conditions. Therefore, performance varies not only with average system properties but also with any variations of these properties encountered in practice. Most factors influencing image quality are variable in nature, and interactions between different factors are frequent and often complex in nature. A standard mathematical approach to modeling such situations is Monte Carlo simulation, which provides an efficient mechanism for integrating information regarding distributions of, and relationships between, pertinent factors.

To model the image quality distribution resulting from a photographic system, one identifies a minimal set of independent factors called primitives that, if known for a particular image, are sufficient to predict its quality through a series of deterministic relationships. In the Monte Carlo simulation, a random sample is drawn from each primitive distribution, and the resulting values are propagated through the deterministic relationships to predict the quality of an individual image. The last few stages of this propagation involve computation of objective metrics, use of the IHIF equation to generate JNDs of change arising from individual attributes, and application of the multivariate formalism to predict overall image quality. The process is then repeated thousands of times to build up the desired distribution of image quality.

29

Interpreting Quality Distributions

29.1 Introduction

The previous two chapters have addressed the motivation for undertaking performance analyses and the mechanics of Monte Carlo calculations. The output of a Monte Carlo model of imaging system performance consists of at least the distribution of image quality resulting from the use of the system under selected conditions. In addition, the distributions of other quantities may be provided, such as: (1) quality changes arising from individual attributes; (2) objective metrics associated with such attributes; and (3) parameters that aid in system diagnosis. Examples of the latter include exposure and camera settings such as aperture, shutter, and focus. Interpretation of the distributions of diagnostic system parameters by an engineer is usually straightforward, because the parameters primarily characterize how the system responds to various situations. In contrast, understanding the shapes of quality distributions may be quite difficult because of the number of contributing factors. In this chapter, we will focus on understanding, interpreting, and comparing distributions of overall image quality; however, many of the points raised also apply to distributions of quality attributes and their corresponding objective metrics.

This chapter is organized as follows. Section 29.2 reviews the mathematical functions that may be used to represent distributions of quality, and defines capability in their terms. In Sect. 29.3, a general discussion of the nature of factors affecting distribution shape is presented, and a few remarks are made regarding the comparison of differently shaped distributions.

29.2 Describing Quality Distributions Mathematically

In Sect. 2.3, two types of functions useful in describing distributions were introduced: the probability density function (PDF) and the cumulative distribution function (CDF). Figure 29.1 depicts PDFs of quality produced by three hypothetical imaging systems. The x-axis is quality in JNDs, with higher quality to the right. The y-axis is the relative probability of occurrence of quality over a small interval centered on the corresponding x-axis value of quality. Better performing systems yield higher quality images more frequently, and so their PDFs plot further right. The peak of the PDF corresponds to the mode of the distribution, which is the quality most likely to be produced by the system. Although a single PDF is very easy to understand, because it intuitively

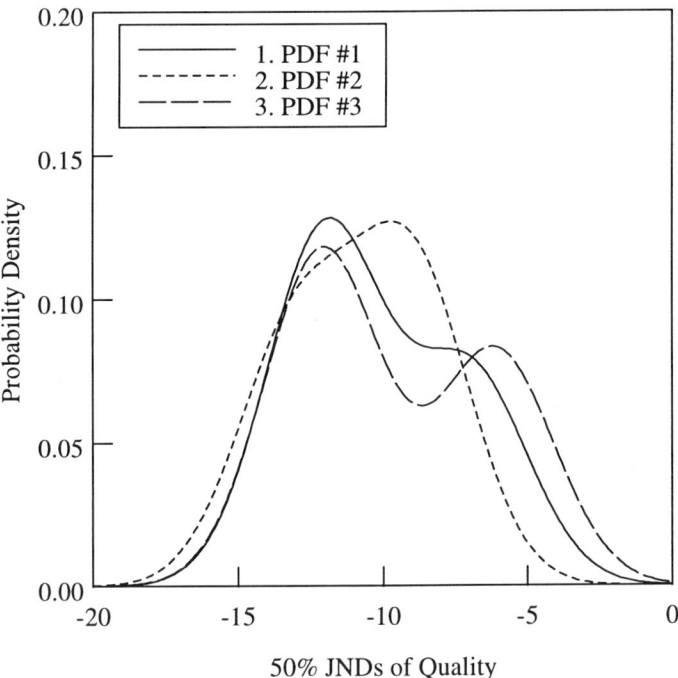

Fig. 29.1 Probability density functions (PDFs) of image quality. The number of potential curve crossings can make this type of plot rather difficult to interpret as the number of curves increases.

Interpreting Quality Distributions

corresponds to what we would normally regard as the shape of the distribution, inter-comparison of several PDFs is sometimes difficult because of curve crossings. For example, it is not easy at a glance to determine the ordering of the three distributions in Fig. 29.1 for overall quality.

Figure 29.2 shows the CDFs corresponding to the three PDFs of Fig. 29.1. The x-axis is as in the preceding figure, but the y-axis is the probability of the quality being less than, or equal to, the corresponding x-axis value. As in the case of PDFs, better performing systems plot farther right. The CDF is the integral of the PDF from minus infinity to x (as exemplified by Eq. 2.5), and therefore, ranges from zero to one inclusive, because the PDF has unit area. The median quality of the system is easily determined graphically; it is the x-axis value of

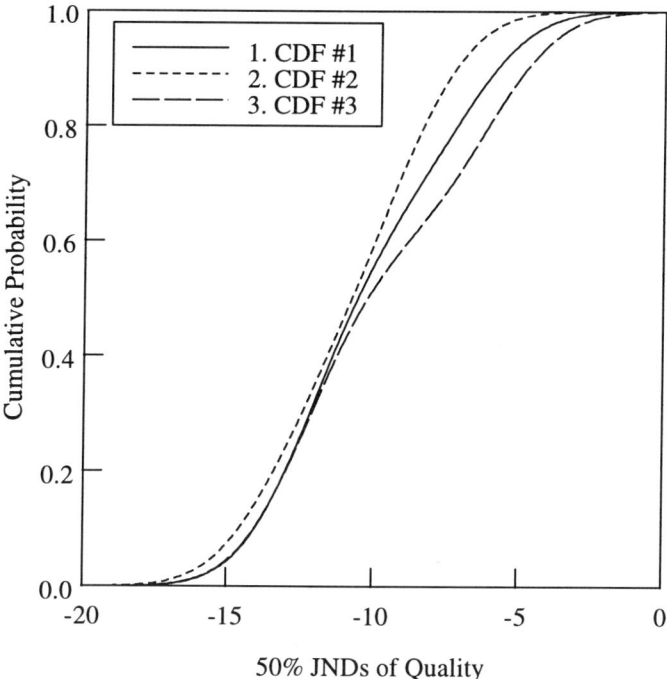

Fig. 29.2 Cumulative distribution functions (CDFs) of image quality. These three CDFs are derived from the PDFs of the preceding figure, but are more easily visually compared by virtue of having been integrated, reducing the number of curve crossings.

quality producing a y-axis cumulative probability of 0.5. Other percentile positions are found in a similar manner; e.g., the 90th percentile quality is the x value yielding y = 0.9. The capability of a system in loose terms is the peak quality it produces; more precisely, it may be defined as the quality of some high percentile position, such as the 99th percentile.

The information contained in the PDF and CDF is identical, but the CDF is often more convenient because the operation of integration frequently removes curve crossings and so makes visual inter-comparisons of CDFs easier than those of PDFs. For example, it is evident at a glance how the three CDFs of Fig. 29.2 are ordered for overall quality, yet such was not the case for the corresponding PDFs of Fig. 29.1. In addition, the simple graphical determination of median quality and capability are advantageous aspects of using CDFs in performance analyses, a convention we will adopt hereafter.

29.3 Quality Distribution Shapes

A simple capability analysis might involve comparison of the peak qualities of two imaging systems. If the two system CDFs were "parallel", having the same shape but merely being shifted from one another, the peak quality difference would adequately describe the quality difference between the two systems. However, if the two CDFs had different shapes, then they would not differ just by a constant shift, and a capability analysis would reflect their difference at only one point on the distribution. Therefore, it is of interest to understand how different types of factors affect the shape of the resulting quality distributions. If a particular change in a system property (including manufacturing and usage factors) were known to simply shift the resulting quality distribution, without significantly altering its shape, then a specification relating to that property could safely be based upon a capability analysis.

Evidently, if a change in some factor affects nearly all images by an approximately equal amount, it will shift the quality CDF without changing its shape appreciably. Although this condition is sufficient to produce such an effect, it is not necessary. Even if only some fraction of images are affected by a change, and even if the degree to which the images are affected varies widely, the quality distribution shape will not be altered substantially unless the probability of an image being affected and/or the extent to which it is affected is correlated with its image quality before or after the change. This requirement for change in shape of the quality distribution will be referred to as the correlation criterion; it will be discussed in greater detail later in this section.

Interpreting Quality Distributions

Examples of changes that do and do not affect distribution shape, and their more complex net effect, are shown in the next three figures. In each of these figures, the same reference system is shown for comparison. This reference system has a camera with fair lens MTF and continuous autofocusing. The effect on performance of changes in these two characteristics will be examined. In Fig. 29.3, the fair lens MTF of the reference system is significantly improved in a modified system. This MTF improvement should increase the sharpness of essentially all images, leading to a commensurate quality increase, at least in images the quality of which is not limited by other attributes (which thereby multivariately suppress the sharpness improvement). Therefore, we might expect this MTF change primarily to shift the quality distribution rather than change its shape. This is borne out by the data shown in Fig. 29.3, where the two CDFs are approximately parallel (equally spaced apart in JNDs).

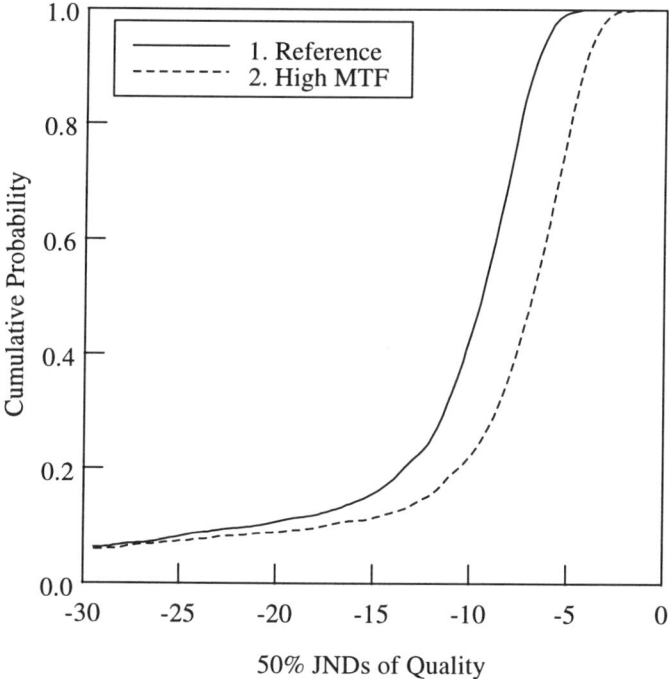

Fig. 29.3 Quality CDFs for two systems differing only in camera lens MTF. The higher MTF improves the quality of most images by a similar amount, so the CDF is approximately shifted, producing nearly parallel curves.

In Fig. 29.4, the reference system is modified in a different fashion. The continuous autofocusing subsystem is omitted, leading to a fixed focus arrangement. The quality of images having the primary subject located approximately at the distance corresponding to the fixed focus point will be relatively unaffected, so the capability of the reference and modified systems are expected to be similar. However, images with primary subjects located farther from the fixed focus position, in terms of reciprocal distance space (see discussion of Fig. 26.6), will be degraded by unsharpness to a greater degree. The reduction in quality of a number of images will populate a low quality tail in the PDF and is expected to lead to a reduction in gradient of the CDF. As shown in Fig. 29.4, our expectations regarding capability and CDF gradient are

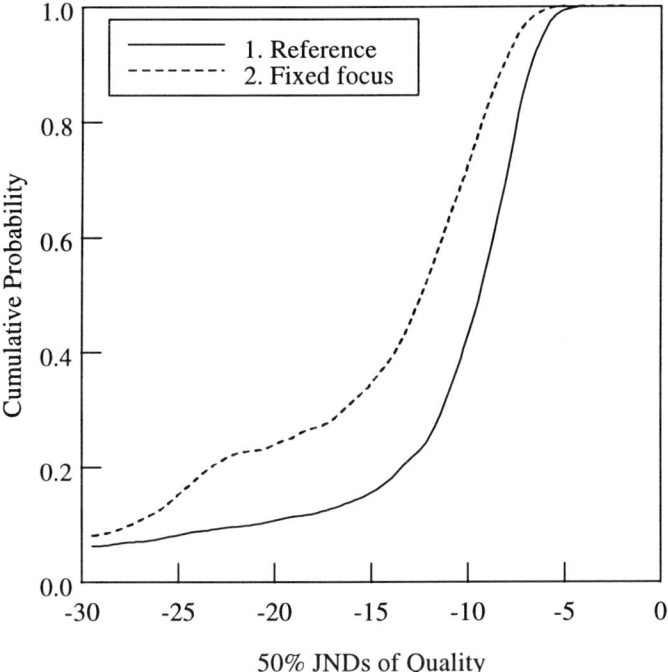

Fig. 29.4 Quality CDFs for two systems differing only in focusing behavior (continuous autofocus versus fixed focus). On average, the relative quality gain from the autofocus is greater in images that were lower in quality with fixed focus; this correlation leads to a gradient change in the CDFs.

approximately borne out, although the change in shape of the CDF is more complicated than just a reduction in gradient.

It is instructive to consider whether the correlation criterion has been met in this instance of a dramatic CDF shape change. Recall that the correlation criterion requires that the quality difference between corresponding images from a pair produced by two systems be correlated with the quality of one or both of the images. It is not obvious than any significant correlation would exist between the quality in the reference system and the quality change between the reference and modified systems. The quality difference has already been stated to depend mostly upon camera-subject distance, relative to the fixed focus position. As noted earlier, subject distance influences image quality in several ways (e.g., by affecting visible detail, depth of field, flash exposure, and redeye severity), so a correlation between quality and subject distance in the reference system is certainly possible. However, it seems unlikely to be a strong correlation, because the different ways in which quality are affected by distance partially cancel one another (e.g., at greater distance there is less evident detail but better depth of field). Therefore, the correlation criterion probably is only weakly met with respect to reference system quality.

In contrast, there is a strong expected correlation between modified system quality and the quality difference between the two systems. Modified system images experiencing a larger loss of quality arising from unsharpness will, on average, have lower quality than other images produced by the modified system. Thus, larger quality differences between the systems will tend to be associated with lower relative quality in the modified system. Consequently, the correlation criterion should be strongly met with respect to modified system quality, thereby explaining the large shape change in the CDF of Fig. 29.4.

As a complementary example involving a system improvement, consider a case in which a reference system has a frequent occurrence of a problem that significantly degrades quality, but is not markedly correlated with other attributes. Under such circumstances, on average, images that possess the problem will be lower in quality than those that are free from the problem. If a factor is changed that reduces the frequency of occurrence of the problem and/or its severity when present, without having other quality ramifications, the performance of the modified system will be improved. In particular, images that possessed the problem in the reference system, and therefore, were on average of lower quality, will be preferentially improved in quality in the modified system. This introduces a correlation between starting quality, in the reference system, and the quality change between the reference and modified systems. The correlation criterion is therefore met, and a shape change in the CDF may be anticipated. By improving lower quality images more, the quality PDF is narrowed (by reduction of the low quality tail), and the gradient of the quality CDF is correspondingly increased.

Finally, let us consider the net effect of simultaneously introducing both of the changes from Figs. 29.3 and 29.4 into the reference system, i.e., improving lens MTF but omitting autofocusing. The former change primarily shifts the CDF to the right (higher quality), without changing its shape much. In contrast, the latter change has relatively little impact on capability, but reduces the gradient of the CDF. The combination of improved capability from higher lens MTF, but increased frequency of lower quality because of misfocus, would be predicted to lead to a crossing of the CDF curves, as is in fact observed in Fig. 29.5, which shows the relevant comparison. Particularly in systems differing in a number of

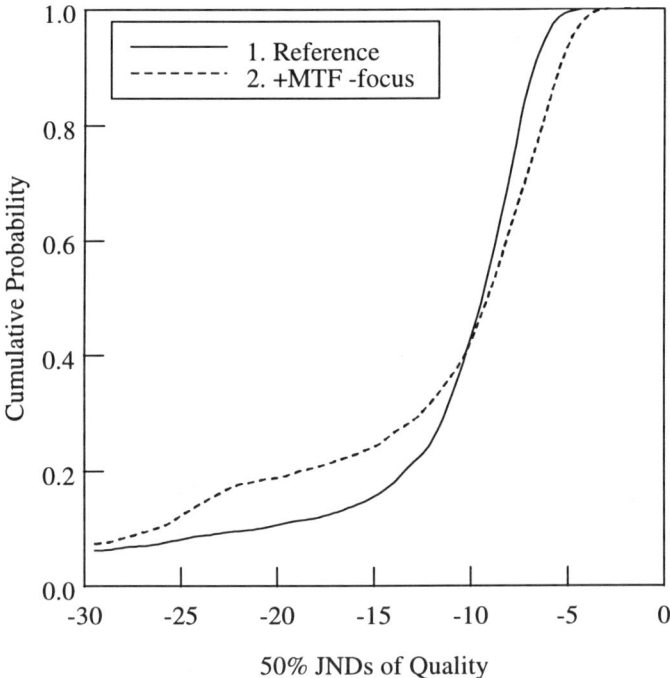

Fig. 29.5 Quality CDFs for two systems, the second with a superior camera lens MTF but without autofocus. The former factor shifts the CDF to the right, whereas the latter decreases its gradient; the net effect is to produce a curve crossing.

Interpreting Quality Distributions 431

factors, more complex behavior such as multiple curve crossings may be encountered.

Associating a single number of JNDs with the difference between two distributions of quality requires assumptions about customer reactions unless the cumulative distribution functions are approximately parallel. In fact, in a case like that of Fig. 29.5, it may not even be obvious which of two distributions corresponds to superior performance as perceived by the customer. Do some especially good images outweigh some particularly bad images (favoring the modified system), or is consistency of quality more desirable (favoring the reference system)? In practice, consistency does seem to confer a small advantage. The median quality is a good first-order measure of performance, but it can be improved by slightly crediting CDFs with higher gradient, corresponding to narrower ranges of quality.

29.4 Summary

The performance of a system is the full distribution of quality resulting from the use of a system under specified conditions, such as representative manufacturing and usage by customers. Performance is most conveniently described as a cumulative distribution function (CDF) of quality. The capability of a system is its peak quality, corresponding to a single point on the CDF, such as the 99th percentile quality. Capability analyses may be misleading when comparing systems having differently shaped CDFs. A shape change between quality distributions is anticipated when the correlation criterion is met, i.e., when the difference in quality between corresponding images from the two systems is correlated with the quality of the image from at least one of the systems. A system change that eliminates or introduces differing levels of a degradation, which is a very common case in practice, usually meets the correlation criterion.

Combinations of system differences producing various shifts and shape changes in the quality distribution commonly lead to curve crossings between the CDFs. In such cases, the median quality difference between the CDFs provides a good single-number estimate of relative performance. A further refinement is to credit CDFs with higher gradients for their more consistent (i.e., narrower) quality distributions.

30

Examples of Performance Calculations

with Richard B. Wheeler
Eastman Kodak Company
Rochester, New York

30.1 Introduction

In this chapter, we provide some representative examples of performance analyses based upon Monte Carlo simulations. All results shown are in the form of CDFs, mostly of JNDs of quality, but distributions of some objective quantities are also shown to demonstrate how they can be used for diagnostic purposes. Sections 30.2–30.6 each present analyses pertaining to different aspects of system performance.

30.2 Photospace-Specific Camera Design

When a system design is tailored to a specific application that is differentiated through marketing, a better optimization of image quality, cost, etc. can often be made. Our first example, concerning the design of a simple camera specifically intended for photography at sporting events, demonstrates this principle. The performance of two camera designs in two photospace distributions will be considered.

Camera #1 is a typical compact 3× zoom point-and-shoot model with autofocus and exposure control. Photospace #1 is a representative consumer photospace for such cameras, like that shown in Fig. 27.1. Curve #1 in Fig. 30.1 shows the reasonable performance of Camera #1, loaded with ISO 400 color negative film, in representative photospace, for which it is designed. Photospace #2 describes consumer picture taking at sporting events, where long distances predominate (especially in a stadium environment) and low artificial light levels are frequently encountered. Curve #2 depicts the very unsatisfactory performance of the same camera in the sporting event photospace, where the yield is low and the coverage inadequate.

Camera #2 differs from Camera #1 in having: (1) a lens with 1.5 stops lower minimum f-number, providing additional light gathering potential; and (2) an

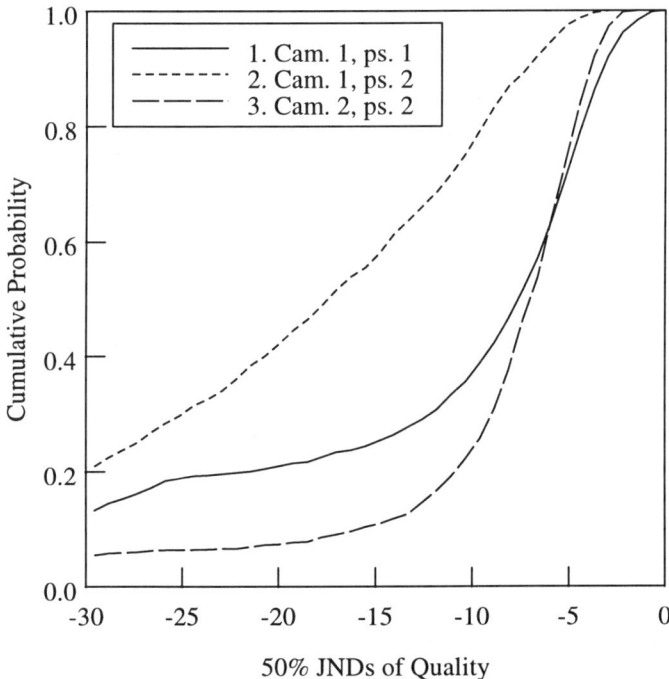

Fig. 30.1 Performance of different camera designs under different usage conditions. Camera #1, designed for use in Photospace #1, performs very poorly in more demanding Photospace #2; however, modification of the design as in Camera #2 permits more than full recovery of performance (but at increased cost).

Examples of Performance Calculations 435

improved shape and shutter actuation mechanism, providing twice as high a stability factor. In standard photospace, these features, while improving performance, would not have made a sufficient difference to warrant the associated increase in camera cost. In contrast, as shown in Curve #3, they make a dramatic improvement in performance in the more demanding sports event photospace. If specialized marketing could target the sale of Camera #2 particularly for use at sporting events, the incremental cost would be justified. This type of niche marketing is particularly practical in the rapidly growing one-time-use camera segment, where distribution at the photographic event or location is often feasible.

30.3 Digital Still Camera Sensor Size

Our second example involves the performance of two systems that are identical except that they incorporate digital still cameras having sensors with the numbers of pixels and the area differing by a factor of two. In this comparison the sensor architecture (size and shape of the active area, spacing of pixel centers, signal and noise characteristics, etc.) is held constant, so the situation is analogous to that of changing film format. The field of view and final displayed image size of the systems are specified to be equal, so their angular magnifications match, but the focal length is $2^{1/2}$ higher and the printing magnification $2^{1/2}$ lower in the system with twice as many pixels (for review, see the discussion of Table 27.1).

As discussed previously (Sects. 12.5, 22.6, and 22.7), systems having higher printing magnification typically have higher noise and lower sharpness because higher capture plane frequencies are visible to the observer viewing the final image. This effect increases the bandwidth and therefore the amount of visible noise, and causes the perceived sharpness to be affected by frequencies at which imaging components usually have lower modulation transfer. Therefore, in the present example, we expect the system with more pixels to have higher capability.

We also anticipate that this system may have lesser photospace coverage arising from lower format speed (Table 27.1), which is caused by the more stringent aperture requirements for adequate depth of field. The degree of loss of photospace coverage depends upon the sophistication of the camera. In the present example, the extreme case of a camera with fixed focus and aperture is assumed, maximizing the effect for demonstrative purposes. Therefore, we expect that the system with more pixels may exhibit a higher number of capture failures, which, in combination with its higher capability, could lead to a crossing of the quality distributions.

This expectation is confirmed in Fig. 30.2, which compares the performance of the two systems described. Although the system with more capture pixels has a higher median quality, it also has a lower distribution slope, which partially offsets its advantage (see Sect. 29.3). With increasing camera sophistication, the frequency of failures in the system with more pixels would decrease, shifting the curve crossing to lower quality and probability, and thereby diminishing its importance.

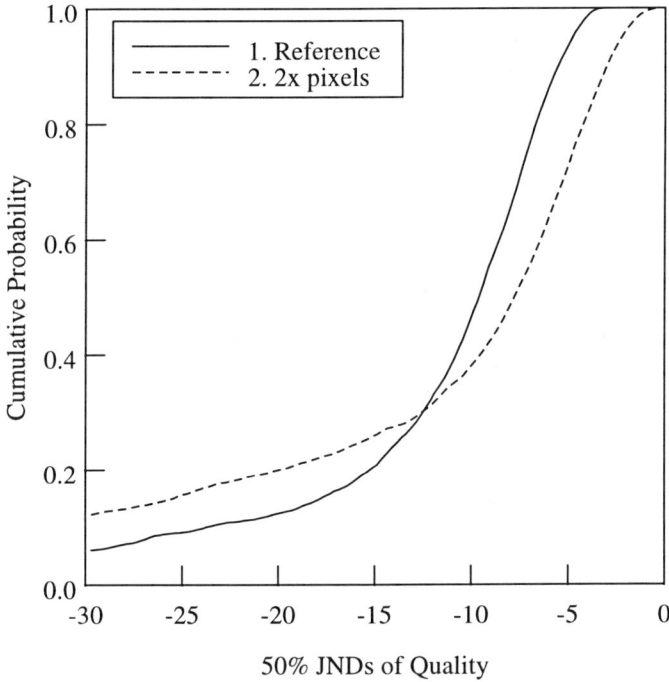

Fig. 30.2 Performance of two systems differing in number of sensor pixels at constant pixel size, field of view, and final image size. The larger capture format has higher capability arising from lower printing magnification, but exhibits a higher frequency of failures associated with poorer photospace coverage.

Examples of Performance Calculations *437*

30.4 Camera Metering

Our third example pertains to the nature of the exposure distribution produced by two different methods of metering light with a camera. Later examples regarding color and density balancing algorithms and film scanner noise will build upon different aspects of the present example. In this case, we will consider the distribution of exposure rather than quality to exemplify the value of intermediate diagnostic factors in understanding system performance.

The impact of exposure on quality varies greatly depending upon the system properties; for example, quality drops more quickly with overexposure than underexposure in digital still cameras and film cameras loaded with reversal (slide) film, but the converse is true with color negative film. Consequently, to understand how exposure-control subcomponent properties influence system performance, it is helpful to break the analysis into two sequential parts. The first part deals with the effect of subcomponent properties on the exposure distribution, and the second part addresses the impact of exposure on quality. In this example, only the exposure distribution will be considered, but the impact of exposure on quality will be addressed in a subsequent example involving scanner noise.

In the integrated metering method a single luminance measurement is made from light reflected (or transmitted or emitted) by the scene. This measurement is then used to estimate what combination of camera settings (aperture and exposure time) will yield the best exposure (usually assumed to be an ISO normal exposure). The luminance measurement covers an acceptance angle that may, or may not, correspond to the field of view being captured in the image, but it is normally fairly wide and is centered somewhere within the field of view. Typically, the reading is more heavily influenced by light arising from the center of the acceptance angle, i.e., the response is center-weighted.

Integrated metering leads to a reasonable exposure when the weighted mean reflectance of the scene over the meter acceptance angle is close to the reflectance of an average scene, i.e., when the scene integrates to a mid-tone. To the extent that scene reflectance varies as a function of scene content, imprecision of exposure will result from integrated metering. Some types of scenes that may cause particularly serious underexposures include backlit scenes as well as those with overcast sky, snow, or white beach sand. Overexposures commonly result from scenes with large shaded areas or flash photographs with distant, poorly illuminated backgrounds.

In segmented metering, multiple luminance measurements are made from within the field of view. By comparing the measurements obtained from different field positions, inferences can be made regarding the type of scene represented, and in some cases, improved estimates of camera settings can be made. For example, if readings from the top portion of the field are very high compared to other readings, overcast sky is likely to be included within the field of view, and a better exposure will probably result if the camera settings are based on the readings from the lower portion of the field. Examples of the other problematic scene types mentioned above can be inferred in a similar manner and appropriate compensation made.

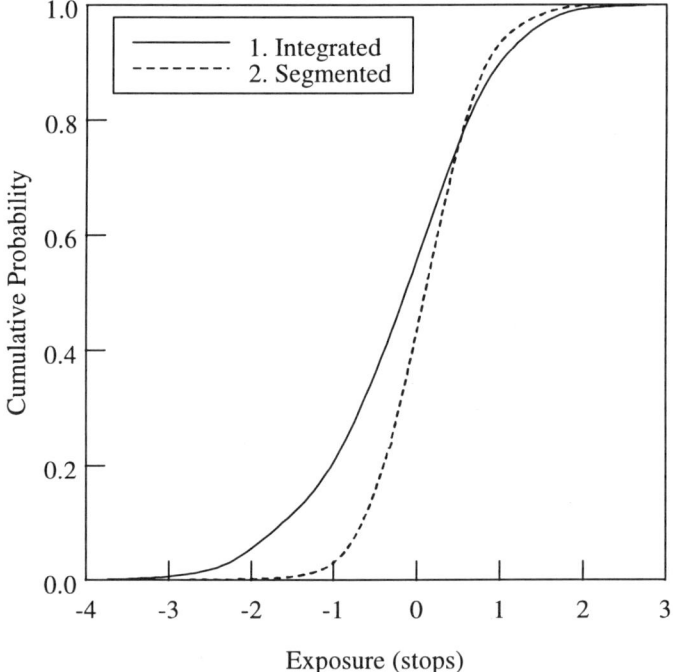

Fig. 30.3 Distribution of exposures (e.g., −1 = one stop underexposure) resulting from two camera metering methods. Integrated metering measures a single field-weighted luminance, whereas segmented metering measures a number of field positions, thereby improving robustness and narrowing the exposure distribution.

Figure 30.3 compares the exposure CDFs resulting from particular implementations of integrated and segmented metering. The x-axis is exposure in stops relative to the aim; e.g., -1 corresponds to a one-stop underexposure. A perfect meter would always produce aim exposures, so the CDF would plot as a vertical line at $x = 0$. The segmented metering subsystem closely approximates this ideal; the integrated metering system exhibits more overexposures and many more significant underexposures. If excessive quality loss were to arise from the significant underexposures, the entire exposure distribution could be shifted to higher exposure by biasing the rule for deriving camera settings from the meter reading. This could be achieved by choosing a higher value for the reflected light metering constant K (see Eq. 27.6), leading to longer exposure times and/or smaller f-numbers. Use of higher K values is a common practice in point-and-shoot cameras, in which color negative film is used almost exclusively. The effect of such an overexposure bias is to shift the photospace coverage by exchanging unsharp images for underexposed ones.

30.5 Output Color and Density Balancing Methods

Our fourth example involves output and is analogous to the metering example just described. Color balance, relating to the appearance of neutrality, and density balance, describing the extent to which an image looks too dark or too light overall, were defined in Sect. 20.2. A balancing method is an optical or digital mechanism for adjusting these attributes. For example, in a digital still camera, there is usually a white balance operation, in which color-specific analog or digital gains are adjusted to try to produce images with neutral appearance. The image may also be lightened or darkened through the equivalent of concerted changes in these gains. Equivalent operations are available in film scanners and also in digital image processing software. Similarly, in an optical color negative-positive system, the red, green, and blue printing exposures to the photographic paper may be adjusted by changing exposure time or filtration. If these three exposures are changed in proportion with one another, the effect is primarily that of lightening or darkening the image, whereas if they are changed disproportionately, the color balance of the final print is altered.

A perfectly calibrated output process would require no balancing operation if camera metering systems were perfect, films and sensors always had responsivities equaling their aims, and all images were captured in the illuminant for which the sensor or film were balanced. Because none of these conditions are met in practice, different schemes are employed to estimate the values of the printing exposures or gains that will cause a particular image to have optimal color and density balance.

In general, scenes that are difficult to meter in the camera will similarly be difficult to print for overall density; e.g., snow scenes are likely to be rendered too dark and flash photographs too light. In the estimation of color balance correction, chromatic analogues of the camera metering challenges are encountered. For example, most films and sensors have responsivities approximately balanced for typical daylight illuminant. Tungsten illumination, as provided by incandescent light, is relatively enriched in red light and deficient in blue compared to daylight. Consequently, an image captured in tungsten illumination with daylight-balanced responsivity and printed without color balance correction will be rendered with an orange cast. Although this cast could readily be corrected by adjusting gains or exposures, the real problem is distinguishing such an image from one of an orange-colored object in a daylight illuminant. The resulting channels' exposures may be identical in these two cases, but color correction is desired for the illuminant imbalance, whereas it would cause the image of the orange object to take on a bluish cyan cast.

If throughput is not critical, a human can visually assess images as they are printed (perhaps using a calibrated softcopy display, as discussed later in this chapter). This approach is taken in many professional operations and some minilabs as well. Higher speed and/or lower cost operations require a fully automated approach. In such cases, overall density corrections are handled in a fashion similar to that of camera metering. The degree of color bias correction is determined from an empirical, statistical algorithm. Higher levels of correction are applied when the apparent color bias is in a direction of expected illuminant variation (e.g., orange from tungsten), and lesser adjustments are made when the bias is towards common object colors (such as the green of foliage). In optical printing, just as in camera meters, either integrated or segmented measurements may be made. Integrated (large area transmission density) measurements dominated in photofinishing operations through the 1980s because of their simplicity, but by the end of the decade low resolution sensors became available at low cost and now most systems are segmented. In digital systems, the pixel values themselves can serve as high-resolution segmented readings, without the need for specialized hardware.

The advantages of segmented measurement for overall density adjustment are the same as those already discussed in connection with camera metering. The advantage of segmented measurements for color bias estimation is more subtle. If integrated measurements indicated an orange cast, a high level of correction would typically be applied because the most likely source of that bias would be tungsten illumination. However, segmented measurements might show that, rather than an overall orange bias, there is one large image area with a red bias and another with a yellow bias, which happen to average to an orange bias like that of tungsten illumination. In this case, it is more likely that the red and yellow biases are object colors and that a low degree of color correction should be applied. Figure 30.4 compares representative performance of color and density balancing based on integrated and segmented measurements; the latter leads to 1.5 JNDs of median improvement. Also shown is the quality distribution for custom printing, in which each image is optimized by human assessment and iterative trials. This yields more than 2 JNDs of additional median advantage, indicating the potential for future improvement as more advanced algorithms, using higher resolution image data, become available.

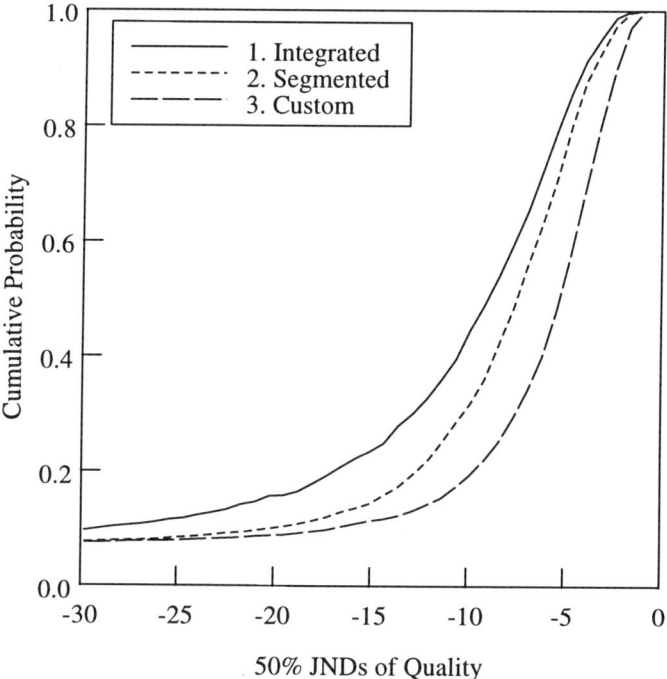

Fig. 30.4 Performance of three color and density balancing methods used in photofinishing operations. Measuring densities at many field positions, rather than determining only an average value, leads to significant improvement, but custom prints made by experienced operators are still substantially superior.

Ironically, at the time of this writing, color and density balancing of consumer digital images, both commercially and at home, is frequently inferior even to that of the optical printing using the integrated balancing method. One reason is that there is not always enough time and computing power in the camera to support inclusion of the best balancing algorithms. Furthermore, there are so many pathways by which digital images may be produced that, in a commercial photofinishing operation, which must print images of unknown pedigree, the resultant source variability may become comparable to that of illuminant and object color, thereby confounding their distinction and compromising the effectiveness of the balancing algorithms. This is one reason why there is currently considerable interest in the definition and use of metadata (auxiliary

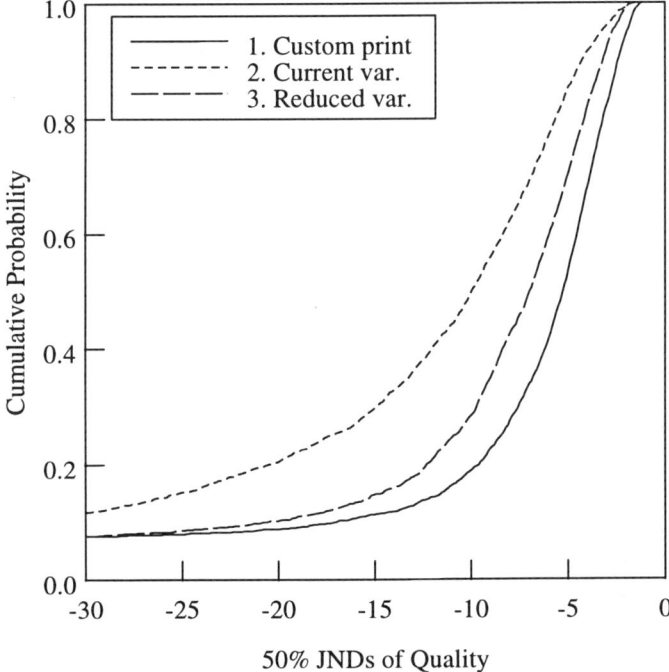

Fig. 30.5 Quality distributions of reflection prints made from digital images adjusted for color balance and contrast on a softcopy display. Current levels of variability in monitor gamma and white-point color temperature cause substantial quality loss (Curve #2 versus #1), most of which could be recovered if the dispersions were halved (Curve #3).

data associated with an image), which could provide a mechanism for tracking the history of a digital image, thereby enabling more effective digital image processing and higher final image quality.

Although custom printing is expensive when done by professionals because of the labor involved, consumers might choose to spend their own time optimizing their digital images using any of a number of commercially available image manipulation software packages. This optimization could include adjustments of density, color balance and even contrast so that the image appeared most pleasing on their computer monitors. After optimization, the adjusted images could be printed at home or transmitted to an online service for remote printing.

In practice, consumer monitors vary considerably in their luminance, the color temperature of their white point, and in their gamma (change in the logarithm of display luminance with respect to input voltage, which is proportional to code value). For a given monitor viewing environment and particular print viewing conditions, there is a monitor luminance, white-point balance and gamma that lead to the best match in appearance of density balance, color balance and contrast between the softcopy and hardcopy images. A deviation of monitor properties from the values leading to the best visual match causes images that are optimized on the monitor to have less preferred density and color balances and contrasts when printed. In this analysis, density balance errors are neglected because it is fairly easy, with a minimum of experience, to adjust for a constant difference in the apparent density balance of softcopy and hardcopy images.

Figure 30.5 shows the impact of monitor calibration on the quality distribution of reflection prints made from digital images optimized on the monitor. Curve #1, represents prints from perfectly calibrated monitors. Curve #2 shows the quality distribution that results from variability of white-point color temperature and gamma that is representative of that found in a sample of consumer computer monitors at the time of this writing. The median quality loss compared to the case of perfect monitor calibration is ≈4.7 JNDs. Curve #3 shows the quality distribution that would result if the standard deviations of white-point color temperature and gamma were both halved. The median improvement, compared to current levels of variability, is ≈3 JNDs, and large improvements are made in the low-quality tail.

30.6 Film Scanner Noise

Our fifth example demonstrates how camera exposure distributions affect the required specifications of film scanner dynamic range. In Fig. 30.3 we saw that the width of the exposure distribution produced by a camera can vary as a function of metering method. It was shown in Sect. 27.6 that cameras with fixed aperture and shutter speed, if loaded with color negative film (with its wide latitude), can have reasonable photospace coverage. Such cameras will produce wide distributions of exposure that directly reflect the ambient light levels and flash distances at which photographers take pictures.

The camera exposure distribution shape can have significant effects on later system components such as film scanners. In particular, the greater the exposure to the negative, the higher its resultant density. A film scanner typically is calibrated to give a nearly saturated sensor signal in each of the color channels at

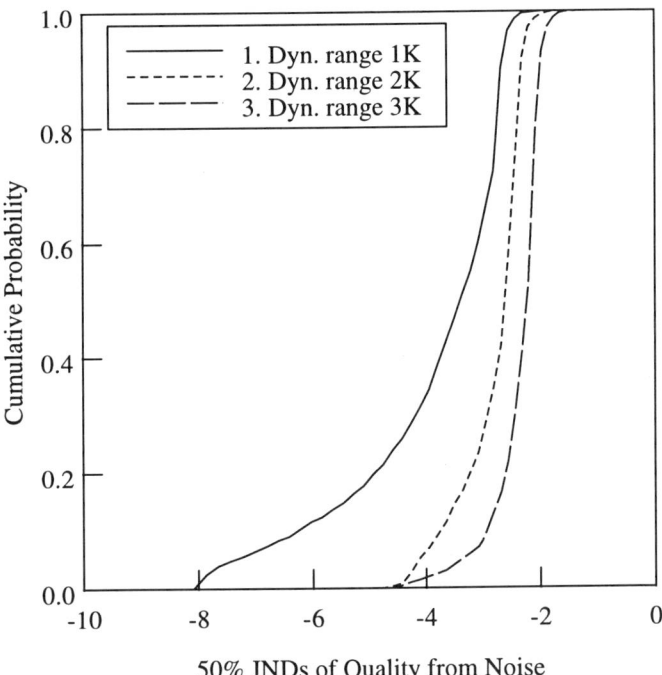

Fig. 30.6 Distribution of quality loss from noise in images made from digital scans of negatives. In this system, which includes a fixed exposure camera, three scanner dynamic ranges (with higher dynamic range corresponding to lower noise) lead to distinctly different quality losses arising from noise.

Examples of Performance Calculations

the expected minimum film densities. In an ISO normal exposure of a uniformly lit scene on color negative film, a black scene object will map slightly above the minimum densities and a white scene object will typically produce densities about one density unit higher. A five-stop overexposure, which is still within the latitude of most color negative films, will shift the densities roughly one density unit higher (because the sensitometric gradient of color negative film is usually approximately 0.65, and $0.65 \cdot \log_{10}(2^5) \approx 1.0$).

Scanner noise arises from a number of sources (see discussion of Fig. 23.1) but generally signal-to-noise improves with increasing signal. A single-point indicator of scanner noise is dynamic range, which reflects primarily signal-independent (dark) noise level. Dynamic range is usually defined as the ratio of the exposure to the sensor that just produces a saturation signal, to that producing a signal-to-noise ratio of one. If the sensor response in photoelectrons is proportional to exposure, as is often approximately true, the dynamic range equals the ratio of the saturation signal to dark noise level, with both expressed in terms of photoelectrons. A signal-to-noise ratio of one is unsatisfactory for pictorial imaging, except perhaps in extremely dark tones; for comparison, the preferred ISO speed of a digital camera is based on a signal-to-noise ratio of 40 (ISO 14524, 1997). Consequently, the dynamic range measure overestimates the useful range of the device for pictorial imaging, but it may still be used as a crude noise correlate.

As discussed above, as negative exposure increases, film density increases, exposure to the sensor decreases, and scanner signal-to-noise decreases. Therefore, we expect that wider camera exposure distributions will place greater demands on scanner noise characteristics. This behavior is confirmed in Figs. 30.6 and 30.7, which show distributions of quality change arising from final image noise for three scanner dynamic ranges (1000:1, 2000:1, and 3000:1). Both the camera film and the scanner contribute to the final image noise, but their relative impact varies with camera exposure, as will be seen.

Figure 30.6 depicts the final image noise performance resulting from the exposure distribution from a camera with a fixed aperture and shutter speed. The three scanner dynamic ranges yield distinctly different performances. Although the median quality shifts by only a little over one JND between the 3000:1 and 1000:1 cases, the latter has a higher frequency of occurrence of significantly degraded images. In about one-quarter of images the noise quality is at least two JNDs lower in the 1000:1 case than in the 3000:1 case, and in 10% of cases at least three JNDs lower.

Figure 30.7 depicts the analogous results for a camera with full exposure control and segmented metering, which produces a very narrow exposure distribution. The noise performance of the three scanner dynamic ranges is now nearly identical because high negative densities are not produced, and so even in the 1000:1 case, operation is in a region of favorable signal-to-noise. Consequently, the scanner contribution to final image noise is diminished, and the film contribution dominates, leading to insensitivity of final image noise to scanner dynamic range.

Interactions among components, such as this one between camera exposure control and scanner noise characteristics, are common in imaging systems but

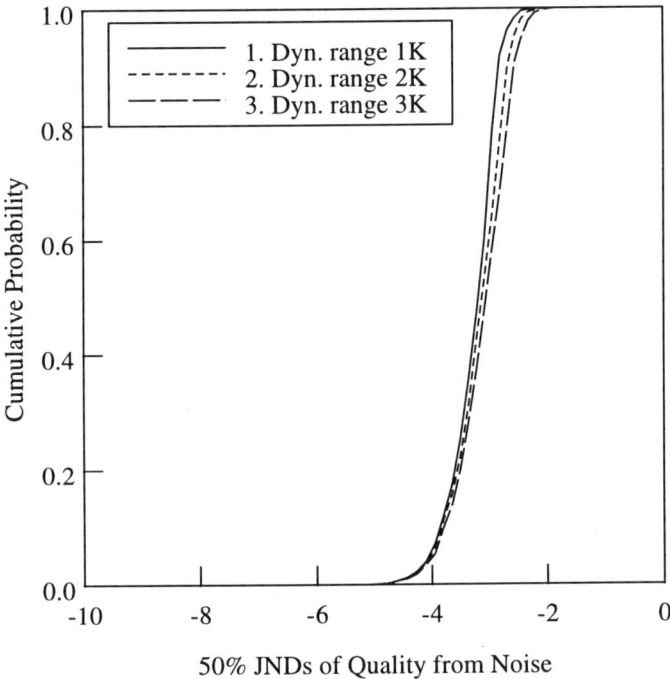

Fig. 30.7 Analogue of the previous figure but for a system having a camera with exposure control. Because overexposures have been curtailed, scanner noise at high negative densities is less important, and the quality loss arising from noise is scarcely affected by scanner dynamic range, being instead dependent primarily on film grain.

Examples of Performance Calculations 447

are difficult to assess in a relevant and timely fashion except through performance modeling.

30.7 Summary

In this chapter we have considered a number of examples of the application of performance modeling to product design. Some of the insights obtained include those listed below.

1. A property of one component can significantly affect the requirements for another component.

2. Camera requirements for a particular application depend upon the nature of the pertinent photospace distribution.

3. Analyzing distributions of intermediate objective quantities, such as camera exposure, is helpful for diagnosing the origin of changes in quality distributions.

4. Like an increase in film format size, higher numbers of equally-sized capture pixels in a digital camera can lead to an increase in capability at the expense of a more dispersed quality distribution.

5. The impact of color and density balancing and contrast adjustment on system performance is substantial. Constraints related to photofinishing throughput, computing power in digital cameras, capture source variability, and output device calibration limit how closely a system can approach the optimal state.

In general, performance analyses are likely to yield valuable insights when influential factors exhibit significant variation, particularly if they affect the impact of multiple components on image quality. These conditions are met in many of the practical problems facing system designers.

31

Verification of Performance Predictions

with Richard B. Wheeler
Eastman Kodak Company
Rochester, New York

31.1 Introduction

Because we have defined image quality in terms of a calibrated mathematical scale with associated physical standards, it is possible to verify the accuracy of modeling predictions by comparing them to independent assessments made against the physical standards. A particularly inclusive and rigorous test of modeling accuracy might have the following characteristics.

1. Predictions of the performance of a new imaging system would be made at the time of its definition, so that parametric estimation played a significant role.

2. The new system would involve a number of differences from existing systems, thereby testing many aspects of the modeling.

3. The new system would be constructed according to plan and then used by actual customers, so that performance under realistic usage conditions would be sampled.

4. A large sample of images generated by customers would be evaluated against the physical image quality standards by independent, trained

observers, thereby accurately and impartially measuring the system performance.

A test meeting all these criteria was made during the development of the Advanced Photo System, a film imaging system introduced in 1996 by a consortium of five photographic manufacturers. This chapter, which describes the use of image quality modeling in the Advanced Photo System effort, is organized as follows. Section 31.2 describes some aspects of the performance modeling carried out during the project. Section 31.3 discusses the definition of image quality aims and establishment of component specifications to meet those aims. Finally, in Sect. 31.4, predicted and measured system performance are compared to verify the accuracy of the modeling results.

31.2 Performance Modeling

Table 31.1 provides a partial listing of factors included in the Monte Carlo modeling that supported the development of the Advanced Photo System. The factors are grouped into categories by the primary aspects of performance that they affect. A somewhat greater number of factors are associated with capture than output because capture occurs under less controlled conditions, is accomplished by less knowledgeable users, and is subject to greater cost constraints.

Most image quality modeling analyses involve comparison of a test system with a reference system, which is often based on products in the marketplace, but may also incorporate prototype or even hypothetical components. For example, as discussed in the next section, the Advanced Photo System was compared to a 35-mm format reference to establish image quality aims. To reliably predict the difference in performance between test and reference systems, the effects of characteristics differing between the two cases must be modeled quite accurately, whereas the effects of the shared characteristics usually require considerably less accuracy. The Advanced Photo System presented a particularly rigorous test of the validity of the image quality modeling because so many of its aspects differed from those of its predecessors. The films, format size, cameras, magnetically encoded information, printer lenses, printing magnifications, and print sizes all departed substantially from existing photographic products. The ramifications of the print size differences are particularly interesting from a perceptual point of view and so are considered further now.

Usage (Distances/Light Levels)	Color/Tone Reproduction
photospace distribution	capture/printing/viewing illuminants
	film/paper spectral sensitivities
Film Exposure	film/paper interimage effects
	film/paper sensitometry
camera metering variability	film/paper image dye spectra
flash activation level	chemical processing variability
flash guide number and variability	printer calibration variability
flash recharging time	rendering algorithm variability
shutter type (e.g., slow-opening)	
flash shutter time	**Redeye**
lens aperture variability	
shutter time variability	subject demographics (age/race)
camera exposure program*	pre-flash power
ISO film speed and variability	flash-lens separation
Camera Motion	**Other Problems**
camera lens focal length	inadequate depth of field
camera stability ratio	subject or photographer motion
optical printing magnification	frame tilt (horizon not level)
viewing distance	frame translation (subject cut off)
	accidental exposure
Camera Lens MTF	double exposure
	blank frame
autofocus error	fogged frame
lens focus vs. subject distance program	dirt/scratches
lens/capture plane separation variability	finger over lens
MTF vs. defocus/aperture	finger over flash
	non-uniformity of flash illumination
Other MTF/NPS	environmental (haze, flare, etc.)
film MTF and NPS vs. exposure	
printer lens to film separation variability	
printer MTF vs. defocus/magnification	*shutter/aperture/flash output vs.
paper MTF and NPS	subject distance/light level

Table 31.1 Factors included in Advanced Photo System modeling.

Three standard print sizes were to be provided by the Advanced Photo System: (1) 4R (4 × 6 inches), the most common print size made from 35-mm format film in the United States and elsewhere; (2) HDTV (high definition television; 4 × 7 inches), having the same short dimension but approximating the higher 16/9 aspect ratio of HDTV; and (3) panoramic (4 × 11.4 inches), again having the same short dimension but a very high aspect ratio of ≈2.85. The choice of different aspect ratios on an image-by-image basis allows photographers to match the image format to the shape of the important elements of the scene being captured, thereby providing the opportunity for improved composition. Notably, many scenic photographs lend themselves to the panoramic format, which seems to invite the eye to scan across the horizon.

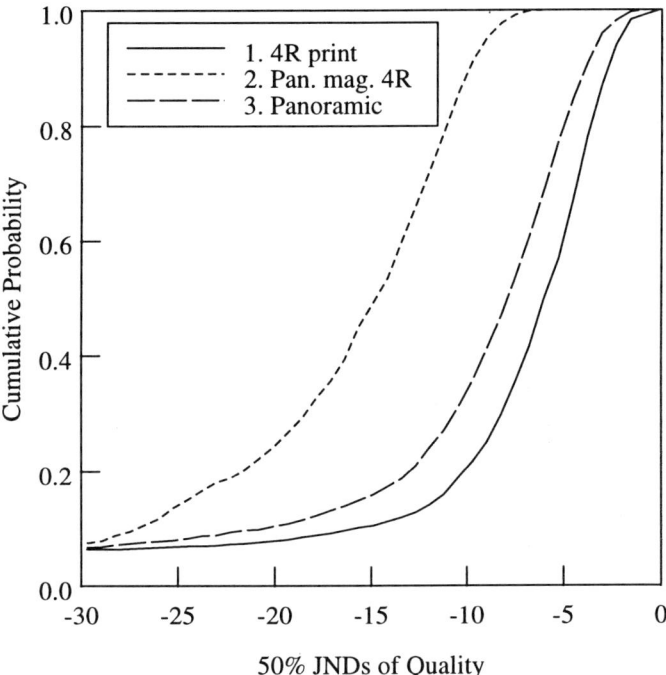

Fig. 31.1 Predicted quality distributions of Advanced Photo System 4R (4 × 6 inch) prints, panoramic (4 × 11.4 inch) prints, and 4R prints made at panoramic (≈1.6× higher) magnification. Most of the quality loss arising from the higher panoramic magnification is compensated by the longer viewing distance and greater "impact" of the larger prints.

Figure 31.1 shows the predicted performance of the Advanced Photo System at three combinations of print size and printing magnification. Curve #1, the reference case, shows the modeled distribution for 4R prints. Curve #2 is the prediction of what would result if 4R prints were made at the printing magnification of a panoramic print. This ≈1.6× higher magnification leads to lower sharpness and higher noisiness, hence the quality is predicted to be significantly lower than in the reference case (about nine JNDs at the median quality). Curve #3 shows the prediction for panoramic prints, which lie only about two JNDs below the reference case. Although the only thing ostensibly changing between Curves #2 and #3 is the print size (not the magnification), two factors mitigate the effects of the panoramic printing magnification. First, on average, people hold larger prints farther away from their eyes (Eq. 26.1). The increased viewing distance associated with the near doubling of the long print dimension partially offsets the increased printing magnification, making it like an increase of ≈1.3×, rather than ≈1.6×. Second, all else being equal, larger prints have greater impact and are assessed to be higher in overall quality than smaller prints. The difference in quality between 4R and panoramic prints having equal perceived image structure is ≈2 JNDs. The combination of increased viewing distance and impact compensates for much of the quality loss that would otherwise be incurred by the panoramic printing magnification.

31.3 Advanced Photo System Aims

The intent of the Advanced Photo System was to provide a number of new features that, compared to the dominant 35-mm format, improved ease of capture and utilization of images, with greater flexibility, while maintaining comparable image quality. These features included: (1) choice of three print formats on an image-by-image basis; (2) true drop-in film loading with a leaderless film cassette; (3) smaller camera size; (4) film return in the cassette; (5) thumbnail index prints cross-referenced to the film cassette for efficient retrieval of negatives; (6) magnetic and latent image information recording on the film; (7) enhanced information printed on back of prints; (8) ability to reload partially exposed film with automatic advance to the proper frame; and (9) a number of incremental improvements in film and camera technology. In this section we will describe how aims were set to meet the overall image quality requirement of parity with 35-mm format.

There are an infinite number of combinations of system properties that could yield performance comparable to that of a stated reference position, but some will be more practical or better support desired features than others. This may be regarded as a design problem in which there are excess degrees of freedom (like a set of N equations in M unknowns, where M > N). Such problems are often best approached by imposing additional constraints based on practical experience and intuition. One general constraint arises from the observation that system specifications that require substantial technological advances in some components, but little change in others, are often less practical than alternatives that distribute the required improvement more uniformly over the system components. This principle of apportioning the burden of responsibility equitably across the system can lead to a higher probability of success in meeting specified aims. For example, if a given improvement in sharpness is needed, it is more likely to be obtained by some combination of modest improvements in autofocus, film MTF, camera lens MTF, printer lens MTF, optical positioning tolerances, etc. than by a dramatic improvement in just one of these factors. There are certainly exceptions to this rule as stated, particularly when the various factors are contingent upon technologies that are at different stages of maturity. These exceptions may be accommodated by refining the definition of equitable apportionment to mean that the burden imposed is commensurate with the ability to deliver improvement.

Some of the desired features of the Advanced Photo System provided constraints of importance in setting aims based on image quality considerations. As mentioned previously, a choice from three standard print sizes with very different aspect ratios (1.5, ≈1.8, and ≈2.9) would be made for each image at the time of capture. Although three print aspect ratios had to be supported, many practical problems in photofinishing would arise if the film format size were variable (i.e., if the spacing between frame centers were not constant). If the aspect ratio of the film frame were set equal to that of the panoramic print, the panoramic printing magnification would be minimized (because no cropping would be required), but more film area would be required per frame. The intermediate HDTV aspect ratio was chosen for the film format to provide a good compromise between efficient usage of film area (for cost) and avoidance of high printing magnifications (for quality).

With the film format aspect ratio chosen, the next critical decision was the actual frame size (it being sufficient to specify one dimension, the other being determined by the aspect ratio). By analogy with the sensor size example of Sect. 30.3, we expect that, all else being equal, as film format size is increased, capability will increase, but photospace coverage will decrease, leading to less steeply sloped quality distributions. The distributions can be reshaped to some degree by optimizing system properties for a particular format size, as discussed further below. The partial cancellation of many compensating effects that vary as format size is changed, leads to a range of format sizes that yield reasonably similar performance at comparable cost.

Verification of Performance Predictions 455

One of the chief contributions of image quality modeling to the development of the Advanced Photo System was to map out this range of advantageous format sizes, which was found to extend from slightly larger than 35-mm format to about one-half its area. At larger format sizes, despite higher capability, poorer performance was obtained at equal cost because of the lower photospace coverage. At smaller format sizes, capability was reduced because of the higher printing magnifications required, and performance suffered accordingly. Thus, a plot of some univariate measure of performance against format size would look like a plateau. Given the strong desire to produce more compact cameras for greater portability and convenience, the final format size was chosen to be as small as possible while still lying within the advantageous range, i.e,. its position was selected to fall near the edge of the plateau, roughly opposite that of 35-mm format. This choice helped mitigate the larger size of leaderless film cassettes, which enabled the drop-in loading feature. The time and resources required to prototype and empirically optimize systems of different format sizes, and then reliably measure their performance in the hands of customers, would have been excessive, so the ability to predict the results of such an experiment through Monte Carlo modeling was extremely valuable.

The resulting format size, about 70% as large as 35-mm format in the short dimension, required ≈40% greater printing magnification to a 4R print, but only ≈20% higher printing magnification to HDTV or panoramic aspect ratio prints because of the higher format aspect ratio. The angular magnification of a 4R print from the Advanced Photo System was stipulated to equal that from 35-mm format so that comparable compositions would be obtained from the two systems in their common print size. A further constraint applied was that the photospace coverage should approximately match that of 35-mm format, so that the system could be used in a fashion similar to that of existing products with which consumers were already familiar. To meet this criterion, camera autofocus and exposure control systems were carefully optimized within cost constraints using Monte Carlo modeling, which was particularly helpful in choosing: (1) the exposure bias that optimally balanced quality loss from underexposure and unsharpness (Sect. 27.6); and (2) the light level for flash activation that best balanced quality loss from near and distant ambient underexposures.

With the system properly optimized, the photospace coverage of the new system was actually greater than that of 35-mm because of the roughly one stop greater relative format speed (see Table 27.1). Because photospace coverage was only required to match 35-mm format, this extra stop of system speed could be expended in various ways (as suggested by Eqs. 27.5, 27.8 and 27.10), such as reducing the size and cost of flash components by selecting a lower flash guide number. However, the higher printing magnifications of the new format pose challenges from the standpoint of image quality, as discussed previously (see Sects. 12.5, 22.6, 22.7, 27.7, and 30.3). Because the new system lay near the edge of the performance plateau, it was necessary to expend the format speed advantage directly on recovery of image quality losses that would otherwise arise from the higher printing magnifications.

One effective means of utilizing format speed benefit to mitigate the increased printing magnification is to reduce film speed, which, depending upon how the film is designed, can lead to reduced film grain, increased film MTF response, improved color reproduction, or a combination thereof. A number of other system properties also had to be improved to offset the impact of higher printing magnification. As discussed in connection with Eq. 27.11, smaller formats require tighter tolerances for the separation of the lens and film planes and for film plane flatness (both in the camera and in the optical printer); fortunately, the latter is somewhat facilitated by a smaller format size. Some lens MTF response is typically sacrificed to cover larger formats, so modest improvement in lens MTF may be anticipated as format size is decreased. All of these factors were taken into account in a Monte Carlo performance analysis to determine the optimum partitioning of needed improvements in image structure, both between attributes (noisiness and unsharpness) and across components. With a global and unbiased analysis of this type, teams assigned to develop individual components of the system had confidence that the aims assigned for their component properties reflected a fair disposition of the required advances in technology, thereby maximizing the probability of meeting the overarching image quality goal of performance parity with 35-mm format.

31.4 Verification of Predictions

Shortly before the introduction of the Advanced Photo System, a four-city trade trial was conducted, providing an excellent opportunity to verify modeling predictions in a particularly rigorous fashion. A total of 8500 consumer images were assessed for overall quality, against physical standards, by trained experts who were not associated with the modeling effort. Because our image quality models are calibrated to the same physical standards used by the trained experts, the predictions made at the initiation of the Advanced Photo System project, two years earlier, could be directly compared to the distribution of quality assessments from the trade trial. No data transformations, assumptions, or interpretations were required to compare the predictions and measurements, because both were expressed in terms of the same numerical scale of quality, which was, in turn, anchored against the physical standards used in the

Verification of Performance Predictions

measurements. Of particular note, this comparison tests an absolute prediction of performance, which is more rigorous than testing predictions of the performance difference between two systems, or of an absolute capability position.

Figure 31.2 depicts the quality distributions predicted by Monte Carlo modeling and measured by expert assessment of Advanced Photo System trade trial images. The agreement is good to one JND and/or 5% cumulative probability over the entire quality distribution, which is a remarkable result! This level of agreement provides a compelling confirmation of the accuracy of the image quality modeling.

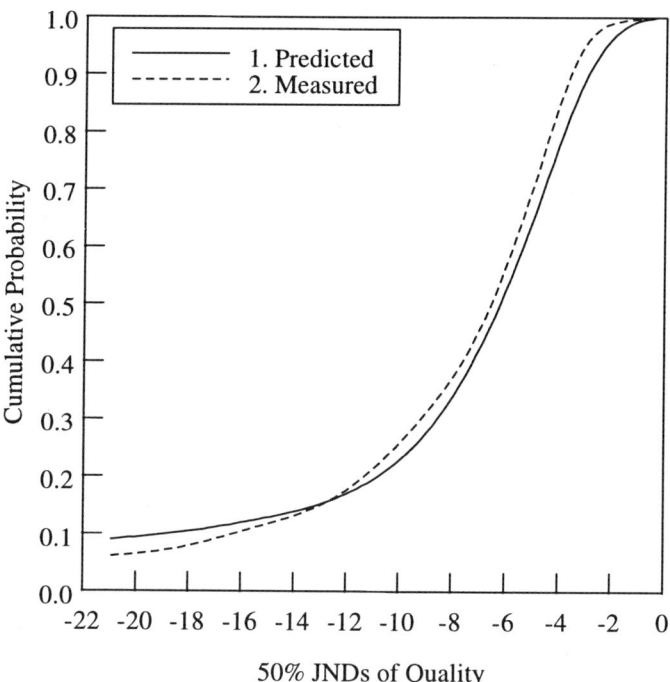

Fig. 31.2 Comparison of performance predictions, made at Advanced Photographic System inception, with expert assessment of customer trade trial orders, collected two years later. The agreement is to within one JND and/or 5% cumulative probability over the entire distribution, dramatically demonstrating the accuracy of the image quality modeling.

31.5 Summary

The Advanced Photo System design and development were strongly influenced by system modeling. The image quality goal of this system was to match that of 35-mm format quality at equal angular magnification and photospace coverage. With image quality at parity, the Advanced Photo System could then be evaluated by customers based on its many new features. System modeling was used to set specifications for format size, film grain, component MTFs, optical positioning tolerances, etc. in a way that equitably partitioned the needed improvements across the different system components and maximized the probability of meeting the image quality goal.

Prior to launch, a large trade trial allowed rigorous verification of modeling predictions made years earlier. Because the new system differed in so many ways from existing systems, this comparison provided a particularly definitive test. Superb agreement was observed between independent assessments of the quality distribution of consumer images by expert observers and the performance predicted by the system models during design phase, strongly substantiating the validity of the image quality modeling.

Conclusion

Although the emphasis of Part III has been on the use of image quality models in photographic system design, we have also employed such models in many other applications, including those following.

1. Research portfolio management: modeling can serve to objectively summarize the state of the art in understanding image quality and to indicate which pursuits have the greatest potential to improve image quality, thereby providing guidance regarding future research topics.

2. Technical intelligence: in addition to providing uniform benchmarking capabilities, modeling can be used to augment tear-down studies to understand how and why a competitor's product produces a particular distribution of image quality.

3. Product planning: modeling can provide input to technology substitution analyses and can make predictions of where significant leverage for improvement still exists, or when diminishing returns are expected.

4. Marketing: modeling can substantiate advertising claims, generate explanatory material, and identify potential application-based niche markets.

5. Intellectual property: modeling can suggest new patent opportunities, identify preferred embodiments, and support claims.

6. Standards activities: modeling can be used to guide recommendations or requirements, and to define the nature of diagnostic criteria employed.

This list of applications gives some idea of the power and versatility of system image quality modeling.

Thinking back upon the question raised in the first chapter, namely, whether image quality can even be quantified, it is apparent how far we have come. Not only can image quality be quantified, but it can be understood in detail, and can even be predicted from the fundamental properties of an imaging system. With more open industrial discussion and increasing academic involvement, further progress in understanding and predicting image quality may be anticipated in the near future. These developments should strengthen the imaging business and lead to greater customer satisfaction.

Appendix 1

Definition of Acronyms and Symbols

A1.1 Acronym Definitions

Acronym	Definition
CDF	cumulative distribution function
CFA	color filter array
CI	confidence interval
CIE	Commission Internationale de l'Eclairage
CIELAB	1976 CIE L* a* b* system (CIE Publication 15.2, 1986)
DVF	detail visibility function
IHIF	integrated hyperbolic increment function
ISO	International Standards Organization
IWS	image wavelength spectrum
JND	just noticeable difference
JPEG	Joint Photographic Experts Group
MCS	mean channel signal
MTF	modulation transfer function
NPS	noise power (Wiener) spectrum
PDF	probability density function
RGB	red, green, blue
RMS	root-mean-square
TDF	tonal distribution function
TIF	tonal importance function

A1.2 Symbol Definitions

Symbols used in equations, other than standard mathematical notation, are tabulated below. Where appropriate, units of quantities are given parenthetically. The first equation in each chapter using the symbol is listed in square brackets. The subscripts i, j, and k refer to integer values unless otherwise noted. Primed quantities have the same definition as the corresponding unprimed quantities but have been subject to some type of transformation or are dummy variables of integration. A bar above a given quantity denotes the mean value thereof. The notation $[\ ]_{jk}$ indicates the matrix element in the jth row and k^{th} column.

Symbol	Definition
A	lens aperture (f-number) [27.1, A5.2]
$a(z_a)$	angular distribution PDF [2.9]
Δa^*	1976 CIE a* difference (CIE Publication 15.2, 1986) [20.2]
Δb^*	1976 CIE b* difference (CIE Publication 15.2, 1986) [20.2]
c_i	i^{th} arbitrary constant (numbering restarted in each chapter)
D_s	scene density [15.7]
$D_{s,i}$	scene density at i^{th} quantization step (code value change) [15.9]
$D_{s,+}$	scene density at which the highlights effectively clip [15.7]
$D_{s,-}$	scene density at which the shadows effectively clip [15.7]
D_v	visual density [15.1]
d_a	aperture dimension (diameter or edge) (mm) [A4.1]
d_b	diameter of geometrical optics blur circle (mm) [A5.3]
d_c	circle of confusion (allowed value of d_b) (mm) [27.1, A5.8]
d_D	display format (e.g., print) diagonal (mm) [26.1]
d_d	capture format (film or sensor) diagonal (mm) [27.2, A5.14]
d_f	depth of focus (mm) [27.11, A5.8]
d_h	hyperfocal distance (mm) [27.1, A5.9]
d_i	distance of image from optical center of thin lens (mm) [A5.1]
d_l	diameter of clear aperture of thin lens (mm) [A5.2]
$d_{\max}$	maximum distance of normal flash exposure (m) [27.10, A5.23]
$d_{\min}$	minimum depth-of-field limit (mm) [27.1, A5.10]
d_o	distance of object from camera (mm) [A5.1]
d_n	distance yielding a normal flash exposure (m) [27.9, A5.22]
d_p	distance between the centers of pixels in a sensor (mm) [A4.3]
d_v	viewing distance (mm) [14.4, 26.1, A5.13]
$[E_i]$	column vector of exposures to i^{th} system component (lux·s) [22.8]

Definition of Acronyms and Symbols 463

Symbol	Definition
E_0	exposure at zero image density and flare (lux·s) [22.8]
F	lens focal length (mm) [27.1, A5.1]
$f([\mu_i])$	function transforming MCS from one component to the next [22.1]
G	flash (strobe) guide number (m) [27.9, A5.22]
$g(z_g)$	Gaussian function (normal distribution) [2.3]
$H(u)$	Heaviside unit function ($H(<0) = 0$; $H(0) = \frac{1}{2}$; $H(>0) = 1$) [A3.5]
$h(u)$	probability density function (PDF) [2.4]
$h_h(\rho_h)$	PDF of head size ratio [21.3]
$h_p(O)$	PDF of objective metric preference [4.1]
$h_t(D_s)$	PDF of scene density tonal importance [15.7]
$I(\lambda)$	illuminant radiance (wavelength spectrum) (ergs/nm·cm^2) [22.7]
$I_a(\lambda)$	ambient flare radiance (wavelength spectrum) (ergs/nm·cm^2) [22.7]
$I_e(\lambda)$	total exposing radiance (ergs/nm·cm^2) [22.7]
$[J]$	inverse Jacobian matrix for transforming MCS differences [22.2]
$[J']$	row-sum-normalized inverse Jacobian matrix [22.10]
K	reflected light metering constant (cd·s/m^2) [27.6, A5.20]
L^*	1976 CIE lightness L* (CIE Publication 15.2, 1986) [15.3]
$[M(v)]$	column vector of MTF at frequency v [22.10]
$[M'(v)]$	column vector of the MTF transformed to new MCS space [22.10]
$[M_{c,i}(v)]$	column vector of the MTF of the ith system component [22.12]
$[M_{s,i}(v)]$	column vector of system MTF through the ith component [22.11]
m_a	angular magnification [27.3, A5.15]
m_p	printing magnification [22.9, A5.13]
n_c	power of one or two in objective metric of color/tone quality [20.1]
n_m	variable power in multivariate formalism Minkowski metric [11.1]
O	flash output (beam candlepower seconds) [27.9, A5.22]
P	number of patches used in color/tone objective metric [20.1]
p	probability of occurrence ($0 \leq p \leq 1$) [2.4]
$p_a(z_a)$	angular distribution CDF [2.7]
p_c	fraction of correct responses in a paired comparison [2.8, 3.1]
p_d	fraction of times difference is detected in a paired comparison [3.1]
$p_g(z_g)$	normal distribution CDF [2.5]
p_p	fraction of times sample is chosen in a paired comparison [2.8, 3.2]
Q	quality scale value (JNDs) [9.1, A3.1]
Q_r	quality at reference threshold/optimal position (JNDs) [9.1, A3.5]
ΔQ	quality difference (JNDs) [4.1, 9.5, 15.7, 21.3]
$\Delta Q(\sigma_e)$	quality loss from noise equally visible at all image densities [15.8]

Symbol	Definition
ΔQ_c	quality difference under critical viewing conditions [21.4]
ΔQ_d	quality difference in JNDs with probability of detection p_d [3.2]
ΔQ_i	quality difference due to i^{th} attribute [11.1]
ΔQ_m	overall multivariate quality difference [11.1]
ΔQ_{max}	maximum quality difference from skin-tone reproduction [21.1]
$(-\Delta Q)_{max}$	magnitude of maximum quality degradation in an image [11.2]
ΔQ_p	quality loss function of preference [4.1]
ΔQ_u	quality difference under uncritical viewing conditions [21.4]
ΔQ_1	quality difference one objective metric unit from optimum [4.3]
q	number of quantization levels; e.g., for 8 bits, $q = 2^8 = 256$ [15.9]
R_i	misregistration radius of the ith color (see w_i) (mm) [14.2]
R_r	radius of curvature of quality loss function at Ω_r [9.2, A3.1]
S	ISO capture (film or digital still camera) speed [27.6, A5.20]
s	rating on an arbitrary (not necessarily interval or ratio) scale [5.8]
s_r	reference value of rating scale s, corresponding to t_r [5.8]
Δs_J	JND increment of a rating scale s [5.8]
T	camera exposure (shutter or integration) time (s) [27.6, A5.11]
T_{max}	maximum exposure time freezing camera motion (s) [27.7, A5.19]
$t(D_s)$	tonal importance function versus scene density [15.7]
u	arbitrary variable or substitution variable for integration
V	number of variable dimensions in color/tone objective metric [20.1]
$\Delta v_{i,j}$	distance of jth variable in i^{th} color patch from its optimum [20.1]
$[W_{c,i}(v)]$	square NPS array of ith system component (mm^2) [22.14]
$[W_{s,i}(v)]$	square system NPS array through i^{th} component (mm^2) [22.14]
w_{a*}	color balance weighting of the a* CIELAB dimension [20.2]
w_i	visual weighting of ith color record (red/green/blue = 1/2/3) [14.1]
$w_{i,p}$	weighting of ith of P patches in color/tone objective metric [20.1]
$w_{j,v}$	weighting of jth of V variables in color/tone objective metric [20.1]
x_c	misregistration center of mass x coordinate (mm) [14.1]
x_i	misregistration x coordinate of the ith color (see w_i) (mm) [14.1]
Y	luminance (cd/m^2) [15.3, 27.6, A5.20]
Y_{min}	minimum luminance (cd/m^2) for normal ambient exposure [A5.21]
Y_0	reference luminance (cd/m^2) [15.3]
y_c	misregistration center of mass y coordinate (mm) [14.1]
y_i	misregistration y coordinate of the ith color (see w_i) (mm) [14.1]
z_a	angular distribution deviate value [2.7, 3.2]
z_g	Gaussian (normal) distribution deviate value [2.2]

Definition of Acronyms and Symbols

Symbol	Definition
α	substitution parameter in derivatives of IHIF [A3.3]
β	substitution parameter in derivatives of IHIF [A3.4]
γ_i	sensitometric tone scale gradient of the ith component [22.14]
θ_c	angle of rotation of camera during exposure (radians) [A5.11]
θ_d	diagonal field of view of capture (radians) [27.2, A5.14]
θ_{max}	maximum acceptable visual angle of blur (radians) [27.4, A5.16]
θ_v	viewing angle of camera motion during exposure (radians) [A5.13]
ι	interval scale of perception [5.1]
ι_r	reference value on interval scale ι, corresponding to s_r [5.8]
$\Delta\iota_J$	JND increment of interval scale ι [5.8]
λ	wavelength of light (nm) [22.7]
$[\mu_i]$	column vector of ith component mean channel signal (MCS) [22.1]
$[\mu]_{R/G/B}$	red, green, or blue channel signal of one pixel [22.5]
$[\delta\mu]$	column vector of small MCS differences [22.3]
ν	spatial frequency (cycles/mm, cycles/sample, cycles/degree) [22.9]
ν_{i-1}	spatial frequency in plane of previous system component [22.9]
ρ	ratio scale of perception [5.4]
ρ_h	ratio of head height to geometric mean image dimension [21.1]
ρ_s	camera stability ratio; >1 more stable than average [27.7, A5.12]
σ_D	RMS granularity calculated in density space [15.1]
σ_e	RMS granularity, corrected for visibility of noise vs. density [15.2]
σ_g	standard deviation of a Gaussian (normal) distribution [2.1]
σ_{L^*}	RMS granularity calculated in CIE lightness space [15.6]
σ_p	standard deviation of a preference distribution [4.2]
$[\tau_i(\nu)]$	system transfer function matrix to i^{th} component [22.13]
$[\tau'_i(\nu)]$	normalized system transfer matrix to i^{th} component [22.11]
Φ	number of phases in a time delay/integrate sensor [A4.3]
$\phi(\lambda)$	fractional reflectance/transmittance of image vs. wavelength [22.7]
$\phi'(\lambda)$	fractional image reflectance/transmittance including flare [22.7]
ϕ_c	fraction of quality loss compared to critical comparison [21.4]
ϕ_h	fraction of maximum skin-tone reproduction quality loss [21.1]
ϕ_s	fraction of exposing illumination that becomes stray light [22.7]
ϕ_v	fraction of image-modified light that becomes veiling flare [22.7]
χ	severity of contouring (quantization) in presence of noise [15.9]
$\Psi_j(\lambda)$	spectral sensitivity of jth channel vs. wavelength (cm^2/erg) [22.8]
Ω	objective metric (units variable) [4.1, 20.1, A3.1]

Symbol	Definition
Ω_a	angular misregistration objective metric (arc-seconds) [14.4]
Ω_d	misregistration objective metric in distance units (mm) [14.3]
Ω_p	observer- and scene-specific preferred objective metric value [4.2]
Ω_r	objective metric at reference threshold position [9.1, A3.1]
Ω_s	preference switch point from one choice (Ω_1) to another (Ω_2) [4.7]
$\Delta\Omega$	objective metric difference [3.4]
$\Delta\Omega_J$	objective metric increment producing one JND of change [3.4, 9.1]
$\Delta\Omega_\infty$	asymptotic objective metric JND increment [9.2, A3.1]
ω	rotation rate of camera during exposure (radians/s) [A5.11]
ω_c	critical value of ω for setting T_{max} (radians/s) [27.7, A5.12]

Appendix 2

Sample Quality Ruler Instructions

A2.1 Introduction

The following is an example of the instructions that are read by the test administrator to the subject in a quality ruler experiment. These sample instructions are taken from a study of misregistration. Text in italics directs the administrator to perform certain actions and so is not read aloud.

A2.2 Instructions for Misregistration Psychophysical Test

Display the Quality Ruler depicting Scene #1 for demonstration purposes.

First, I would like to thank you for participating in this study. Please put on the lab coat and gloves and make yourself comfortable in the chair in front of the viewing table.

In this experiment, you will be evaluating the overall quality of prints made from images that have one or more of the three color planes shifted out of register. This misregistration may affect the image sharpness and may cause various image artifacts. Let me show you some examples of these images.

Give the subject Preview Print #1.

Some images you will see may exhibit only small or even unnoticeable levels of unsharpness and color fringing around edges. For example, in this image, look at the windows in the building on the right.

Give the subject Preview Print #2.

Some images may appear as two or more sharp, colored images offset from each other as in this scene. Notice the horizontal edges in the lipstick cases in this image.

Do you have any questions about the attribute that you will be judging today?

You will be evaluating these samples using a quality ruler like the one in front of you now. The ruler provides you with a series of prints at different levels of quality, produced by variations in sharpness. The ruler print quality decreases from left to right; however, the ruler numerical values increase from left to right, to reflect quality degradation. The numerical values are defined so that one unit is approximately one just noticeable difference. Here is how you use the ruler:

1. Place the test image flat in the holder above the ruler. Slide the ruler right or left to permit comparison of the test sample with different ruler images, which must be underneath the test sample for a valid assessment.

2. Locate the position of equality on the ruler such that each print farther right is lower in quality than the test image, and each image farther left is higher in quality.

3. Read off the number of the position of equality. If this position falls in between two ruler prints, as it often will, please select an intermediate number from the ruler scale. For example, if the point of equality is in between the ruler prints "12" and "15", but is closer to "12", you might assign the test print a value of "13".

4. If you feel a test print is higher in quality than the "3" print on the ruler, you may assign it an integer value less than 3. If the test print is lower in quality than the "21" print you may assign it an integer value greater than 21.

When you are finding the point of equality on the ruler, it is important to consider overall quality. It may help to imagine that the image is one that you treasure, and you have a choice of two prints of that image: one is misregistered, whereas the other is unsharp. Compare the test print with different ruler prints

Sample Quality Ruler Instructions

and decide which one you would choose to keep. From these comparisons, identify the position on the ruler from which prints to the right are less desirable than the test sample, and those to the left are more desirable.

Note: although some misregistered images may simply appear unsharp, please don't just match the sharpness of the test and ruler images. It is important to concentrate on assessing overall quality, because most of the misregistered images will show artifacts in addition to unsharpness.

Let me show you how I would evaluate one print.

Evaluate Preview Print #3 for the observer.

Finally, please disregard any physical damage such as scratches on the test prints during your evaluations.

Do you have any questions at this time?

Answer questions, then begin test.

Appendix 3

The Integrated Hyperbolic Increment Function

A3.1 Continuity and Curvature

Equation 9.4, reproduced below, defines the integrated hyperbolic increment function (IHIF) above the reference (threshold or optimal) position.

$$Q(\Omega) = Q_r + \frac{R_r}{\Delta\Omega_\infty^2} \cdot \ln\left(1 + \frac{\Delta\Omega_\infty \cdot (\Omega - \Omega_r)}{R_r}\right) - \frac{\Omega - \Omega_r}{\Delta\Omega_\infty} \quad (9.4)$$

In this equation, Ω is an objective metric value; Ω_r is the threshold of detection (or distinction from the optimum of preference); $\Delta\Omega_\infty$ is the asymptotic JND increment well above threshold; R_r is a curvature parameter that determines how rapidly above threshold the asymptotic behavior is reached; Q is quality in JNDs, and Q_r is the quality at the reference position.

It was stated in Ch. 9 that Eq. 9.4, which is valid when $\Omega > \Omega_r$, and the equation $Q = Q_r$, which is valid elsewhere, are equal and have equal first derivatives at $\Omega = \Omega_r$, and therefore, join smoothly to first order. Obviously, the latter equation has a value of Q_r at $\Omega = \Omega_r$, and its first derivative at this point is zero. By inspection, Eq. 9.4 also has a value of Q_r at $\Omega = \Omega_r$, so the functions match at that point, as claimed. The first derivative of Eq. 9.4 with respect to the objective metric Ω is given by Eq. A3.1.

$$\frac{\partial Q}{\partial \Omega} = \frac{-(\Omega - \Omega_r)}{R_r + \Delta\Omega_\infty \cdot (\Omega - \Omega_r)} \qquad (A3.1)$$

This likewise has a value of zero at $\Omega = \Omega_r$, and so the slopes of the two functions match at this point as well, as claimed.

It is instructive to take the derivative of Eq. A3.1 with respect to the objective metric Ω, which yields the following result.

$$\frac{\partial^2 Q}{\partial \Omega^2} = \frac{-1}{R_r \cdot (1 + \Delta\Omega_\infty \cdot (\Omega - \Omega_r)/R_r)^2} \qquad (A3.2)$$

At $\Omega = \Omega_r$ this second derivative has a value of $-1/R_r$, allowing R_r to be identified as the radius of curvature of the IHIF at the reference position.

A3.2 Derivatives

It is sometimes helpful in nonlinear regression applications to specify the derivatives of the assumed functional form with respect to each of the fit parameters. This can lead to more robust and faster convergence, and facilitate calculation of confidence intervals associated with the regression equation. The derivatives of the IHIF with respect to R_r, $\Delta\Omega_\infty$, Ω_r, and Q_r will be given below. First, it is helpful to define two parameters to make the equations more compact.

$$\alpha \equiv 1 + \frac{\Delta\Omega_\infty \cdot |\Omega - \Omega_r|}{R_r} \qquad (A3.3)$$

$$\beta \equiv |\Omega - \Omega_r| \qquad (A3.4)$$

As discussed in the previous section, Eq. 9.4 is valid for $\Omega > \Omega_r$; otherwise $Q = Q_r$. It can simplify nonlinear regression programming to combine these into a single equation. This can be done using the Heaviside step function $H(u)$, which is zero for $u < 0$, one for $u > 0$, and one-half at $u = 0$. If the Heaviside function is unavailable in a programming language, $(1 + \text{sign}(u))/2$ may usually be substituted for it, or logical branching used instead. Employing the parameters in Eqs. A3.3 and A3.4 and the Heaviside function, Eq. 9.4 may be rewritten as:

The Integrated Hyperbolic Increment Function

$$Q = Q_r - H(\Omega - \Omega_r) \cdot \left(\frac{\beta}{\Delta\Omega_\infty} - \frac{R_r \cdot \ln\alpha}{\Delta\Omega_\infty^2} \right) \quad \text{(A3.5)}$$

which is now valid at all values of Ω. This is the most general form of the IHIF. The derivatives of Eq. A3.5 with respect to the four fit parameters may now be written compactly as follows.

$$\frac{\partial Q}{\partial Q_r} = 1 \quad \text{(A3.6)}$$

$$\frac{\partial Q}{\partial R_r} = H(\Omega - \Omega_r) \cdot \left(\frac{\ln\alpha}{\Delta\Omega_\infty^2} - \frac{\beta}{R_r \cdot \alpha \cdot \Delta\Omega_\infty} \right) \quad \text{(A3.7)}$$

$$\frac{\partial Q}{\partial \Omega_r} = H(\Omega - \Omega_r) \cdot \left(\frac{\beta}{R_r \cdot \alpha} \right) \quad \text{(A3.8)}$$

$$\frac{\partial Q}{\partial \Delta\Omega_\infty} = H(\Omega - \Omega_r) \cdot \left(\frac{\beta \cdot (1+\alpha)}{\alpha \cdot \Delta\Omega_\infty^2} - \frac{2 \cdot R_r \cdot \ln\alpha}{\Delta\Omega_\infty^3} \right) \quad \text{(A3.9)}$$

Appendix 4

Sample Help Screen

A4.1 Introduction

In addition to containing standard software usage information, it is valuable if help screens provide some scientific background and make specific process recommendations (e.g., regarding measurement protocols). The following is the text of the help screen associated with the data entry screen shown in Fig. 25.2.

A4.2 Linear Sensor Help Screen Text

The sensor parametric model provided includes aperture dimensions, sampling frequency, motion during sampling, charge carrier diffusion, and color filter array interpolation. The first three effects are described below, whereas the latter two effects are addressed in separate help screens (see links at bottom of screen). Velocity tracking errors and charge transfer inefficiency are neglected in this parametric model.

In some treatments (e.g., J. C. Feltz and M. A. Karim, *Appl. Opt.* **29**(5), 1990, pp. 717–722), phase effects are included as part of the sensor MTF based on an assumed reconstruction/interpolation scheme. In contrast, in this full system model, phase effects are treated at the reconstruction stage. If a sensor samples sparsely, the system MTF will eventually be penalized in terms of more severe interpolation MTF loss and/or greater required blur (spot size) in the output stage to produce a visually rasterless output. Consequently, caution is needed when assessing which component in a digital system is responsible for the greatest degradation.

The aperture dimensions may be specified directly or they may be calculated from the pixel pitch and fill factor, the latter being the ratio of the dimension to the pixel pitch expressed as a length. The MTF of a rectangular aperture depends on the dimension of the aperture, d_a, in that direction:

$$M(v) = \left| \frac{\sin(\pi \cdot d_a \cdot v)}{\pi \cdot d_a \cdot v} \right| \quad (A4.1)$$

where d_a is in inverse units of frequency v. (Note that the MTF is the modulus of the optical transfer function, which may be negative, hence the absolute value operation.) The comparable equation for a circular aperture of diameter d_a is:

$$M(v) = \left| \frac{2 \cdot J_1(\pi \cdot d_a \cdot v)}{\pi \cdot d_a \cdot v} \right| \quad (A4.2)$$

where J_1 is a Bessel function of the first kind. Sampling frequency (i.e., pixel spacing, rather than active area) determines the Nyquist frequency and thereby affects aliasing.

There is often relative motion between the original and the sensor during the sampling period, especially in scanners with linear arrays. Frequently, this entails motion of the scanned object at a constant rate past the fixed sensor. This introduces blur in the direction of motion because a given pixel value is affected by a larger area on the original. By far the most common situation is that the motion during one sampling period is exactly equal to the spacing between the pixel centers, corresponding to essentially continuous integration. In special applications such as aerial photography, time delay and integration (TDI) sensors based on multiphase CCDs (charge coupled devices) are used to mitigate the effect of motion on MTF [D. F. Barbe, *Proc. IEEE*, Vol. **63**(1), 1975]. These allow reduction of effective motion during the sampling period to be reduced to 1/Φ times the pixel spacing where Φ is the number of phases. The MTF associated with such motion is given by:

$$M(v) = \left| \frac{\sin(\pi \cdot d_p \cdot v / \Phi)}{(\pi \cdot d_p \cdot v / \Phi)} \right| \quad (A4.3)$$

where d_p is distance between pixel centers in inverse units of frequency v. In the case of linear arrays the MTF caused by motion is applied in the scan direction; in the case of point sensors, it is applied in the fast direction.

The user either may provide a path to a file containing MTF data, or may choose to model the sensor MTF parametrically. Some care is required to ensure that the measurements are specified at the plane expected by the program. The MTF of a sensor following a Camera Lens (as in digital still cameras) or a General Lens (as in a scanner) in a block diagram must have frequencies specified in the image (sensor) plane. By convention, the Camera Lens MTF is also specified in the image plane, but a General Lens MTF is specified in the object plane. Frequently, composite lens plus sensor MTF measurements are made for scanners, in which case there is no lens (or a "null lens"; see below) preceding the sensor component in the block diagram. In such cases, the scanner MTF must instead be expressed in terms of frequencies in the object (scanned) plane. It is important that the pixel pitch be specified in the same plane as the MTF measurement.

To facilitate comparison of composite measured data with parametric models, where the lens effect must be explicitly specified, it may be convenient to include in your block diagram a "null" preceding lens (Camera Lens for digital still camera sensors, General Lens for scanners) with unit MTF and (for scanners) unit magnification.

For further information, click on:

Recommended MTF Measurement Protocols
Charge Carrier Diffusion
Color Filter Array Interpolation

Appendix 5

Useful Optical Formulas and Photospace Coverage

A5.1 Depth of Field and Focus

In this appendix, several relationships found useful in Sect. 26.4 and Ch. 27 are derived primarily from consideration of simple optical properties of image capture and display. Suppose that an object at a distance d_o from a simple lens produces a sharply focused image at the capture (film or sensor) plane located a distance d_i behind the lens. According to the thin lens equation, derived in any standard optics text (or see Jacobson (1978) for a photographically oriented discussion), the distances d_o, d_i, and the lens focal length F, expressed in the same units, are related by:

$$d_o^{-1} + d_i^{-1} = F^{-1} \tag{A5.1}$$

provided that the imaging is done in media with indices of refraction sufficiently close to one (as is the case for air). Even with complex lenses containing many elements, there exists a unique plane from which measurements of d_o and d_i obey the thin-lens equation, so this equation may be taken to be generally applicable. From Eq. A5.1, if an object were infinitely far away, it would be imaged in focus exactly one focal length behind the lens. Except in extreme close-up (macro) photography, the main subject of the composition is many camera lens focal lengths away, and so the first term of the left hand side of Eq. 5.1 is small, and the image focus distance is very close to F (usually within a few percent).

Figure A5.1 depicts an object at d_o being imaged in sharp focus at the capture plane a distance d_i behind a thin lens. Also shown in Fig. A5.1 is a second, nearer object at distance $d_o - \Delta d_o$, which would be imaged in sharp focus at a plane located at $d_i + \Delta d_i$, but in the actual capture plane at d_i, is defocused by an amount Δd_i. This figure is not to scale; as noted above, if the image distances were shown to scale for a typical photographic capture, the image distances would be too small and crowded together to readily distinguish. Therefore, for clarity, the image distances have been greatly exaggerated in Fig. A5.1.

The blur circle (geometrical optics) approximation discussed in Chs. 23 and 26 is sufficiently accurate for the analysis that follows. According to this approximation, the image of a defocused point is a solid disk, called the blur circle, the diameter of which is denoted d_b. If the clear lens aperture diameter is denoted d_l, then the lens f-number, denoted A for aperture, is defined by Eq. A5.2.

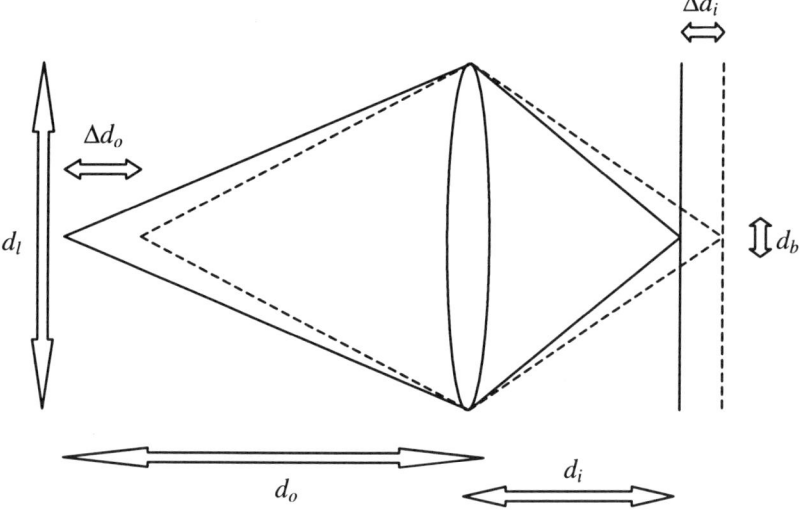

Fig. A5.1 Focusing of light by a thin lens. An object d_o from the lens, which has a clear aperture of diameter d_l, is focused d_i behind the lens, whereas a closer object at $d_o - \Delta d_o$ is out of focus by Δd_i, producing a disk with diameter d_b.

Useful Optical Formulas and Photospace Coverage

$$A = \frac{d_i}{d_l} \tag{A5.2}$$

As discussed in any optics text (or see Jacobson, 1978), the illuminance at the image plane is inversely proportional to the f-number squared. In imprecise terms, this is because: (1) the amount of light collected by the lens is proportional to its area and therefore to d_l^2; and (2) the area of the image, over which the collected light must be spread, is proportional to d_i^2.

Recalling that the image distances have been greatly exaggerated in Fig. A5.1 (with Δd_i typically a few percent or less of d_i), it may be recognized that the cone angles at the focused and defocused image points are approximately equal. Consequently, from A5.2:

$$\frac{\Delta d_i}{d_b} \approx \frac{d_i}{d_l} \Rightarrow d_b \approx \frac{\Delta d_i}{A} \tag{A5.3}$$

We will now show that the blur circle diameter is approximately proportional to focus error expressed in reciprocal object distance units, a result referenced in the discussion of Fig. 26.6. If we rewrite Eq. A5.1, the thin-lens equation, for the nearer object in Fig. A5.1, we obtain:

$$(d_o - \Delta d_o)^{-1} + (d_i + \Delta d_i)^{-1} = F^{-1} \tag{A5.4}$$

Equating the left hand sides of Eq. A5.1 and Eq. A5.4 yields:

$$(d_o - \Delta d_o)^{-1} - d_o^{-1} = \frac{\Delta d_i}{d_i \cdot (d_i + \Delta d_i)} \tag{A5.5}$$

Denoting the difference in reciprocal object distances by $\Delta(d_o^{-1})$ and recalling that typically $d_i \approx F$ and $\Delta d_i \ll F$ allows simplification to:

$$\Delta(d_o^{-1}) \approx \frac{\Delta d_i}{F^2} \tag{A5.6}$$

and further substitution of Eq. A5.3 with rearrangement yields Eq. A5.7.

$$d_b \approx \frac{F^2 \cdot \Delta(d_o^{-1})}{A} \qquad (A5.7)$$

This equation explains why the plots in Fig. 26.6 consist of line segments. The blur circle diameter is proportional to the defocus in reciprocal object distance units, with the slope of the line being the focal length squared over the f-number.

The remainder of this appendix is devoted to substantiating results needed in Ch. 27. As discussed in connection with Fig. 26.6, it is commonly assumed that if the diameter of the blur circle is sufficiently small, the image, although slightly defocused, will still appear to be sharp or at least adequately so. This maximum permissible value of d_b is called the circle of confusion and will be denoted by d_c. The depth of focus d_f is defined to be the range of capture plane positions yielding a sharp image, i.e., one for which the blur circle diameter does not exceed the circle of confusion. Setting $d_b = d_c$ in Eq. A5.3, the region of depth of focus extends in front of and behind the best focus position by an amount $A \cdot d_c$, yielding a total depth-of-focus range of:

$$d_f = 2 \cdot A \cdot d_c \qquad (A5.8)$$

Depth of field is analogous to depth of focus except that it is defined in terms of object, rather than image, distances. Suppose we wish to focus a lens so that the depth of field is maximized, as is desirable in a fixed focus camera. Evidently the maximum "limit" of the depth of field should be at infinity, which, as noted earlier, is focused at $d_i = F$. Based on the logic leading to Eq. A5.8, for infinity to fall at the depth-of-field limit, the capture plane must be at $d_i \approx F + A \cdot d_c$, and the image of the other depth-of-field limit must be at $d_i \approx F + 2 \cdot A \cdot d_c$. Setting d_i equal to these values in Eq. A5.1 gives the corresponding object distances, which are called the hyperfocal distance d_h and the minimum focus distance d_{min}, respectively.

$$d_h = \frac{F \cdot (F + A \cdot d_c)}{A \cdot d_c} \approx \frac{F^2}{A \cdot d_c} \qquad (A5.9)$$

$$d_{min} = \frac{F \cdot (F + 2 \cdot A \cdot d_c)}{2 \cdot A \cdot d_c} \approx \frac{F^2}{2 \cdot A \cdot d_c} \approx \frac{d_h}{2} \qquad (A5.10)$$

Thus, a lens focused at its hyperfocal distance, given by Eq. A5.9, has depth of field extending from half its hyperfocal distance to infinity.

A5.2 Angular Magnification

One of the most useful parameters for characterizing an imaging system is its angular magnification, which relates the size of elements as viewed in an image to the size of those elements as seen by the photographer in life. In this section, the formula for angular magnification will be derived in the context of an analysis of camera motion during exposure.

In hand-held ambient light photography, camera motion during the exposure, arising from unsteadiness of the photographer, may cause noticeable blur. The motion of the camera may be broken down into six components, each referenced to one of three mutually orthogonal axes meeting at the optical center of the lens, one of which is coincident with the optical axis of the lens. Three components correspond to translations along the axes, and three to rotations about these axes. Translational motions slightly shift the region of object space that falls within the field of view and depth of field (which we will call the captured volume), but the shifts are small compared to the dimensions of the captured volume (except in macro photography), and so may be neglected. In contrast, if the optical center of the lens were stationary, a small motion made by the photographer, acting at a distance comparable to a camera dimension, could produce a substantial angular deviation in the field of view. Whereas the impact of a pure translation depends on its magnitude compared to the corresponding dimension of the normally large captured volume, the effect of a rotational motion is scaled relative to a much smaller camera dimension, and so rotational motions are of far greater consequence.

In this analysis we assume that rotations about the optical axis, which is usually approximately perpendicular to the plane of the body of a standing person, are likely to be relatively small, given the stability of a standing person. In combination, rotations about the remaining two axes during an exposure trace out an arc in the capture plane. Although, in general, this arc may be complex and irregular in shape, we are concerned with the case in which the exposure time is sufficiently short that the motion does not unduly degrade the sharpness of the resulting image. On such a short time scale (typically 1/30 second or less), it is plausible that the arc may be approximated by a line segment that is traversed at a constant angular rate of ω radians per second. In this linear smear approximation, the angle subtended by the composite rotation at capture (in either image or object space) is then given by:

$$\theta_c = \omega \cdot T \qquad (A5.11)$$

where T is the duration of the exposure in seconds, which is controlled by either the shutter speed or a sensor integration time. The rotation rate ω is a function of the steadiness of the photographer and the stability inherent in the camera design, the latter of which depends primarily on mass, shape, and dynamics of shutter button actuation. It will be convenient to express ω as the ratio of two factors, one of which is camera-dependent, and the other of which reflects the stability of an average photographer.

$$\omega = \frac{\omega_c}{\rho_s} \qquad (A5.12)$$

Here ω_c is a critical rotation rate that, if assumed in the calculation of the maximum allowable exposure time for an average camera, leads to tolerably small and infrequent degradation arising from camera motion (a reasonable value of ω_c is supplied below). The camera-specific stability ratio ρ_s has a value of one for a typical camera and is larger for cameras with greater than average stability.

We ultimately seek to map the rotation θ_c of Eq. A5.11 to the viewer's eye to determine its perceptual significance and to thereby establish a criterion for maximum permissible exposure time to avoid undue blur from camera motion during exposure. Figure A5.2 shows a schematic of the propagation of an arc of camera rotation through a two-stage photographic system.

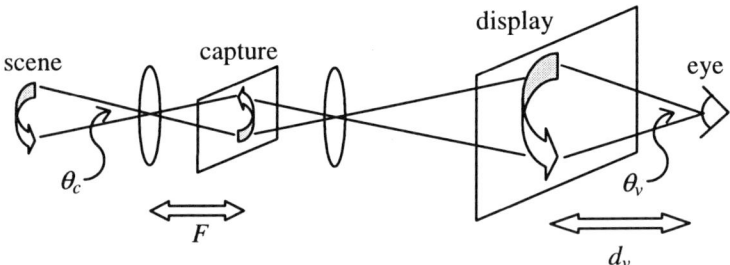

Fig. A5.2 Relationship between capture and viewing angles. The ratio of the latter to the former is the angular magnification of the system, given by Eq. A5.15; it relates the size of a feature in the viewed image to the size of that feature as seen by the naked eye of the photographer.

If this system were a conventional color negative-positive system, the capture plane would be color negative film, the final display would be a reflection print, and the intervening lens would belong to an optical printer. If the system instead involved electronic camera and display devices, the capture plane would be a sensor, the display plane would be a monitor, and the intervening lens would symbolize digital image processing.

Let us now translate the rotation angle at capture θ_c to that subtended at the eye of the observer viewing the final displayed image θ_v. Referring to Fig. A5.2, to achieve this, θ_c is:

1. converted to arc length in the capture plane by multiplying by the lens focal length F (assuming units of radians and invoking a small angle approximation);

2. scaled to the final displayed image plane by multiplying by the printing magnification m_p (the linear ratio of display and capture format sizes); and

3. divided by the final image viewing distance d_v to convert back to radians (again using a small angle approximation).

This procedure yields the following relationship.

$$\theta_v = \frac{\theta_c \cdot F \cdot m_p}{d_v} \qquad (A5.13)$$

The viewing angle θ_v and the capture angle θ_c as defined here pertain to an arbitrary arc and should not be confused with the field of view, which is the angular subtense of the imaged portion of the scene. The angular field of view of capture is determined by the camera lens focal length and the format dimensions of the film or sensor at the capture plane. Like display sizes, the field of view is usually defined along the diagonal orientation, and so will be denoted by θ_d; analogously, the dimension of the format diagonal will be represented by d_d. Suppose in Fig. A5.2 that the arc traced in the capture plane were a line segment extending from the center of the format to one of its corners; this arc would then subtend one-half the field of view and have a length equal to the format semi-diagonal. Furthermore, this arc, in combination with line segments joining its endpoints to the center of the camera lens, would form a right triangle, from which simple trigonometry would yield Eq. A5.14.

$$\tan(\theta_d / 2) = \frac{d_d / 2}{F} \Rightarrow F = \frac{d_d}{2 \cdot \tan(\theta_d / 2)} \quad \text{(A5.14)}$$

Once format dimensions have been specified, Eq. A5.14 allows computation of the necessary focal length to achieve a desired field of view; larger formats require lenses of longer focal length to obtain a given field of view.

We now define the quantity angular magnification to be the ratio of the subtense of a feature in a final displayed image, as seen by the observer viewing that image, to the subtense of that same feature in the original scene, as seen by the photographer, looking naked eye from the camera position. A unit angular magnification implies that the image as viewed by the observer and the corresponding portion of the original scene as viewed by the photographer appear to be the same "size". Angular magnification is an important system parameter, with higher angular magnifications (e.g., from zoom lenses) being compositionally desirable in many scenes.

Although the capture and viewing angles of Eq. A5.13 have been defined in terms of camera rotation during exposure, they could equally well pertain to the angles subtended by some feature in the original scene and final image, respectively. Consequently, we may identify the system's angular magnification m_a as being the ratio of θ_v to θ_c. This ratio may be derived from Eq. A5.13.

$$m_a = \frac{F \cdot m_p}{d_v} \quad \text{(A5.15)}$$

By analogy with the circle of confusion, we now define a subtense of confusion θ_{max} that depends primarily upon the angular resolution of the human visual system. If the visual subtense of motion, θ_v, exceeds θ_{max}, noticeable degradation of image sharpness will result. Although the frequency dependence of MTF loss arising from defocus and linear smear differ somewhat, it is still anticipated that the circle of confusion diameter should be approximately related to the subtense of confusion by:

$$d_c = \frac{\theta_{max} \cdot d_v}{m_p} \quad \text{(A5.16)}$$

which is obtained by logic like that leading to Eq. A5.13, with $\theta_v = \theta_{max}$.

The quantity θ_{max} is expected to be close to the angular resolution of the eye. The Snellen eye chart defines 20/20 vision based on the ability to resolve letters composed of segments subtending one arc-minute (Bartleson and Grum, 1984). In 35-mm format professional systems a circle of confusion $d_c = 0.033$ mm is commonly assumed in computing the positions of engraved depth-of-field markings on lenses. If we assume this tolerance to be appropriate for 8 × 10-inch prints having an optical printing magnification of $m_p = 8.88$ and viewed at a typical hand-held distance of $d_v \approx 412$ mm (Eq. 26.1), the above equation yields $\theta_{max} = (0.033 \text{ mm}) \cdot (8.88)/(412 \text{ mm}) \approx 7 \cdot 10^{-4}$ radians ≈ 2.4 minutes of arc, which is $\approx 2.4\times$ higher than the Snellen resolution of one arc-minute.

Equation A5.10 gave the minimum depth-of-field limit (for a maximum limit of infinity) in terms of the format-dependent properties of focal length and circle of confusion. Now that angular magnification, field of view, and subtense of confusion have been defined, it is desirable, for completeness and for use in Ch. 27, to express the minimum depth-of-field limit in a way that the format dependence is more explicit. Substituting Eqs. A5.14–A5.16 into Eq. A5.10 yields Eq. A5.17.

$$d_{min} = \frac{d_d \cdot m_a}{4 \cdot \tan(\theta_d/2) \cdot A \cdot \theta_{max}} \quad (A5.17)$$

Likewise, depth of focus can be rewritten in two other useful forms via substitution of A5.10 and then A5.14 into Eq. A5.8.

$$d_f = \frac{F^2}{d_{min}} = \frac{d_d^2}{4 \cdot d_{min} \cdot \tan^2(\theta_d/2)} \quad (A5.18)$$

We now can define an expression for the maximum permissible hand-held exposure time T_{max} that does not cause undue loss of sharpness arising from camera motion during exposure. The exposure time has units of seconds and can be controlled by shutter time or by sensor integration time. Setting θ_v equal to θ_{max} in Eq. A5.13, substituting in Eqs. A5.11 and A5.12, and rearranging yields:

$$T_{max} = \frac{\rho_s \cdot d_v \cdot \theta_{max}/(\omega_c \cdot m_p)}{F} = \frac{\rho_s \cdot \theta_{max}}{\omega_c \cdot m_a} \quad (A5.19)$$

where the second equality follows from Eq. A5.15.

The oft-quoted "one over focal length rule" states that hand-held exposure time in seconds should not exceed the reciprocal of the lens focal length in mm. This rule is a special case of the first equality of Eq. A5.19, in which the values of all parameters except focal length are fixed at values typical in consumer photography. Coincidentally, these typical values combine to yield an overall constant in the numerator of approximately one second-millimeter, leading to the mnemonic value of "one" in the one over focal length rule. A value of $d_c = 0.033$ mm is commonly assumed in professional 35-mm format systems. Using this value in the left hand equality of Eq. A5.16, substituting the result into the first equality of Eq. A5.19, and requiring that the new result reduce to $1/F$ for a camera of average stability, implies that $\omega_c = 0.033$ radians per second, which is ≈ 1.9 degrees per second.

A5.3 Photospace Coverage

In this final section, we consider two system photospace coverage metrics, the minimum light level at which a normal ambient exposure can be obtained, and the maximum distance at which a normal flash exposure is possible. Equations relating these metrics to fundamental quantities are derived for a simple photographic system containing a camera with fixed focus and aperture. Such systems can yield surprisingly good performance, notwithstanding the simplicity of the camera and their resultant low cost. Consequently, they represent an excellent value, as demonstrated by the fact that about half of the point-and-shoot cameras sold in the United States at the time of this writing were of this type.

While it is possible to define more sophisticated photospace coverage metrics and apply them to more advanced photographic systems, as argued in Sect. 27.7, the value of such metrics is primarily in developing intuition regarding system design interactions, and the simple case considered here is sufficient for that purpose. Characterization of real systems is better accomplished through analysis of system performance via Monte Carlo modeling.

To quantify the minimum practical ambient light level, we start with the standard reflected light metering equation from ANSI/ISO 2720 (1994).

$$Y = \frac{K \cdot A^2}{S \cdot T} \qquad (A5.20)$$

In this equation Y is the average scene luminance in cd/m^2 that yields an ISO normal exposure; K is the reflected light metering constant, having a currently recommended range of values of 10.6–13.4 cd·s/m^2 (ANSI/ISO 2720, 1994); S is the ISO capture (film or sensor) speed; and A and T are the f-number (aperture) and exposure times, as previously stated.

The plausibility of the recommended range of values of K may be confirmed as follows. The commonly used "sunny 16" rule of thumb for exposure in bright sunlight states that at f/16 aperture, the shutter time in seconds equals the reciprocal of the ISO speed. The illuminance in a plane oriented perpendicular to the sun's rays at the earth's surface typically is about 10,000 foot-candles ($\approx$110,000 lux) in bright sunlight. However, to obtain a pleasing rendition of three-dimensional subjects not specifically oriented to maximize illuminance, an average exposure increase of two-thirds of a stop is needed (ANSI PH2.7, 1986), corresponding to an equivalent illuminance of $110,000 \cdot 2^{-2/3} = 69,000$ lux. Although a range of average scene reflectances have been suggested in the literature, we favor a value of 14% for outdoor scenes, which implies a corresponding mean luminance of $69,000 \cdot 0.14/\pi = 3100$ cd/m^2. Substituting this value for Y in Eq. A5.20 and setting $T = 1/S$ and $A = 16$ yields $K \approx 12$, which is in the middle of the recommended range.

Equation A5.20 shows that capture is possible at lower light levels if the ISO capture speed is higher, the exposure time is longer, or the lens f-number is lower (thereby increasing illuminance at the capture plane). Therefore, to reach the minimum light level at which a normal exposure may be obtained at a given ISO speed, denoted Y_{min}, the f-number must be minimized and the exposure time maximized.

Equation A5.19 already gives us a maximum permissible hand-held exposure time. The minimum permissible f-number is that providing adequate depth of field to span those distances at which consumers frequently take photographs. There are many outdoor photographs depicting distant objects (e.g., scenic shots taken on vacations), so the depth of field must include infinity. Given some cut-off criterion, a lower required depth-of-field limit could be established based on photospace distribution data (see Fig. 27.1 and associated discussion). Equating this limit to d_{min} of Eq. A5.17 allows solution for the fixed aperture required to provide adequate depth of field. Using this expression for A in Eq. A5.20, setting $Y = Y_{min}$, $T = T_{max}$, and substituting in Eq. A5.19 (right hand equality) finally yields the desired relationship, given by Eq. A5.21.

$$Y_{min} = \frac{K \cdot \omega_c \cdot d_d^2 \cdot m_a^3}{16 \cdot \rho_s \cdot S \cdot d_{min}^2 \cdot \theta_{max}^3 \cdot \tan^2(\theta_d/2)}$$

$$= \frac{K \cdot \omega_c \cdot F^5 \cdot m_p^3}{4 \cdot \rho_s \cdot S \cdot d_{min}^2 \cdot \theta_{max}^3 \cdot d_v^3}$$

(A5.21)

The first equality has format size dependence expressed explicitly; the second equality shows focal length dependence and results from further substitution of Eqs. A5.14 and A5.15. The minimum light level that can be captured in an ambient exposure is a very sensitive function of the system angular magnification (cubed) and the camera lens focal length (fifth power). These limitations pose significant challenges in designing systems with zoom camera lenses and/or zoom and crop features in the printing stage.

At lower light levels, where ambient photography becomes difficult, if the subject is sufficiently close to the camera (as is often the case indoors), flash illumination provides an alternative means of achieving adequate exposure. The "output" O of the flash unit may be quantified by its luminous energy expressed in units of beam candlepower seconds, which depends upon capacitance and various efficiencies (Ray, 1983). The duration of the electronic flash pulse determines the effective exposure time, which is normally so short that camera motion during exposure is insignificant. Neglecting the effect of secondary reflections off walls, ceilings, etc., the illuminance provided at the subject falls off with the square of the distance from the camera, because the flash tube is effectively a point source under most practical conditions. Consequently, the photographic exposure in lux-seconds at the capture plane depends only on camera lens aperture, flash output, and subject distance. Because the capture plane exposure required for an ISO normal exposure is inversely proportional to capture speed (Eq. A5.20), and flash exposure is proportional to flash output and inversely proportional to the squares of camera lens aperture and object distance, the product of the distance d_n and the aperture A yielding a normal exposure is proportional to the square root of capture speed times flash output. This product is referred to as a guide number G; it has units of distance and is quoted in connection with a particular ISO capture speed.

$$G = A \cdot d_n \propto \sqrt{S \cdot O} \quad (A5.22)$$

The output of a flash may be reduced by only partially discharging the capacitor, either by manually setting an output fraction or by quenching the flash during exposure, the latter usually in response to real-time measurement of light return from the subject. Such reductions in output are not important in the present discussion because we seek the maximum flash exposure range, which will occur at full output.

The maximum distance at which a normal exposure can be achieved is obtained by substituting A, from Eq. A5.17, into Eq. A5.22, and equating d_n and d_{max}:

$$d_{max} = \frac{G}{A} = \frac{4 \cdot d_{min} \cdot \theta_{max} \cdot G \cdot \tan(\theta_d/2)}{d_d \cdot m_a}$$

$$= \frac{2 \cdot d_{min} \cdot \theta_{max} \cdot G \cdot d_v}{F^2 \cdot m_p} \quad (A5.23)$$

where the final equality results from further substitution of Eqs. A5.14 and A5.15. As with Eq. 5.21, one equality explicitly demonstrates the format size dependence and the other equality shows the focal length dependence. As in the ambient case, photospace coverage by flash photography decreases with increasing focal length and angular magnification, and so greater challenges are encountered in systems with zoom camera lenses and/or zoom and crop capability in the printing stage.

References

Adams, J. (1997). Design of Practical Color Filter Array Interpolation Algorithms for Digital Cameras, *Proc. SPIE 3028*, pp. 117–125.

Adams, J. (1998). Design of Practical Color Filter Array Interpolation Algorithms for Digital Cameras, Part 2, *Proc. ICIP 1*, pp. 488–492.

Adams, J., Parulski, K., and Spaulding, K. (1998). Color Processing in Digital Cameras, *IEEE Micro. 18(6)*, pp. 20–30.

ANSI PH2.7 (1986). *Photography – Photographic exposure guide*, American National Standards Institute, New York, pp. 18–26.

ANSI/ISO 2720 (1994). *General Purpose Photographic Exposure Meters (Photoelectric Type) – Guide to Product Specification*, American National Standards Institute, New York, Section 6. More complete designations are ANSI/ISO 2720-1974 (R1994) and ANSI/NAPM IT3.302–1994.

Barten, P. G. J. (1990). Evaluation of Subjective Image Quality with the Square-Root Integral Method, *J. Opt. Soc. Am. A7*, pp. 2024–2031.

Barten, P. G. J. (1999). *Contrast Sensitivity of the Human Eye and Its Effects on Image Quality*, Society of Photo-Optical Instrumentation Engineers, Bellingham, Washington, Ch. 8.

Bartleson, C. J. (1982). The Combined Influence of Sharpness and Graininess on the Quality of Color Prints, *J. Photogr. Sci. 30*, pp. 33–38.

Bartleson, C. J. (1985). Predicting Graininess from Granularity, *J. Photogr. Sci. 33*, pp. 117–126.

Bartleson, C. J. and Breneman, E. J. (1967a). Brightness Perception in Complex Fields, *J. Opt. Soc. Am. 57*, pp. 953–957.

Bartleson, C. J. and Breneman, E. J. (1967b). Brightness Reproduction in the Photographic Process, *Photogr. Sci. Eng. 11*, pp. 254–262.

Bartleson, C. J. and Grum, F. (1984). *Optical Radiation Measurements*, Vol. 5, Academic Press, New York, Chs. 6–10 (psychophysics), p. 259 (Snellen chart), and pp. 377–382 (Weber's Law).

Bendat, J. S. and Piersol, A. G. (1980). *Random Data: Analysis and Measurement Procedures*, Wiley, New York, pp. 120–126 (cross-spectra), pp. 240–244 (noise propagation), and pp. 339–341 (quantization noise).

Berns, R. S. (2000). *Billmeyer and Saltzman's Principles of Color Technology*, 3rd ed., John Wiley and Sons, New York, Ch. 4.

Biberman, L. M., ed. (2000). *Electro-Optical Imaging: System Performance and Modeling*, Society of Photo-Optical Instrumentation Engineers, Bellingham, Washington.

Bock, R. D. and Jones, L. V. (1968). *The Measurement and Prediction of Judgment and Choice*, Holden-Day, San Francisco, pp. 71–75 and 134–136.

Bovick, A. C. (2000). *Handbook of Image and Video Processing*, Academic Press, New York.

Buhr, J. D. and Franchino, H. D. (1994). Color Image Reproduction of Scenes with Preferential Tone Mapping, U. S. Patent #5,300,381.

Buhr, J. D. and Franchino, H. D. (1995). Color Image Reproduction of Scenes with Preferential Tone Mapping, U. S. Patent #5,447,811.

Burningham, N. and Dalal, E. (2000). Status of the Development of International Standards of Image Quality, *Proc. IS&T's PICS 2000 Conference*, Portland, Oregon, Society for Imaging Science and Technology, Springfield, Virginia, pp. 121–123.

Carlson, C. R. and Cohen, R. W. (1980). A Simple Psychophysical Model for Predicting the Visibility of Displayed Information, *Proc. SID 21(3)*, 229–246.

Castleman, K. R. (1996). *Digital Image Processing*, Prentice Hall, Upper Saddle River, New Jersey, Sect. 7.2.1.

CIE Publication 15.2 (1986). *Colorimetry*, 2nd ed., Commission Internationale de l'Eclairage, Vienna.

Crane, E. M. (1964). An Objective Method for Rating Picture Sharpness: SMT Acutance, *J. Soc. Mot. Pict. Tel. Eng. 73*, pp. 643–647.

Dainty, J. C. and Shaw, R. (1974). *Image Science*, Academic Press, London.

Dalal, E., Rasmussen, D. R., Nakaya, F., Crean, P. A., and Sato, M. (1998). Evaluating the Overall Image Quality of Hardcopy Output, *Proc. IS&T's PICS 1998 Conference*, Portland, Oregon, Society for Imaging Science and Technology, Springfield, Virginia, pp. 169–173.

de Ridder, H. (1992). Minkowski-metrics as a combination rule for digital image coding impairments, *Human Vision, Visual Processing, and Digital Display III*, SPIE Vol. 1666, Society of Photo-Optical Instrumentation Engineers, Bellingham, Washington, pp. 16–26.

de Ridder, H. (1996). Saturation and Lightness Variation in Color Images of Natural Scenes, *J. Imaging Sci. Technol. 40(6)*, 487–493.

DeMarsh, L. E. (1972). Optimum Telecine Transfer Characteristics, *J. Soc. Mot. Pict. Tel. Eng. 81*, pp. 784–787.

Doerner, E. C. (1962). Wiener Spectrum Analysis of Photographic Granularity, *J. Opt. Soc. Am. 52(6)*, pp. 669–672.

Draper, N. and Smith, H. (1981). *Applied Regression Analysis*, Wiley and Sons, New York, Sect. 2.7.

Engeldrum, P. G. (1999). Image Quality Modeling: Where Are We?, *Proc. IS&T's PICS 1999 Conference*, Savannah, Georgia, Society for Imaging Science and Technology, Springfield, Virginia, pp. 251–255.

Engeldrum, P. G. (2000). *Psychometric Scaling: A Toolkit for Imaging Systems Development*, Imcotek Press, Winchester, Mass., Ch. 6.

Fairchild, M. D. (1998). *Color Appearance Models*, Addison-Wesley, Reading, Massachusetts.

Frieser, H. (1975). *Photographic Information Recording*, Focal Press, London. In German with English summaries.

Gescheider, G. A. (1985). *Psychophysics, Method, Theory, and Application*, 2nd ed., Lawrence Erlbaum Associates, Publishers, New Jersey.

Giorgianni, E. J. and Madden, T. E. (1998). *Digital Color Management: Encoding Solutions*, Addison-Wesley, Reading, Mass., Chs. 4–6 (tone reproduction) and Ch. 7 and Appendices B, C, F, and G (color matrixing).

Gonzalez, R. C. and Woods, R. E. (1992). *Digital Image Processing*, 3rd ed., Addison-Wesley, Reading, Mass.

Görgens, E. (1987). Methoden der objectiven Bewertung der Bildgüte (Methods of Objective Image Quality Evaluation), *Bild und Ton 40*, pp. 69–74. In German.

Greivenkamp, J. E. (1990). Color-Dependent Optical Prefilter for the Suppression of Aliasing Artifacts, *Appl. Opt. 29(5)*, pp. 676–684.

Guilford, J. P. (1954). *Psychometric Methods*, McGraw-Hill, New York.

Hanson, W. T., Jr., and Horton, C. A. (1952). Subtractive Color Reproduction: Interimage Effects, *J. Opt. Soc. Am. 42*, 663–669.

Hanson, W. T., Jr. and Vittum, P. W. (1947). Color Dye-Forming Couplers in Subtractive Color Photography, *J. Photogr. Soc. Am. 13*, 94–96.

Harris, F. J. (1978). On the Use of Windows for Harmonic Analysis with the Discrete Fourier Transform, *Proc. IEEE 66(1)*, pp. 51–83.

Hecht, E. and Zajac, A. (1976). *Optics*, Addison-Wesley, Reading, Mass., Ch. 8.

Hultgren, B. O. (1990). Subjective Quality Factor Revisited, *Proc. Human Vision and Electronic Imaging: Models, Methods, and Applications*, Santa Clara, Calif., SPIE Vol. 1249, Society of Photo-Optical Instrumentation Engineers, Bellingham, Washington, pp. 12–22.

Hunt, R. W. G. (1976). *The Reproduction of Color in Photography, Printing, and Television*, 4th ed., Fountain Press, Tolworth, England.

Hunt, R. W. G. (1991). *Measuring Color*, 2nd ed., Ellis Horwood Ltd., London.

Hunt, R. W. G., Pitt, I. T., and Winter, L. M. (1974). The Preferred Reproduction of Blue Sky, Green Grass, and Caucasian Skin in Color Photography, *J. Photogr. Sci. 22*, pp. 144–149.

ISO 12233 (1998). *Photography – Electronic still picture cameras – Resolution measurements*, International Standards Organization, Switzerland.

ISO 14524 (1997). *Photography – Electronic still picture cameras – Methods for measuring opto-electronic conversion functions (OECF's)*, International Standards Organization, Switzerland.

Jacobson, R. E. (ed.) (1978). *The Manual of Photography*, 7th ed., Focal Press, London, Ch. 4.

James, T. H. (1977). *The Theory of the Photographic Process*, MacMillan Publishing Co., Inc., New York.

Janssen, R. (2001). *Computational Image Quality*, Society of Photo-Optical Instrumentation Engineers, Bellingham, Washington.

Jin, E. W., Feng, X.-F., and Newell, J. (1998). The Development of a Color Visual Difference Model (CVDM), *Proc. IS&T's PICS 1998 Conference*, Portland, Oregon, Society for Imaging Science and Technology, Springfield, Virginia, pp. 154–158.

Johnson, G. M. and Fairchild, M. D. (2000). Sharpness Rules, *Proc. IS&T/SID Eighth Color Imaging Conference*, Scottsdale, Arizona, Society for Imaging Science and Technology, Springfield, Virginia, pp. 24–30.

Jones, L. A. (1920). On the Theory of Tone Reproduction, with a Graphic Method for the Solution of Problems, *J. Franklin Inst. 190*, pp. 39–90.

Jones, L. A. and Nelson, C. N. (1942). The Control of Photographic Printing by Measured Characteristics of the Negative, *J. Opt. Soc. Am. 32*, pp. 558–619.

Jones, R. C. (1955). New Method of Describing and Measuring the Granularity of Photographic Materials, *J. Opt. Soc. Am. 45*, pp. 799–808.

Jones, R. C. (1961). Information Capacity of Photographic Films, *J. Opt. Soc. Am. 51*, pp. 1159–1171.

Kane, P. J., Bouk, T. F., Burns, P. D., and Thompson, A. D. (2000). Quantification of Banding, Streaking, and Grain in Flat Field Images, *Proc. IS&T's PICS 2000 Conference*, Portland, Oregon, Society for Imaging Science and Technology, Springfield, Virginia, pp. 79–83.

Kang, H. R. (1999). *Digital Color Halftoning*, IEEE Press, Piscataway, New Jersey.

Kendall, M. and Stuart, A. (1967). *The Advanced Theory of Statistics*, 4th ed., Vol. 2, Macmillan, New York, Ch. 29.

Kodak Publication E-58 (1994). *Print Grain Index*, Eastman Kodak Company, Rochester, New York.

Linfoot, E. H. (1964). *Fourier Methods in Optical Image Evaluation*, Focal Press, London.

Montgomery, D. C. (1984). *Design and Analysis of Experiments*, 2nd ed., John Wiley and Sons, New York.

Nijenhuis, M., Hamberg, R., Teunissen, C., Bech, S., Looren de Jong, H., Houben, P., and Pramanik, S. K. (1997). Sharpness, Sharpness Related Attributes, and their Physical Correlates, *Proc. Very High Resolution and Quality Imaging II Conference*, San Jose, California, SPIE Vol. 3025, Society of Photo-Optical Instrumentation Engineers, Bellingham, Washington, pp. 173–184.

Nunnaly, J. C. and Bernstein, I. H. (1994). *Psychometric Theory*, 3rd ed., McGraw Hill, New York.

Pennebaker, W. B. and Mitchell, J. L. (1992). *JPEG Still Image Compression Standard*, Van Nostrand Reinhold, New York.

PIMA IT2.37 (2001). *Print Grain Index – Assessment of Print Graininess from Color Negative Films*, Photographic and Imaging Manufacturers Association, Harrison, New York, Working Draft 4.

Prosser, R. D., Allnatt, J. W., and Lewis, N. W. (1964). Quality Grading of Impaired Television Images, *Proc. IEEE 111(3)*, pp. 491–502. See also Lewis, N. W. and Allnatt, J. W. (1965). Subjective Quality of Television Images with Multiple Impairments, *Electr. Letters 1(7)*, pp. 188–189.

Rabbani, M. and Jones, P. W. (1991). *Digital Image Compression Techniques*, SPIE Tutorial Text Series, Vol. TT7, Society of Photo-Optical Instrumentation Engineers, Bellingham, Washington.

Ray, S. (1983). *Camera Systems*, Focal Press, London, pp. 152–172.

Rice, T. M. and Faulkner, T. W. (1983). The Use of Photographic Space in the Development of the Disc Photographic System, *Proc. J. Appl. Photogr. Eng. 9(2)*, pp. 52–57.

SAS Institute (1989). *SAS/STAT$^©$ User's Guide*, Ver. 6, 4th ed., Vol. 1, SAS Institute, Inc., Cary, North Carolina, Ch. 18.

Searle, S. S. (1982). *Matrix Algebra Useful for Statistics*, John Wiley, New York, pp. 338–340.

Shannon, C. E. (1948). A Mathematical Theory of Communication, *Bell System Tech. J. 27*, pp. 379–423.

Simonds, J. L. (1961). A Quantitative Study of the Influence of Tone-Reproduction Factors on Picture Quality, *Photogr. Sci. Eng. 5(5)*, pp. 270–277.

Spaulding, K. E., Miller, R. and Schildkraut, J. (1997). Methods for Generating Blue-Noise Dither Matrices for Digital Halftoning, *J. Electr. Imaging 6(2)*, pp. 208–230.

Stevens, S. S. (1946). On the Theory of Scales of Measurement, *Science 103*, pp. 677–680.

Stokseth, P. A. (1969). Properties of a Defocused Optical System, *J. Opt. Soc. Am. 59(10)*, pp. 1314–1321.

Stultz, K. F. and Zweig, H. J. (1959). Relationship between Graininess and Granularity for Black and White Samples with Nonuniform Granularity Spectra, *J. Opt. Soc. Am. 49*, pp. 693–702.

Taubman, D. and Marcellin, M. (2001). *JPEG 2000: Image Compression Fundamentals, Practice, and Standards*, Kluwer Academic Publishers, Boston, Mass.

Thurstone, L. L. (1927). A Law of Comparative Judgment, *Psych. Rev. 34*, pp. 273–286.

Töpfer, K., Adams, J. E., and Keelan, B. W. (1998). Modulation Transfer Functions and Aliasing Patterns of CFA Interpolation Algorithms, *Proc. IS&T's PICS 1998 Conference*, Portland, Oregon, Society for Imaging Science and Technology, Springfield, Virginia, pp. 367–370.

Töpfer, K. and Jacobson, R. E. (1993). The Relationship between Objective and Subjective Image Quality Criteria, *J. Inf. Rec. Mats. 21*, pp. 5–27.

Torgerson, W. S. (1958). *Theory and Methods of Scaling*, Wiley, New York.

Velleman, P. F. and Wilkinson, L. (1993). Nominal, Ordinal, Interval, and Ratio Typologies are Misleading, *Am. Statistician 47(1)*, pp. 65–72.

Wheeler, R. B. and Keelan, B. W. (1994). Automatic Optimization of Photographic Exposure Parameters for Non-standard Display Sizes and/or Different Focal Length Photographic Modes through Determination and Utilization of Extra System Speed, U. S. Patent #5,323,204.

Wolberg, G. (1990). *Digital Image Warping*, IEEE Computer Society Press, Washington DC.

Yendrikhovskij, S., MacDonald, L., Bech, S., and Jensen, K. (2000), Enhancing Color Image Quality in Television Displays, *Imaging. Sci. J. 47(4)*, pp. 197–211.

Zwick, D. F. (1984). Psychometric Scaling of Terms Used in Category Scales of Image Quality Attributes, *Proc. Symposium on Photogr. and Electr. Image Quality*, Royal Photographic Society, Cambridge, England, pp. 46–55.

Zwick, D. F. and Brothers, D. (1975). RMS Granularity: Determination of Just Noticeable Differences, *Photogr. Sci. Eng. 19*, pp. 235–238.

Index

Aberrations, lens, 351–353
Adaptation
   color balance influencing, 311, 315
   direct, 13, 294
   lateral, 13
   pupil size, effect on, 277–278
Addressability, device, 390, 392–393
Adjectives, quality, 81–83, 92
Advanced Photo System
   aims, 453–456
   modeling, 450–453, 456–457
   scanning, 384–386
Age, effect on redeye, 277–279
Algorithms
   color balance, 440–442
   contrast customization, 59–60
   interpolation, 359–361
   research emphases, 15
Aliasing
   anti-aliasing filter
      digital, 258, 270
      optical, 257–258, 388–390
   appearance, 263, 266
   color filter array, 388–390
   origins, 254–257
   pictorial examples, 156–157, 266

Ambient light level (*see* Light level, ambient)
Anchor (reference image), 90–92
Angular distribution, 29–32, 40–41
Angular magnification (m_a)
   definition, 402, 483–484, 486
   effects on
      ambient light level limit, 406, 489–490
      depth of field, 403, 487
      flash range, 407, 491
      maximum hand-held exposure time, 405, 487
      redeye, 278–279
Aperture (*see* F-number)
Arcsine (angular) distribution, 29–32, 40–41
Artifacts (*see* Aliasing; Attribute, artifactual; Banding; Blocking; CFA interpolation artifacts; Clipping, tonal; Misregistration; Noise, isotropic; Oversharpening; Pixelization; Reconstruction infidelity; Redeye; Ringing; Sharpness; Streaking)

501

Assessment
  attribute, 228–230
  first/third party, 4
  reference image, effect of, 313–314
  singe/double-stimulus, 311–312
  uncertainty, 116, 130–131
Attributes
  aesthetic, 5, 8
  artifactual, 6, 8, 52–53 (*see also* Artifacts)
  assessment, 228–230
  classification
    general types, 4–8
    intergrading complexes, 270–273, 288–289
  color/tone, 287–289, 315–317
  dimensions, 188–193
  interactions (*see* Interactions, attribute)
  personal, 4, 8
  preferential, 6, 8, 51–52 (*see also* Preference)
Autofocus
  ranging requirements, 386–388
  system performance, effect on, 401, 409, 428–429

Balance
  color
    algorithm performance, 439–443
    contour plots, 302
    definition, 287
    objective metric, 292–293
    reference image effect, 311–315
    softcopy optimization, 442–443
  density
    algorithm performance, 439–443
    definition, 287
    objective metric, 293

Banding artifact
  definition, 185–188
  NPS, 185, 188
  objective metric, 193–196
  pictorial example, 183
  raster frequency, 348–349
  signal dependence, 208
Bias
  color (*see* Balance, color)
  correction, 229
  exposure, 439, 455
  quality ruler, 115–116
Bit depth (*see* Quantization)
Blocking artifact, 394–395
Blur (*see* Cameras, motion during exposure; Circle, blur; Defocus; Sharpness)
Boosting (*see* Unsharp masking)
Boxcar integration (*see* Interpolation, pixel replication)

Camera (*see also* Focal length; Format dimensions)
  compact (point-and-shoot)
    fixed focus/aperture, 401, 407, 488
    photospace, 398–399
  digital still (DSC)
    anti-aliasing filter, 257–258, 388–390
    noise, 347–348, 410–411
    resolution and quality, 390–393
    sensor size and quality, 435–436
    tonal clipping, 217–218, 410–411
  motion during exposure, 405, 491–493, 487–488

Index

Camera-subject distance
  photospace, dimension of, 398–399, 420–421
  redeye, effect on, 276–280
  system quality, effect on, 379, 429
Capability
  analyses, examples of
    anti-aliasing filter, 388–390
    autofocus ranging, 386–388
    capture/display resolution, 390–393
    image compression, 393–395
    optical tolerances, 384–386
    unsharp masking gain, 380–384
  definition, 379, 426
  vs. performance, 379, 413–414
Cascading MTFs, 260, 262, 337
Categorical sort
  limitations, 90, 93
  method, 89–90
  range effects, 81–83
  uncertainty, 116
CCD (charge coupled device) (*see* Sensor)
CDF (cumulative distribution function) (*see also* Distribution)
  definition, 25–26
  of quality
    advantages over PDF, 424–426
    interpreting shapes, 426–431
    sampling, 419–422
Central limit theorem, 23–24
CFA (color filter array)
  artifacts, 359, 388–390
  interpolation, 360–364
  patterns, 359–360
Chroma
  boosting, 331
  channel, 360–364, 394
  definition, 287–289
  ratios and colorfulness, 293–294

CIE (Commission Internationale de l'Eclairage)
  color matching functions, 332, 335
  lightness (L*)
    attribute correlates, 287–288
    definition, 216
    objective metrics using, 289–293
    quantization steps, 223, 248
  other metrics, 287, 289–293, 301
Circle
  blur (d_b)
    autofocus design using, 386–388
    definition, 351, 480–482
    model, 350, 385
  of confusion (d_c) (*see also* Subtense of confusion)
    autofocus design using, 386–388
    definition, 386, 482
    depth of field, effect on, 401, 487
    depth of focus, effect on, 410, 482
    values, 403, 488
Classification
  attribute
    general types, 4–8
    intergrading complexes, 270–273, 288–289
  observer (*see also* Observer sensitivity)
    by expertise, 76, 78, 140
    by preference, 295–299
  scene, 295–297 (*see also* Scene susceptibility)
Clipping, tonal, 217–219, 410–411
CMOS (complementary metal oxide semiconductor) (*see* Sensor)

Color
  balance (*see* Balance, color)
  filter array (*see* CFA)
  gamut, 287, 293
  hue, 287–289, 293, 300
  memory
    experimentation, 289–292
    foliage, 296, 298–299, 302
    objective metrics, 293–295
    skin, 298, 301–302, 307–310
    sky, 296, 298, 300, 302
  (and tone) quality
    compositional impact, 307–310
    contours, 298–302
    multivariate, 315–318
    reference image effect, 311–315
  (and tone) reproduction (*see also* Contrast)
    attributes, 287–289, 315–317
    definition, 285
    historical overview, 13–14
    objective metrics, 292–295
  saturation, 287, 301
Colorfulness, 287–288, 293–294, 331
Components, 324, 344–346, 454–456
Compression, image, 393–395
Contouring artifact
  definition, 136
  masking by noise, 241, 243–247
  objective metric, 222–224, 247–249
  pictorial examples, 134, 155, 242
Contours, iso-quality, 165, 298–302
Contrast
  amplification, 330–331
  customization, 54–57
  definition, 287–288
  interaction with sharpness, 316
  mismatch, 293
  objective metric, 293
  pictorial examples, 48–49
  preference, 50–52

[Contrast, *cont.*]
  sensitivity function, 191–194, 211, 402, 488
  softcopy optimization, 442–443
  surround effects, 294
Convolution
  definition, 26–27
  kernel, 232, 375–376
  preference modeling, 52–54
Correlation
  attribute, 151, 227
  color channel, 360, 393
  criterion for shape of quality CDFs, 426, 429–430
  noise, 328, 340
  preference vs. sensitivity, 52
  spatial signal, 393
Cropping (electronic zoom) and
  capture resolution, 390–392
  enlargeability factor, 177–179
  photospace coverage, 403, 406, 491
Customer
  expectations over time, 82
  influence by experts, 313
  profile, 59–60, 282
  response, vs. surrogates, 140
  satisfaction, 7, 9, 80
  usage distribution (*see* Photospace)
Customization, 57–60
Cycle, product development, 344–346

D log E/H curve (*see* Sensitometry)
Decimation (*see* Down-sampling)
Defects, image (*see* Artifacts; Attributes, artifactual)
Defocus (*see also* Tolerances, optical positioning)
  blur circle, effect on, 480–481
  limits in autofocus design, 386–388
  MTF, effect on, 350–352, 356–357
  sources, 350

Demographics
    color/tone effects, 290, 296, 298
    redeye effects, 277–279
Density balance
    algorithm performance, 439–443
    definition, 287
    objective metric, 293
Depth of field
    definition, 401, 482
    minimum (d_{min})
        and ambient light level limit,
            406, 489–490
        computation, 401, 403, 482, 487
        and depth of focus, 410, 487
        and flash range, 407, 491
    photospace coverage requirements,
        400–403, 408–409
Depth of focus (d_f), 410, 482, 487
Design, central composite, 290–291
Detail
    highlight/shadow (tonal clipping),
        217–219, 410–411
    saturated color, 287–288, 293
    visibility function (DVF), 215–216,
        219–221
Development cycle, product, 344–346
Difference
    estimation method, 91–94
    just noticeable (*see* JND)
Distance
    camera to subject (*see* Camera-
        subject distance)
    hyperfocal, 401, 482
    viewing (*see* Viewing distance)
Distribution
    ambient light level (*see*
        Photospace)
    angular (arcsine), 29–32, 40–41
    CDF (cumulative distribution
        function)
        definition, 25–26
        interpreting quality, 425–431
        sampling, 419–422

[Distribution, *cont.*]
    head size, 308
    histogram computation, 55–56
    mathematical definition, 24–26
    normal (Gaussian), 23–27, 30–32
    objective quantity, 437–439
    perceptual, 20–22, 27–29
    Poisson, 347–348
    preference, 52–57
    PDF (probability density function),
        24–25, 55–56
    quality
        CDFs vs. PDFs, 424–426
        interpreting shapes, 426–431
    sampling from, 419–422
    scene tone, 212–214
    subject distance (*see* Photospace)
Doerner equation, 340, 361, 374
Down-sampling (*see also*
        Interpolation)
    aliasing from, 258, 270–272
    definition, 255
    frequency scaling, 337
    limiting pixel subtense scaling, 271
DPI (dots per inch), 175, 390,
        392–393
DQE (detective quantum efficiency),
        12
DSC (*see* Camera, digital still)
DVF (detail visibility function),
        215–216, 219–221
Dynamic range
    sensor, 443–446
    tone scale, 287–288

Edge
    artifacts (*see* Oversharpening;
        Pixelization; Reconstruction
        infidelity; Ringing)
    map, 232–233
    orientation in scenes, 211, 361, 363
Engineering parameter, 175–176

Enlargeability factor, 177–179
Evaluation (*see* Assessment)
Exposure
   camera (ISO exposure)
      distributions, 438–439, 443–436
      film quality, effect on, 221–222
      ISO normal, 405, 488–489
      photospace coverage
         requirements, 400, 409, 411
      optical printing, 332–335
      time (shutter/integration time)
         camera motion during, 483–484
         ISO normal exposure, yielding (T), 405, 488–489
         maximum hand-held (T_{max}), 405, 487–488

Field of view (θ_d)
   ambient light level limit, effect on, 406, 489–490
   definition, 402, 485–486
   depth of field, effect on, 403, 487
   depth of focus, effect on, 410, 487
   flash range, effect on, 407, 491
   sample calculation, 403
Film
   Advanced Photo System, 453–454
   enlargeability, 177–179
   format, 404, 454 (*see also* Format dimensions)
   grain, 186–187, 207
   NPS, 357–359, 374
   optical printing, 332–335
   properties in modeling, 451
   scanning (*see* Scanner, film)
   speed (*see* ISO speed)
   tonal clipping, 221–222, 410–411
Filter
   anti-aliasing
      digital, 258, 270
      optical, 257–258, 388–390

[Filter, *cont.*]
   array, color (*see* CFA)
   kernel, 232, 375–376
   low-pass, 258–259, 263–265
   pass-band, 259–260, 262–263, 270
   spatial (*see* Unsharp masking)
   stop-band, 260, 262–263, 270, 390
Flare, 333–334
Flash
   photospace coverage requirements, 400, 407–409, 490–491
   redeye, 276–279
F-number (aperture), lens (A)
   definition, 480–481
   effects on
      depth of field, 401, 403, 488
      depth of focus, 410, 482
      flash range, 407, 490–491
      photospace coverage, 409
   format dependence, 403–404
   yielding normal exposure, 405, 488
Focal length, camera lens
   definition, 479
   depth of focus, relationship with, 410, 487
   effect on
      ambient light level limit, 406, 489–490
      angular magnification, 402, 486
      depth of field, 401, 482
      field of view, 402, 485–486
      flash range, 407, 491
      maximum hand-held exposure time, 405, 487–488
      photospace coverage, 403
      redeye, 278
   format, dependence on, 404
Foliage reproduction, 296, 298–299, 302 (*see also* Color, memory)
Formalism, multivariate (*see* Multivariate formalism)

Index 507

Format dimensions
  effects on
    ambient light level limit, 406, 489–490
    depth of field, 403, 487
    depth of focus, 410, 487
    field of view, 401–402, 485–486
    flash range, 407, 491
    system speed, 403–404, 410, 455–456
  optimization, 454–455

Gamut, color, 287, 293
Gaussian distribution, 23–27, 30–32
Glare, veiling (flare), 333–334
Gradient, sensitometric
  amplification operation, 330–331
  banding, effect on, 208
  color negative film, 445
  definition, 287–288
  NPS, effect on, 340
  sharpness, effect on, 316
Graininess (*see* Granularity, RMS; Noise, isotropic)
Granularity, RMS (root-mean-square) (*see also* Noise; NPS)
  density profile effect, 217, 219–221
  exposure dependence, 207
  film speed dependence, 178
  JND, 12, 123
  lightness space (σ_{L*}), 216, 219, 223
  measurement aperture size, 12
  printing magnification effect, 177
  quality specification, 146
  visibility vs. density, 215–216
Guessing and detection certainty, 38
Guide number, 407–408, 490–491

Halftone, 15, 211, 364, 392
Head size
  and redeye severity, 279–282
  and skin-tone reproduction impact, 307–310
Highlight detail, 213, 217–218, 220–221
Hue, color, 287–289, 293, 300
Hyperfocal distance, 401, 482

IHIF (integrated hyperbolic increment function)
  continuity of, 471–472
  derivation, 123–124
  derivatives of, 472–473
  fit parameters, 124–127, 472
  fitting variability data, 142–145, 203–205
  Monte Carlo simulation, role in, 415–416
  regression uncertainty, 130–131
Illuminance (*see* Light level, ambient)
Image processing (*see* Algorithms)
Image quality (*see also* Artifact; Attribute; Capability; Color and tone quality; Models, image quality; Multivariate formalism; Performance; Quality loss function; Quality ruler; Software, image quality)
  adjectives, 81–83, 92
  contours, 165, 298–302
  definition, 9
  distributions
    CDFs vs. PDFs, 424–426
    interpreting shapes, 426–431
  frameworks, 15–16

Image size, effect on quality, 452–453
Image structure, historical overview, 12–13 (*see also* Granularity, RMS; MTF; Noise; NPS)
Imaging system, 324
Impairment, 11, 152 (*see also* Artifact)
Instructions to observers
　avoiding appearance matching in quality assessment, 98–99
　magnitude/difference estimation, examples, 91–92
　preview image use, 113–114
　quality ruler, example, 467–468
Interactions, attribute
　color/tone, 315–317
　correlations, 151, 227
　metrocentric/psychocentric viewpoints, 249–251
　strong, 241–244, 247–249
　test for, 150–152
　weak, 239–241
Interimage effect, 14, 333, 357–358
Interpolation, image (*see also* Downsampling; Reconstruction)
　color filter array (*see* CFA interpolation)
　cubic convolution, 262, 265
　definition, 255
　frequency scaling, 227
　pixel replication, 260–262, 264
　sinc, 259, 262, 264
Interval scale, 61–64, 67–69
ISO (International Standards Organization) speed (S)
　ambient light level limit, effect on, 406, 410, 489–490
　digital still camera (DSC), 410, 445
　film, 410
　guide number, effect on, 407, 490
　normal exposure, relationship to (metering equation), 405, 488

Isolation method, 228–229, 233
IWS (image wavelength spectrum)
　definition, 325–327
　propagation, 331–335

Jacobian matrix
　definition, 330–331
　MTF propagation, 337–338
　NPS propagation, 339–341
　sharpness prediction, 335–336
JND (just noticeable difference) (*see also* JND increment)
　adjectives, relationship to, 82–83
　advertising, use in, 80
　definition, 38–39
　deviates, relationship to, 40–41, 44
　granularity, 12, 123
　images spaced by 3 JNDs, 104–109
　observer dependence, 80, 146
　preference, 50–51
　rating scale calibration in, 69–71
　sign convention, 121
　univariate vs. multivariate, 79–80
　utility of, 35–37
JND increment ($\Delta \Omega_j$) (*see also* JND)
　asymptotic ($\Delta \Omega_\infty$), 123, 126, 471–473
　definition, 63
　determination, 41–44
　objective metric, dependence on, 121–122
　observer dependence, 80, 146
　rating scale, 69–71
　univariate vs. multivariate, 79–80
JPEG (Joint Photographic Experts Group), 394–395
Judge (*see* Observer classification; Observer sensitivity)

Key objective measures
  definition, 175–176
  kinds, 325–326
  Monte Carlo calculations, 416–417
  prediction (*see* Parametric estimation)

Lens design software, 351–354
Light level, ambient
  ISO normal exposure, yielding, 405, 488
  minimum (Y_{min}), 406–408, 489–490
  photospace, dimension of, 398–399, 420–421
  redeye, effect on, 276–279
  sunlight, in bright, 489
  system quality, effect on, 399
Lightness, CIE (L*)
  attribute correlates, 287–288
  definition, 216
  objective metrics using, 289–293
  quantization steps, 223, 248
Limen (*see* JND)
Luminance
  channel, 360
  lightness, relationship to, 216
  scene (*see* Light level, ambient)

Magnification
  angular (*see* Angular magnification)
  printing (*see* Printing magnification)
Magnitude estimation method
  calibrating scales from, 70–71, 317
  comparing scales from two experiments, 66
  deficiencies, 93
  description, 90–92
  uncertainty, 116
Market segmentation, 58–59, 433–435

Masking
  by colored couplers, 14
  of contouring by noise
    explanation of, 244–246
    objective treatment of, 247–249
    pictorial example of, 242–243
  of streaking by noise, 240–241
  unsharp
    description, 232–233
    gain optimization, 374–376, 380, 384
    pictorial examples, 381–383
MCS (mean channel response)
  definition, 325–327
  and IWS, 332–335
  propagation, 329–331
  and tone scale, 288
Measurement protocol
  advantages of standardization, 355
  color filter array MTF, 359–364
  development of, 356
  film NPS, 357–359
  through-focus lens MTF, 356–357
Measurement target, 173, 294, 326, 342, 361–363
Measures, key objective
  definition, 175–176
  kinds, 325–326
  Monte Carlo calculations, 416–417
  prediction (*see* Parametric estimation)
Memory color (*see* Color, memory)
Mensuration method, 228–229, 233
Metadata, 59, 442–443
Metering, reflection
  constant, 405–406, 439, 488–490
  integrated vs. segmented, 437–438
Metric
  benchmark, 175–179
  Minkowski, 160–162
  objective (*see* Objective metric)

Minkowski metric, 160–162
Misregistration artifact
  definition, 114
  experimental design, 197–198
  experimental instructions, 467–469
  objective metric, 198–203
  observer sensitivity, 138, 203–205
  quality loss function, 130, 202–203
  scene susceptibility, 137, 203–205
Models, image quality (*see also* Monte Carlo simulation; Software, image quality)
  applications, 323, 459–460
  use during product development cycle, 344–346
Monitor, video
  grayscale function, 329–330
  MTF, 348–349
  white point and gamma variability, 442–443
Monte Carlo simulation
  examples of input factors, 450–451
  organizing calculations, 415–418
  and performance, 414, 423
  random and pseudorandom numbers, 418–419, 421–422
  sampling distributions, 419–422
Motion picture system, 332–335
MTFs (modulation transfer functions)
  cascading, 260, 262
  CFA interpolation, 361–363
  dependencies on
    field position, 208–209
    motion, 208, 375, 476–478
    orientation, 208
  digital boosting, 234
  film scanner, 374–375, 475–477
  imaging applications, 12, 326
  lens, 350–353, 356–357
  measurement,
    CFA interpolation, 361–364
    general, 326–328
    lens, 356–357

[MTFs, *cont.*]
  propagation, 337–338
  quality distribution, effect on, 427
  reconstruction, 263, 270–271
  visual weighting, 335–336
Multivariate formalism
  additivity, 153, 158, 162, 164–165
  assumptions, 152–153
  attribute delineation, 270–271, 288
  color/tone quality, 215–218
  decomposition, 228–231, 236
  historical perspective, 11
  mathematical formulation, 160–162, 164
  Monte Carlo calculations, 415–416
  pictorial demonstration, 154–160
  predictions, 164–167
  psychocentric view, 251
  suppression, 153, 159, 162, 164–165
  verification, 162–163, 215–218

Noise (*see also* Banding; Granularity, RMS; NPS; Streaking)
  dark, 176, 348, 445
  definition of types, 185–186
  digital still camera, 347–348
  fixed pattern, 207, 347–348
  isotropic
    aliasing of, 270
    density profiles, 217, 219–221
    masking of contouring, 243–249
    masking of streaking, 240–241
    NPS, 184, 186
    pictorial example, 182
    quality loss function, 220
    visibility vs. density, 215–216
  shot (Poisson), 207, 347, 348

Index

NPS (noise power spectra)
  banding, 185, 188
  cross-spectra, 328
  film, 357–359, 374
  imaging applications, 12, 326
  isotropic noise, 184, 186
  measurement, 328
  propagation, 339–341
  streaking, 184, 187
Nyquist frequency, 255–256, 362, 476

Objectionability, 83
Objective measure, key
  definition, 175–176
  kinds, 325–326
  Monte Carlo calculations, 416–417
  prediction (*see* Parametric estimation)
Objective metric (*see also* JND increment)
  definition, 171
  design example, 198–201
  determination methods, 173–174
  empirical optimization, 193–194
  extension for interactions
    contouring/noise, 248–249
    psychocentric view, 250–251
    streaking/noise, 240–241, 249
  Monte Carlo calculations, 416–417
  preference-dependent (color/tone), 292–295, 297–298
  scene-dependent (redeye), 275–276, 278–282
  sign convention, 121
  utility, 172
  verification, 194–196, 202–203
  weighting over image, 210–211
Observer classification
  by expertise, 76, 78, 140
  by preference, 295–299

Observer sensitivity
  expert vs. representative, 140
  Monte Carlo sampling, 415–416
  normalization by quality ruler, 139
  pictorial demonstration, 134–136
  use of variability data, 145–146
  variation, 96, 131, 138, 203–205
OECF (opto-electronic conversion function), 329
Optimum
  multivariate, 166–167
  in objective metrics, 292, 294–295, 297–298
  preference position, 121, 471–472
Oversharpening artifact
  balancing against sharpness and noise, 374, 376, 380–384
  definition, 230
  objective metric, 235–237
  origins, 232–234
  pictorial examples, 231, 382–383

Paired comparison
  conversion to interval scale, 89
  experimental methods, 88, 113
  outcome in terms of CDF, 27–29
  probabilistic model, 19–22
  saturation, 27–29, 32, 41, 61
Panoramic images, 403, 406, 452–453
Parameter, engineering, 175–176
Parametric estimation
  definition, 343
  digital still camera noise, 347–348
  lens MTF, 350–354
  usefulness, 345
  video monitor MTF, 348–349
PDF (probability density function), 24–25, 55–56 (*see also* Distribution)
Perception, probabilistic model, 19–22

Performance
    definition, 379, 413–414
    distributions, 424–431
    examples of analyses
        Advanced Photo System
            verification, 456–457
        camera light metering, 437–439
        camera sensor size, 435–436
        color/density/contrast balancing, 439–443
        film scanner noise, 443–446
        panoramic prints, 452–453
        photospace-specific camera design, 433–435
    modeling (*see* Monte Carlo simulation)
PGI (print grain index), 174, 341
Photomotivation, 399–400, 403, 489
Photospace
    coverage requirements
        ambient exposure, 403–406, 488–490
        depth of field, 401–403, 487
        example, 407–409
        flash exposure, 407, 490–491
        limitations, 410–411
    distribution
        examples, 398–399, 408
        sampling from, 419–422
    effect on camera design, 433–435
Pixelization artifact
    definition, 262
    examples of, 264–265, 267
    relationship with ringing, 262–263, 270–271
PPI (pixels per inch), 390, 392–393
Preference
    distribution, 51–56
    JNDs, 50–51
    quality loss function, 51–57
    summary of treatment, 285–286
Preview images, 114, 442–443

Primitive factors, 415–418
Printing magnification (m_p)
    Advanced Photo System, 454–456
    definition, 177
    effects on
        ambient light level limit, 406, 489–490
        angular magnification, 402, 486
        circle of confusion, 402
        flash range, 407, 491
        maximum hand-held exposure time, 487
        redeye, 278
    enlargeability factor, 178–179
    format dimensions, dependence on, 403–404, 454
    frequency scaling, 337, 339–340
Projection method, 98, 228–229
Propagation
    Image wavelength spectrum, 331–335
    Mean channel signal, 329–331
    MTF, 337–338
    Noise power spectrum, 339–341
Prototyping product designs, 344, 346
Psychometrics, 10–11 (*see also* Assessment; Scale, rating; Scaling methods)
Psychophysics, 10
Pupil, human
    origins of redeye, 276–278
    redeye objective metric, 278–282

Quality, image (*see* Image quality)
Quality loss function (*see also* IHIF)
    advantages of universal form, 120
    contours, 298–302
    definition, 51–52
    quadratic approximation, 53, 55–57
    typical, 130

Index

Quality ruler
  assessing individual attributes (projection), 98–99
  attributes varied in, 95–98
  concept, 93–95
  hardcopy implementation
    description, 101–103
    sample images, 104–109
    sample MTFs, 110
  instructions, 113–114, 475–477
  observer sensitivity effects, 139
  performance, 115–118
  softcopy implementation, 111–113
Quality scale (*see* Scale, rating)
Quantization
  in image compression, 393–394
  tonal artifacts (*see* Contouring)

Race, effect on redeye, 277–279
Radius of curvature at threshold (R_r), 123, 126–127, 471–473
Random numbers, 418–419, 421–422
Range, dynamic
  sensor, 443–446
  tone scale, 287–288
Range effect, 81–83, 91
Rank ordering method, 113
Rating (*see* Assessment; Scale, rating; Scaling methods)
Ratio scale, 61, 64–69
Reconstruction
  artifacts (*see* Pixelization, Reconstruction infidelity, Ringing)
  definition, 255
  error, 260–262
  infidelity (*see also* Pixelization; Ringing)
    description, 270–273
    examples of, 265, 269

Redeye artifact
  objective metric, 278–282
  origins, 276–278
Reference image
  in critical assessments, 311–315
  in scaling experiments, 90–92
Reflectance, average scene, 489
Regression fitting (*see* IHIF)
Reproduction, color/tone (*see* Color)
Resampling (*see* Interpolation)
Resolution, 175, 390, 392–393
Resolving power, 12
Ringing artifact
  definition, 258–259
  examples of, 264–265, 268
  relationship with pixelization, 262–263, 270–271
RMS error, 160–161, 201, 203
RMS granularity (*see* Granularity, RMS)
Ruler, quality (*see* Quality ruler)

Sample and hold (*see* Interpolation, pixel replication)
Sampling theory, 254–258
Saturation
  color, 287, 301
  in paired comparisons, 27–29, 32, 41, 61
Scale, rating
  calibration, 69–71, 74–77
  ideal interval, 61–64, 67–69
  ideal ratio, 61, 64–69
  magnitude estimation, 66, 70
  psychometrics, 10–11
  standard, 73–77, 87–88
Scaling methods (*see* Categorical sort; Difference estimation; Magnitude estimation; Paired comparison; Quality ruler; Rank ordering)

Scanner, film
  MTF, 374–375, 475–477
  noise (dynamic range), 443–446
  optical tolerances, 384–386
Scene classification, 295–297
Scene susceptibility
  Monte Carlo sampling, 415–416
  pictorial demonstration, 132–133
  use of variability data, 145–146
  variation, 97, 131, 136–137, 203–205
Seed, pseudorandom number, 419
Segmentation, market, 58–59, 433–435
Selection
  scene, 75, 141, 291–292, 307
  observer, 76–78, 139–141, 290
Sensitivity
  observer (see Observer sensitivity)
  spectral, 334–335
Sensitometry (see also MCS)
  analog transform, 329–330, 333
  gradient, 13, 208, 340
  speed and clipping, 221, 410
  tone reproduction, 13, 288
Sensor
  color filter array (see CFA)
  MTF, 375, 483–485
  noise (dynamic range), 347–348, 443–446
  optical anti-aliasing, 388–389
  resolution and quality, 390–393
  size and quality, 435–436
Shadow detail, 213, 217, 220–221
Shake, camera, 405, 483–485, 487–488
Sharpening (see Unsharp masking)
Sharpness
  anti-aliasing filter effects, 257–258
  boosting (see Unsharp masking)
  digital filter effects, 259–260, 262
  film speed effects, 178
  images spaced by 3 JNDs, 104–109

[Sharpness, cont.]
  interaction with contrast, 316
  observer sensitivity variability, 96
  printing magnification effects, 177
  scene susceptibility variability, 97
  visual weighting, 335–336
Signal-to-noise ratio (SNR), 12–13, 410, 445–446 (see also Noise; Sharpness)
Skin-tone reproduction, 298, 301–302, 307–310 (see also Color, memory)
Sky reproduction, 296, 298, 300, 302 (see also Color, memory)
Softcopy
  hardware (see Video monitor)
  quality ruler, 111–113
Software
  image quality
    advantages of general models, 367–368
    example of use, 373–376
    features, 16–17
    on-line help, 372–373, 475–477
    output, 373
    user interface design, 368–372
  lens design, 351–354
Specification setting
  Advanced Photo System, 453–457
  component (see Capability analyses, examples of)
  general approach, 146, 172–173, 345–346, 454
Spectral truncation, 258–259, 262
Speed
  format, 403–406, 455–456
  ISO (see ISO speed)
SQRI (square-root integral), 13
Standards, image quality
  international, 16
  physical, constructing, 74–78
  quality ruler (see Quality ruler)
  univariate vs. multivariate, 78–81

Index

Streaking artifact
  definition, 131, 185–187
  masking by noise, 340–341
  NPS, 184–185, 187
  objective metric, 189–192
  pictorial examples, 132–133, 135
Strobe (*see* Flash)
Subject
  distance to camera (*see* Autofocus; Camera-subject distance)
  importance and color quality, 296, 298, 307–310
  location in images, 211–212
Subsystem, 324
Subtense
  limiting pixel, 271–272
  of confusion (θ_{max}) (*see also* Circle of confusion)
    definition, 402, 486
    depth of field, effect on, 403
    maximum flash range, effect on, 407, 491
    maximum hand-held exposure time, effect on, 405, 487
    minimum ambient light level, effect on, 406, 489–490
    values, 402, 411, 487
Sunny 16 rule, 405, 489
Susceptibility, scene (*see* Scene susceptibility)
System, imaging, 324
System modeling (*see* Models, image quality; Monte Carlo simulation; Software, image quality)

Target, measurement, 173, 294, 326, 342, 361–363
TDF (tonal distribution function)
  definition, 212–214
  use in contouring metric, 223, 248

Thin-lens equation, 479, 481
Threshold
  placement of experimental levels near, 188–190
  radius of curvature (R_r), 123, 126–127, 471–473
  reference position (Ω_r), 121–125, 471–472
  weighting of artifacts near, 210
Thurstone's Law, 11, 22, 89–90
TIF (tonal importance function)
  definition, 212–241
  determination, 217, 219–221
  use in contouring metric, 223, 248
Tolerances, optical positioning, 384–386, 410, 417
Tonal clipping, 217–219, 410–411
Tone reproduction (*see* Color and tone reproduction; Contrast)
Torgerson's Law, 90
Tristimulus values, 332, 335–336 (*see also* CIE)
Truncation, spectral, 258–259, 262

Uncertainty, assessment, 116, 130–131
Unsharp masking
  description, 232–233
  gain optimization, 374–376, 380, 384
  pictorial examples, 381–383
Unsharpness (*see* Sharpness)
Up-sampling (*see* Interpolation)

Video monitor
  grayscale function, 329–330
  MTF, 348–349
  variability, 442–443

Viewing distance
  effects on
    ambient light level limit, 406, 489–490
    angular magnification, 402, 486
    circle of confusion, 402
    flash range, 407, 491
    maximum hand-held exposure time, 487
    redeye, 278
  of hand-held prints, 390–391, 453
Visibility function, detail (DVF), 215–216, 219–221
Visual system
  appearance modeling, 15
  frequency response, 191–194, 211, 402, 487

Weber's Law, 123
Weighting
  color channel, 202–203, 335–336
  in color metrics, 292–294, 298

[Weighting, *cont.*]
  in objective vs. perceptual space, 209–211
  spatial, 209, 211–212
  tonal (signal-dependent) (*see* TDF, TIF)
Wiener spectrum (*see* NPS)
Windowing function, 259, 262, 264

Yield, 397, 399, 410–411 (*see also* Photospace coverage requirements)

Zoom
  electronic (cropping)
    capture resolution, 390–392
    enlargeability factor, 177–179
    photospace coverage, 403, 406, 491
  optical (lenses), 403, 406, 491